IØØ55341

This report contains the collective views of international groups of experts and does not necessarily represent the decisions or the stated policy of the United Nations Environment Programme, the International Labour Organization, or the World Health Organization.

Environmental Health Criteria 221

ZINC

First draft prepared by Drs B. Simon-Hettich and A. Wibbertmann, Fraunhofer Institute of Toxicology and Aerosol Research, Hanover, Germany, Mr D. Wagner, Department of Health and Family Services, Canberra, Australia, Dr L. Tomaska, Australia New Zealand Food Authority, Canberra, Australia, and Mr H. Malcolm, Institute of Terrestrial Ecology, Monks Wood, England.

Published under the joint sponsorship of the United Nations Environment Programme, the International Labour Organization, and the World Health Organization, and produced within the framework of the Inter-Organization Programme for the Sound Management of Chemicals.

World Health Organization
Geneva, 2001

The **International Programme on Chemical Safety (IPCS)**, established in 1980, is a joint venture of the United Nations Environment Programme (UNEP), the International Labour Organization (ILO), and the World Health Organization (WHO). The overall objectives of the IPCS are to establish the scientific basis for assessment of the risk to human health and the environment from exposure to chemicals, through international peer-review processes, as a prerequisite for the promotion of chemical safety, and to provide technical assistance in strengthening national capacities for the sound management of chemicals.

The **Inter-Organization Programme for the Sound Management of Chemicals (IOMC)** was established in 1995 by UNEP, ILO, the Food and Agriculture Organization of the United Nations, WHO, the United Nations Industrial Development Organization, the United Nations Institute for Training and Research, and the Organisation for Economic Co-operation and Development (Participating Organizations), following recommendations made by the 1992 UN Conference on Environment and Development to strengthen cooperation and increase coordination in the field of chemical safety. The purpose of the IOMC is to promote coordination of the policies and activities pursued by the Participating Organizations, jointly or separately, to achieve the sound management of chemicals in relation to human health and the environment.

WHO Library Cataloguing-in-Publication Data

Zinc.

    (Environmental health criteria ; 221)

    1.Zinc – analysis 2.Zinc – toxicity 3.Occupational exposure
    4.Environmental exposure 5.Risk assessment I.Series

    ISBN 92 4 157221 3   (NLM Classification: QD 181.Z6)
    ISSN 0250-863X

Computer typesetting by I. Xavier Lourduraj, Chennai, India

Printed in Finland
2001/14046 – Vammala – 5000

CONTENTS

ENVIRONMENTAL HEALTH CRITERIA FOR
ZINC

NOTE TO READERS OF THE CRITERIA MONOGRAPHS

Every effort has been made to present information in the criteria monographs as accurately as possible without unduly delaying their publication. In the interest of all users of the Environmental Health Criteria monographs, readers are requested to communicate any errors that may have occurred to the Director of the International Programme on Chemical Safety, World Health Organization, Geneva, Switzerland, in order that they may be included in corrigenda.

* * *

A detailed data profile and a legal file can be obtained from the International Register of Potentially Toxic Chemicals, Case postale 356, 1219 Châtelaine, Geneva, Switzerland (telephone no. + 41 22 - 9799111, fax no. + 41 22 - 7973460, E-mail irptc@unep.ch).

* * *

This publication was made possible by grant number 5 U01 ES02617-15 from the National Institute of Environmental Health Sciences, National Institutes of Health, USA, and by financial support from the European Commission.

Environmental Health Criteria

PREAMBLE

Objectives

In 1973 the WHO Environmental Health Criteria Programme was initiated with the following objectives:

(i)   to assess information on the relationship between exposure to environmental pollutants and human health, and to provide guidelines for setting exposure limits;

(ii)   to identify new or potential pollutants;

(iii)   to identify gaps in knowledge concerning the health effects of pollutants;

(iv)   to promote the harmonization of toxicological and epidemiological methods in order to have internationally comparable results.

The first Environmental Health Criteria (EHC) monograph, on mercury, was published in 1976 and since that time an ever-increasing number of assessments of chemicals and of physical effects have been produced. In addition, many EHC monographs have been devoted to evaluating toxicological methodology, e.g. for genetic, neurotoxic, teratogenic and nephrotoxic effects. Other publications have been concerned with epidemiological guidelines, evaluation of short-term tests for carcinogens, biomarkers, effects on the elderly and so forth.

Since its inauguration the EHC Programme has widened its scope, and the importance of environmental effects, in addition to health effects, has been increasingly emphasized in the total evaluation of chemicals.

The original impetus for the Programme came from World Health Assembly resolutions and the recommendations of the 1972 UN Conference on the Human Environment. Subsequently the work became an integral part of the International Programme on Chemical Safety (IPCS), a cooperative programme of UNEP, ILO and WHO.

In this manner, with the strong support of the new partners, the importance of occupational health and environmental effects was fully recognized. The EHC monographs have become widely established, used and recognized throughout the world.

The recommendations of the 1992 UN Conference on Environment and Development and the subsequent establishment of the Intergovernmental Forum on Chemical Safety with the priorities for action in the six programme areas of Chapter 19, Agenda 21, all lend further weight to the need for EHC assessments of the risks of chemicals.

Scope

The criteria monographs are intended to provide critical reviews on the effect on human health and the environment of chemicals and of combinations of chemicals and physical and biological agents. As such, they include and review studies that are of direct relevance for the evaluation. However, they do not describe *every* study carried out. Worldwide data are used and are quoted from original studies, not from abstracts or reviews. Both published and unpublished reports are considered and it is incumbent on the authors to assess all the articles cited in the references. Preference is always given to published data. Unpublished data are used only when relevant published data are absent or when they are pivotal to the risk assessment. A detailed policy statement is available that describes the procedures used for unpublished proprietary data so that this information can be used in the evaluation without compromising its confidential nature (WHO (1999) Guidelines for the Preparation of Environmental Health Criteria. PCS/99.9, Geneva, World Health Organization).

In the evaluation of human health risks, sound human data, whenever available, are preferred to animal data. Animal and *in vitro* studies provide support and are used mainly to supply evidence missing from human studies. It is mandatory that research on human subjects is conducted in full accord with ethical principles, including the provisions of the Helsinki Declaration.

The EHC monographs are intended to assist national and international authorities in making risk assessments and subsequent risk management decisions. They represent a thorough evaluation of risks and are not, in any sense, recommendations for regulation or standard setting. These latter are the exclusive purview of national and regional governments.

Content

The layout of EHC monographs for chemicals is outlined below.

- Summary – a review of the salient facts and the risk evaluation of the chemical
- Identity – physical and chemical properties, analytical methods
- Sources of exposure
- Environmental transport, distribution and transformation
- Environmental levels and human exposure
- Kinetics and metabolism in laboratory animals and humans
- Effects on laboratory mammals and *in vitro* test systems
- Effects on humans
- Effects on other organisms in the laboratory and field
- Evaluation of human health risks and effects on the environment
- Conclusions and recommendations for protection of human health and the environment
- Further research
- Previous evaluations by international bodies, e.g. IARC, JECFA, JMPR

Selection of chemicals

Since the inception of the EHC Programme, the IPCS has organized meetings of scientists to establish lists of priority chemicals for subsequent evaluation. Such meetings have been held in Ispra, Italy, 1980; Oxford, United Kingdom, 1984; Berlin, Germany, 1987; and North Carolina, USA, 1995. The selection of chemicals has been based on the following criteria: the existence of scientific evidence that the substance presents a hazard to human health and/or the environment; the possible use, persistence, accumulation or degradation of the substance shows that there may be significant human or environmental exposure; the size and nature of populations at risk (both human and other species) and risks for environment; international concern, i.e. the substance is of major interest to several countries; adequate data on the hazards are available.

If an EHC monograph is proposed for a chemical not on the priority list, the IPCS Secretariat consults with the Cooperating Organizations and all the Participating Institutions before embarking on the preparation of the monograph.

Procedures

The order of procedures that result in the publication of an EHC monograph is shown in the flow chart on p. xv. A designated staff member of IPCS, responsible for the scientific quality of the document, serves as Responsible Officer (RO). The IPCS Editor is responsible for layout and language. The first draft, prepared by consultants or, more usually, staff from an IPCS Participating Institution, is based initially on data provided from the International Register of Potentially Toxic Chemicals, and reference data bases such as Medline and Toxline.

The draft document, when received by the RO, may require an initial review by a small panel of experts to determine its scientific quality and objectivity. Once the RO finds the document acceptable as a first draft, it is distributed, in its unedited form, to well over 150 EHC contact points throughout the world who are asked to comment on its completeness and accuracy and, where necessary, provide additional material. The contact points, usually designated by governments, may be Participating Institutions, IPCS Focal Points, or individual scientists known for their particular expertise. Generally some four months are allowed before the comments are considered by the RO and author(s). A second draft incorporating comments received and approved by the Director, IPCS, is then distributed to Task Group members, who carry out the peer review, at least six weeks before their meeting.

The Task Group members serve as individual scientists, not as representatives of any organization, government or industry. Their function is to evaluate the accuracy, significance and relevance of the information in the document and to assess the health and environmental risks from exposure to the chemical. A summary and recommendations for further research and improved safety aspects are also required. The composition of the Task Group is dictated by the range of expertise required for the subject of the meeting and by the need for a balanced geographical distribution.

EHC PREPARATION FLOW CHART

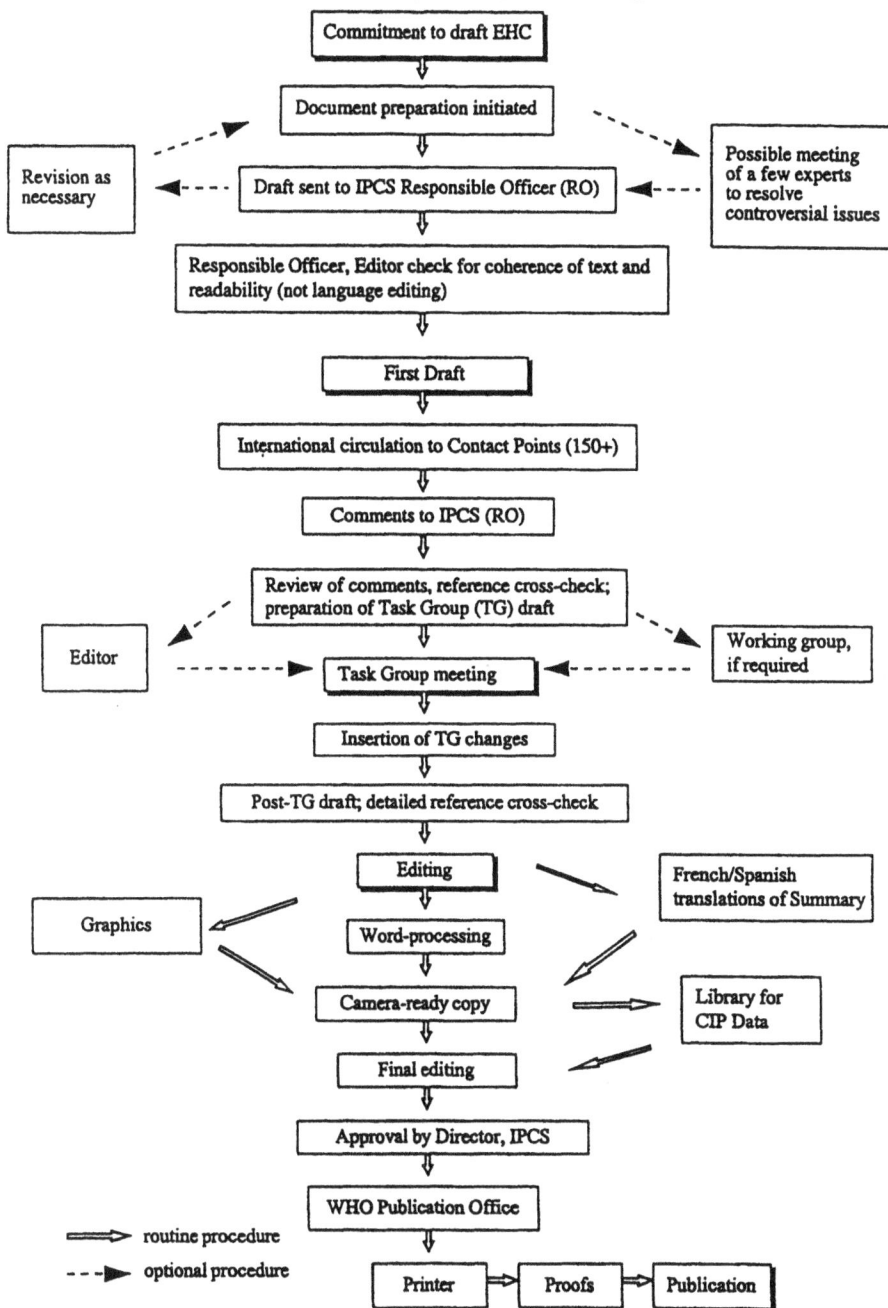

Commitment to draft EHC
⇩
Document preparation initiated
⇩
Revision as necessary ◄ - - - Draft sent to IPCS Responsible Officer (RO) ◄ - - - - Possible meeting of a few experts to resolve controversial issues
⇩
Responsible Officer, Editor check for coherence of text and readability (not language editing)
⇩
First Draft
⇩
International circulation to Contact Points (150+)
⇩
Comments to IPCS (RO)
⇩
Review of comments, reference cross-check; preparation of Task Group (TG) draft
⇩
Editor - - - - - ► Task Group meeting ◄ - - - - - Working group, if required
⇩
Insertion of TG changes
⇩
Post-TG draft; detailed reference cross-check
⇩
Editing → French/Spanish translations of Summary
⇩
Graphics ← Word-processing
⇩
Camera-ready copy ⇒ Library for CIP Data
⇩
Final editing
⇩
Approval by Director, IPCS
⇩
WHO Publication Office
⇩

⇒ routine procedure
- - ► optional procedure

Printer ⇒ Proofs ⇒ Publication

The three cooperating organizations of the IPCS recognize the important role played by nongovernmental organizations. Representatives from relevant national and international associations may be invited to join the Task Group as observers. Although observers may provide a valuable contribution to the process, they can only speak at the invitation of the Chairperson. Observers do not participate in the final evaluation of the chemical; this is the sole responsibility of the Task Group members. When the Task Group considers it to be appropriate, it may meet *in camera.*

All individuals who as authors, consultants or advisers participate in the preparation of the EHC monograph must, in addition to serving in their personal capacity as scientists, inform the RO if at any time a conflict of interest, whether actual or potential, could be perceived in their work. They are required to sign a conflict of interest statement. Such a procedure ensures the transparency and probity of the process.

When the Task Group has completed its review and the RO is satisfied as to the scientific correctness and completeness of the document, it then goes for language editing, reference checking and preparation of camera-ready copy. After approval by the Director, IPCS, the monograph is submitted to the WHO Office of Publications for printing. At this time a copy of the final draft is sent to the Chairperson and Rapporteur of the Task Group to check for any errors.

It is accepted that the following criteria should initiate the updating of an EHC monograph: new data are available that would substantially change the evaluation; there is public concern for health or environmental effects of the agent because of greater exposure; an appreciable time period has elapsed since the last evaluation.

All Participating Institutions are informed, through the EHC progress report, of the authors and institutions proposed for the drafting of the documents. A comprehensive file of all comments received on drafts of each EHC monograph is maintained and is available on request. The Chairpersons of Task Groups are briefed before each meeting on their role and responsibility in ensuring that these rules are followed.

WHO TASK GROUP ON ENVIRONMENTAL HEALTH CRITERIA FOR ZINC

Members

Dr H.E. Allen, Department of Civil and Environmental Engineering, University of Delaware, Newark, Delaware, USA

Dr G. Batley, CSIRO Centre for Advanced Analytical Chemistry, Division of Coal and Energy Technology, Lucas Heights Research Laboratories, Menai, Australia

Dr G. Cherian, Department of Pathology, University of Western Ontario, London, Ontario, Canada (*Vice-Chairman*)

Dr G. Dixon, Department of Biology, University of Waterloo, Waterloo, Ontario, Canada

Professor W.H.O. Ernst, Vrije University, Amsterdam, the Netherlands

Professor R. Gibson, Department of Human Nutrition, University of Otago, Dunedin, New Zealand

Dr C.R. Janssen, University of Ghent, Laboratory for Biological Research in Aquatic Pollution, Ghent, Belgium

Dr L.M. Klevay, US Department of Agriculture, Grand Forks Human Nutrition Research Center, Grand Forks, North Dakota, USA

Mr H. Malcom, Institute of Terrestrial Ecology, Monks Wood, Huntingdon, Cambridgeshire, United Kingdom (*Co-Rapporteur*)

Dr L. Maltby, Department of Animal and Plant Sciences, School of Biological Sciences, University of Sheffield, Sheffield, United Kingdom

Professor M.R. Moore, University of Queensland, National Research Centre for Environmental Toxicology, Coopers Plains, Brisbane, Australia

Dr G. Nordberg, Department of Occupational and Environmental Medicine, Environmental Medicine Unit, Umea University, Umea, Sweden

Dr H.H. Sandstead, University of Texas School of Medicine, Department of Preventive Medicine and Community Health, Galveston, Texas, USA

Dr B. Simon-Hettich, Fraunhofer Institute of Toxicology and Aerosol Research, Hanover, Germany

Dr J.H.M. Temmink, Wageningen Agricultural University, Department of Toxicology, Wageningen, Netherlands (*Chairman*)

Dr J. Vangronsveld, Limburgs University Centre, University Campus, Diepenbeek, Belgium

Dr D. Wagner, Chemicals Safety Unit, Human Services and Health, Canberra, Australia (*Co-Rapporteur*)

Observers/Representatives

Dr K. Bentley, Commonwealth Department of Health and Family Services, Canberra, Australia

Dr C. Boreiko, International Lead Zinc Research Organization, Inc., Research Triangle Park, North Carolina, USA

Dr P. Chapman, EVS Environment Consultants, Ltd., North Vancouver, Canada (*Representing the International Lead Zinc Research Organization*)

Dr T.M. Florence, Centre for Environmental Health Sciences, Oyster Bay, New South Wales, Australia

Dr T.V. O'Donnell, University of Otago, Wellington South, New Zealand

Mr D. Sinclair, Pasminco Ltd., Melbourne, Victoria, Australia

Dr L. Tomaska, Australia New Zealand Food Authority, Canberra, Australian Capital Territory, Australia

Dr F. Van Assche, European Zinc Institute, Brussels, Belgium

Dr W.J.M. Van Tilborg, Rozendaal, Netherlands *(Representing the European Chemical Industry Ecology and Toxicology Centre)*

Mr H. Waeterschoot, Union Minière, Brussels, Belgium
(Representing the International Zinc Association)

Secretariat

Dr G.C. Becking, International Programme on Chemical Safety, World Health Organization, Interregional Research Unit, Research Triangle Park, North Carolina, USA *(Secretary)*

Mr P. Callan, Environmental Health Policy, Department of Health and Family Services, Canberra, Australian Capital Territory, Australia

Dr A. Langley, Hazardous Substances Section, South Australia Health Commission, Adelaide, South Australia, Australia

Mr S. Mangas, Hazardous Substances Section, South Australian Health Commission, Adelaide, South Australia, Australia

WHO TASK GROUP ON ENVIRONMENTAL HEALTH CRITERIA FOR ZINC

A WHO Task Group on Environmental Health Criteria for Zinc met in McLaren Vale, Australia, from 16 to 20 September 1996. The meeting was sponsored by a consortium of Australian Commonwealth and State Governments through a national steering committee chaired by Dr K. Bentley, Commonwealth Department of Health and Family Services, Canberra. The meeting was co-hosted and organized by the South Australian Health Commission, Dr A. Langley and Mr S. Mangas being responsible for the arrangements. Participants were welcomed on behalf of the host organizations by Dr I. Calder, Director, Environmental Health Branch, South Australian Health Commission. Dr G.C. Becking, IPCS, opened the meeting and, on behalf of the Director, IPCS and the three cooperating organizations (UNEP/ILO/WHO), thanked the Australian Commonwealth and State Governments for their funding of the Task Group as well as their financial and in-kind support for the preparation of the first draft of the Environmental Health Criteria for Zinc. He thanked the staff of the Hazardous Substances Section, South Australian Health Commission for their excellent work in organizing the Task Group. The Task Group reviewed and revised the draft criteria monograph, and made an evaluation of the risks to human health and the environment from exposure to zinc.

The first draft of this monograph was prepared by Dr B. Simon-Hettich and Dr A. Wibbertmann, Fraunhofer Institute of Toxicology and Aerosol Research, Hanover, Germany; Mr D. Wagner, Commonwealth Department of Health and Family Services, Canberra, Australia; Dr L. Tomaska, Australia New Zealand Food Authority (ANZFA), Canberra, Australia, and Mr H. Malcolm, Institute of Terrestrial Ecology, Monks Wood, United Kingdom. The draft reviewed by the Task Group, incorporating the comments received from the IPCS Contact Points, was prepared through the cooperative efforts of the Commonwealth Department of Health and Family Services, ANZFA, Institute of Terrestrial Ecology, and the Secretariat.

Dr G.C. Becking (IPCS Central Unit, Interregional Research Unit) and Ms S.M. Poole (Birmingham, England) were responsible for the overall scientific content and technical editing, respectively, of this monograph.

The efforts of all who helped in the preparation and finalization of the monograph are gratefully acknowledged.

ABBREVIATIONS

AAS	atomic absorption spectroscopy
AES	atomic emission spectroscopy
ASV	anodic stripping voltametry
BAF	bioaccumulation factor
BCF	bioconcentration factor
CRIP	cysteine-rich intestinal protein
CSV	cathodic stripping voltametry
DNA	deoxyribonucleic acid
DP-ASV	differential pulse-anodic stripping voltametry
DTPA	diethylenetriamine pentaacetic acid
dw	dry weight
E_h	redox potential
EC_{50}	effective concentration, affecting 50% of test organisms
EDTA	ethylenediaminetetraacetic acid
EPA	Environmental Protection Agency (USA)
ESOD	Cu, Zn erythrocyte superoxide dismutase
FAAS	flame atomic absorption spectroscopy
GF-AAS	graphite furnace atomic absorption spectroscopy
HDL	high-density lipoprotein
ICP-AES	inductively-coupled plasma-atomic emission spectroscopy
ICP-MS	inductively-coupled plasma-mass spectrometry
Ig	immunoglobulin
IGF	insulin-like growth factor
LC_{50}	lethal concentration killing 50% of test organisms
LDL	low-density lipoprotein
LOEC	lowest-observed-effective concentration
$LT_{(50)}$	lethal time$_{(50)}$ for specified concentration of chemical killing 50% of test organisms
MS	mass spectrometry
NAA	neutron activation analysis

NHANES	National Health and Nutrition Examination Survey (USA)
NOEC	no-observed-effect concentration
NOEL	no-observed-effect level
RNA	ribonucleic acid
SEM	standard error of the mean
TFIIIA	transcription factor IIIA
UV	ultraviolet
XRF	X-ray fluorescence

1. SUMMARY AND CONCLUSIONS

1.1 Identity, and physical and chemical properties

Zinc metal does not occur in the natural environment. It is present only in the divalent state Zn(II). Ionic zinc is subject to solvation, and its solubility is pH and anion dependent. Zinc is a transition element and is able to form complexes with a variety of organic ligands. Organometallic zinc compounds do not exist in the environment.

1.2 Analytical methods

Because zinc is ubiquitous in the environment, special care is required during sampling, sample preparation and analysis to avoid sample contamination. Sample preparation for solid samples typically involves microwave-assisted mineralization with concentrated acids. For water samples, solvent extraction in the presence of complexing agents and chelating resin separation have been used to preconcentrate zinc.

Inductively-coupled plasma atomic emission spectrometry (ICP-AES), graphite furnace atomic absorption spectrometry (GF-AAS), anodic stripping voltammetry (ASV) and ICP-mass spectrometry (ICP-MS) are commonly used instrumental techniques for zinc determination. For low-level analyses, GF-AAS, ASV and ICP-MS are preferred.

With special care, zinc concentrations as low as 0.006 µg/litre and 0.01 mg/kg are detectable in water and solid samples, respectively.

Speciation analyses in water require the application of separation techniques with any of the above methods or use of the labile-bound discrimination offered by ASV.

1.3 Sources of human and environmental exposure

Most rocks and many minerals contain zinc in varying amounts. Commercially, sphalerite (ZnS) is the most important ore mineral and the principal source of the metal for the zinc industry. In 1994, world metal production of zinc was 7 089 000 tonnes and zinc metal consumption amounted to 6 895 000 tonnes.

Zinc is widely used as a protective coating of other metals, in dye casting and the construction industry, and for alloys. Inorganic zinc compounds have various applications, e.g., for automotive equipment, storage and dry cell batteries, and dental, medical and household applications. Organo-zinc compounds are used as fungicides, topical antibiotics and lubricants.

Zinc becomes malleable when heated to 100–150 °C and is then readily machined into shapes. It is capable of reducing most other metal states and is therefore used as an electrode in dry cells and in hydrometallurgy.

The largest natural emission of zinc to water results from erosion. Natural inputs to air are mainly due to igneous emissions and forest fires. Anthropogenic and natural sources are of a similar magnitude. The main anthropogenic sources of zinc are mining, zinc production facilities, iron and steel production, corrosion of galvanized structures, coal and fuel combustion, waste disposal and incineration, and the use of zinc-containing fertilizers and pesticides.

1.4 Environmental transport, distribution and transformation

Zinc in the atmosphere is primarily bound to aerosol particles. The size of particle is determined by the source of zinc emission. A major proportion of the zinc released from industrial processes is adsorbed on particles that are small enough to be in the respirable range.

The transport and distribution of atmospheric zinc vary according to the size of particles and the properties of the zinc compounds concerned. Zinc is removed from the atmosphere by dry and wet deposition. Zinc adsorbed on particles with low densities and diameters can be transported over long distances.

The distribution and transport of zinc in water, sediment and soil are dependent upon the species of zinc present and the characteristics

2

of the environment. The solubility of zinc is primarily determined by pH. At acidic pH values, zinc may be present in the aqueous phase in its ionic form. Zinc may precipitate at pH values greater than 8.0. It may also form stable organic complexes, for example, with humic and fulvic acids. The formation of such complexes can increase the mobility and/or solubility of zinc. Zinc is unlikely to be leached from soil owing to its adsorption on clay and organic matter. Acidic soils and sandy soils with a low organic content have a reduced capacity for zinc absorption.

Zinc is an essential element and *in vivo* levels are therefore regulated by most organisms. Zinc is not biomagnified. The absorption of zinc by aquatic animals tends to be from water rather than food. Only dissolved zinc tends to be bioavailable, and bioavailability depends on the physical and chemical characteristics of the environment and biological processes. Consequently, environmental assessment must be conducted on a site-specific basis.

1.5 Environmental concentrations

Zinc occurs ubiquitously in environmental and biological samples. Concentrations in soil sediments and fresh water are strongly determined by local geological and anthropogenic influences and thus vary substantially. Natural background total zinc concentrations are usually < 0.1–50 µg/litre in fresh water, 0.002–0.1 µg/litre in seawater, 10–300 mg/kg dry weight (dw) in soils, up to 100 mg/kg dw in sediments, and up to 300 ng/m^3 in air. Increased levels can be attributed to natural occurrence of zinc-enriched ores, to anthropogenic sources or to abiotic and biotic processes. In anthropogenically contaminated samples, zinc levels of up to 4 mg/litre in water, 35 g/kg in soil, 15 µg/litre in estuarine water, and 8 µg/m^3 in air are found.

Zinc concentrations in representative organisms during exposure to water-borne zinc are in the range 200–2000 mg/kg.

Concentrations in plants and animals are higher near anthropogenic point sources of zinc contamination. Interspecies variations in zinc content are considerable; intraspecies levels vary, for instance, with life stage, sex, season, diet and age. Normal levels of zinc in most crops and pastures are in the range 10–100 mg/kg dw. Some

3

plants are zinc accumulators, but the extent of the accumulation in plant tissues varies with soil and plant properties.

Only negligible quantities of zinc are inhaled from ambient air, but a broad range of exposures to dusts and fumes of zinc and zinc compounds is possible in occupational settings.

1.5.1 Human intakes

Estimated ranges of daily dietary intakes of total zinc are 5.6–10 mg/day for infants and children aged 2 months–11 years, 12.3–13.0 mg/day for children aged 12–19 years, and 8.8–14.4 mg/day for adults aged 20–50 years. Mean daily zinc intake from drinking-water is estimated to be < 0.2 mg/day.

Dietary reference values for zinc vary according to the dietary pattern of the country, assumptions on the bioavailability of dietary zinc, and age, sex and physiological status. Dietary reference values range from 3.3 to 5.6 mg/day for infants aged 0–12 months, 3.8 to 10.0 mg/day for children aged 1–10 years, and 8.7 to 15 mg/day for adolescents aged 11–18 years. Adult values range from 6.7 to 15 mg/day for those aged 19–50 years, 7.3 to 15 mg/day during pregnancy, assuming diets of moderate zinc availability, and 11.7 to 19 mg/day during lactation, depending on the stage.

1.6 Kinetics and metabolism in laboratory animals and humans

For inhalation studies (nose only) in guinea-pigs, rats and rabbits, retention values of 5–20% in the lung were observed after exposure to zinc oxide aerosols at a concentration of 5–12 mg/m^3 for 3–6 h. The intestinal absorption of zinc is controlled by a homeostatic mechanism which is not fully understood but is mainly controlled by pancreatic and intestinal secretion and faecal excretion. Homeostasis may involve metal-binding proteins such as metallothionein and cysteine-rich intestinal protein. Other unknown mechanisms may also exist. The uptake from intestinal mucosa may involve both active and passive transport processes. In animals, absorption can vary in the range 10–40% depending on nutritional status and other ligands in the diet. Dermal absorption of zinc from zinc oxide and zinc chloride can occur and is increased in zinc deficiency. Absorbed zinc is mainly deposited in muscle, bone, liver, pancreas, kidney and other organs. The biological half-life of zinc is about 4–50 days in rats, depending on the administered dose, and about 280 days in humans.

1.7 Effects on laboratory animals

Acute oral toxicity in rodents exposed to zinc is low, with LD_{50} values in the range 30–600 mg/kg body weight, depending on the zinc salt administered. Acute effects in rodents following inhalation or intratracheal instillation of zinc compounds include respiratory distress, pulmonary oedema and infiltration of the lung by leukocytes.

Toxic effects of zinc in rodents following short-term oral exposure include weakness, anorexia, anaemia, diminished growth, loss of hair and lowered food utilization, as well as changes in the levels of liver and serum enzymes, morphological and enzymatic changes in the brain, and histological and functional changes in the kidney. The level at which zinc produces no adverse symptoms in rats has been set at about 160 mg/kg body weight. Pancreatic changes were observed in calves exposed to high levels of dietary zinc. Short-term inhalation exposure of guinea-pigs and rats to zinc oxide at concentrations of ≥ 5.9 mg/m^3 resulted in inflammation and pulmonary damage.

Long-term oral exposure to zinc indicated the target organs of toxicity to be the haematopoietic system in rats, ferrets and rabbits; the kidney in rats and ferrets; and the pancreas in mice and ferrets. The no-observed-effect level (NOEL) with respect to growth and anaemia for zinc sulfate in the diet was reported to be < 100 mg/kg in rats. Increases in zinc concentrations in the bodies of experimental animals exposed to zinc are accompanied by reduced levels of copper, suggesting that some of the signs of toxicity ascribed to exposure to excess levels of zinc may be caused by zinc-induced copper deficiency. Moreover, studies have shown that exposure to zinc alters the levels of other essential metals, including iron, in the bodies of exposed animals. Some signs of toxicity observed in animals exposed to high levels of zinc can be alleviated by the addition of copper or iron to the diet.

Very high levels of zinc are toxic to pregnant mice and hamsters. Rats exposed to zinc at 0.5% and 1% in the diet for 5 months were unable to conceive until the zinc was withdrawn. High levels of zinc in the diet (2000 mg/kg) were also associated with an increase in resorptions and stillbirths in mice and rats; a finding also observed in sheep and hamsters. Resorptions were increased in one study in which rats were exposed, throughout the entire gestation period, to zinc at doses as low as 150 mg/kg. In another rat study, however, no deleterious effects on the developing fetus were observed at doses of 500 mg/kg. Exposure of rats to dietary zinc levels of 4000 mg/kg post coitus was shown to interfere with the implantation of ova. Elevation of zinc levels in rat pups exposed to zinc was accompanied by reductions in the levels of copper and iron.

Genotoxicity studies have been conducted in a variety of systems. Most of the findings have been negative, but a few positive results have been reported.

Zinc deficiency in animals is characterized by reduction in growth, cell replication, adverse reproductive effects, adverse developmental effects, which persist after weaning, and reduced immunoresponsiveness.

1.8 Effects on humans

Poisoning incidents with symptoms of gastrointestinal distress, nausea and diarrhoea have been reported after a single or short-term exposure to concentrations of zinc in water or beverages of 1000–2500 mg/litre. Similar symptoms, occasionally leading to death, have been reported following the inadvertent intravenous administration of large doses of zinc. Kidney dialysis patients exposed to zinc through the use of water stored in galvanized units have developed symptoms of zinc toxicity that were reversible when the water was subjected to activated carbon filtration.

A disproportionate intake of zinc in relation to copper has been shown to induce copper deficiency in humans, resulting in increased copper requirements, increased copper excretion and impaired copper status. Pharmacological intakes of zinc have been associated with effects ranging from leukopenia and/or hypochromic microcytic anaemia to decreases in serum high-density lipoprotein concentrations. These conditions were reversible upon discontinuation of zinc therapy together with copper supplementation.

The human health effects associated with zinc deficiency are numerous, and include neurosensory changes, oligospermia, impaired neuropsychological functions, growth retardation, delayed wound healing, immune disorders and dermatitis. These conditions are generally reversible when corrected by zinc supplementation.

There is no single, specific and sensitive biochemical index of zinc status. The most reliable method for detecting deficiency is to show a positive response to zinc supplementation in controlled double-blind trials (in the absence of other limiting nutrient deficiencies). This approach is time-consuming and often impractical, however, and determination of a combination of dietary, biochemical and functional physiological indices is generally preferred. Several concordant abnormal values are more reliable than a single aberrant value in diagnosing a zinc deficiency state. The inclusion of functional physiological indices, such as growth, taste acuity and dark adaptation with a biochemical test (e.g., plasma or hair zinc concentration) allows the extent of the functional consequences of the zinc deficiency state to be assessed.

Inhalation exposure to zinc chloride following the military use of "smoke bombs" has resulted in effects that include interstitial oedema, interstitial fibrosis, pneumonitis, bronchial mucosal oedema, ulceration and even death under extreme exposure conditions in confined spaces. These effects are possibly attributable to the hygroscopic and astringent nature of the particles released by such devices.

Occupational exposure to finely dispersed particulate matter formed when certain metals, including zinc, are volatilized can lead to an acute illness termed "metal-fume fever", characterized by a variety of symptoms including fever, chills, dyspnoea, nausea and fatigue. The condition is generally severe but transient, and individuals tend to develop tolerance. Exposure of volunteers to zinc concentrations of 77–150 mg/m^3 for 15–30 min gave rise to symptoms in some of the subjects, a marked dose-related inflammatory response with increased polynuclear lymphocytes in broncheoalveolar lavage fluid, and a marked increase in cytokines.

Occupational asthma has been reported among those working with soft solder fluxes, but the evidence was not sufficient to indicate a causative relationship. A rare case suggesting such a relationship has been diagnosed recently in a worker from a hot-dip (zinc) galvanizing plant.

1.9 Effects on other organisms in the laboratory and field

Zinc is important in membrane stability, in over 300 enzymes, and in the metabolism of proteins and nucleic acids. The adverse effects of zinc must be balanced against its essentiality. Zinc deficiency has been reported in a wide variety of cultivated plants and animals, with severe effects on all stages of reproduction, growth and tissue proliferation. Zinc deficiencies in various crops have resulted in large crop losses worldwide. Zinc deficiency is rare in aquatic organisms in the environment, but can be induced under experimental conditions.

The toxicity of zinc can be influenced by both biotic and abiotic factors, such as organism age and size, prior exposure, water hardness, pH, dissolved organic carbon and temperature. The integration of environmental chemistry and toxicology has allowed a better prediction of the effects on organisms in the environment. This has led to the now accepted view that the total concentration of an essential element such as zinc in an environmental compartment is not, taken alone, a good predictor of its bioavailability.

Acute toxicity values of dissolved zinc to freshwater invertebrates range from 0.07 mg/litre for a water flea to 575 mg/litre for an isopod. Acute toxicity values for marine invertebrates range from 0.097 mg/litre for a mysid to 11.3 mg/litre for a grass shrimp. Acutely lethal concentrations for freshwater fish are in the range 0.066–2.6 mg/litre; the range for marine fish is 0.19–17.66 mg/litre.

Zinc has been shown to exert adverse reproductive, biochemical, physiological and behavioural effects on a variety of aquatic organisms. Zinc concentrations of > 20 µg/litre have been shown to have adverse effects on aquatic organisms. However, the toxicity of zinc to such organisms is influenced by many factors, such as the temperature, hardness and pH of the water, and previous zinc exposure.

Zinc toxicity in plants generally causes disturbances in metabolism, which are different from those occurring in zinc deficiency. The critical leaf tissue concentration of zinc for an effect on growth in most species is in the range 200–300 mg/kg dw.

Field studies have revealed adverse effects on aquatic invertebrates, fish and terrestrial plants close to sources of zinc contamination. Zinc tolerances in terrestrial plants, algae, microorganisms and invertebrates have developed in the vicinity of areas with elevated zinc concentrations.

1.10 Conclusions

1.10.1 Human health

- There is a decreasing trend in anthropogenic zinc emissions.

- Many pre-1980 environmental samples, in particular in water samples, may have been subject to contamination with zinc during sampling and analysis and, for this reason, zinc concentration data for such samples should be viewed with extreme caution.

- In countries where staple diets are based on unrefined cereals and legumes, and intakes of flesh foods are low, dietary strategies should be developed to improve the content and bioavailability of zinc.

- Preparations intended to increase the zinc intake above that provided by the diet should not contain zinc levels that exceed dietary reference values, and should contain sufficient copper to ensure a ratio of zinc to copper of approximately 7, as is found in human milk.

- There is a need for better documentation of actual exposures to zinc oxide fume in occupational settings. Workplace concentrations should not result in exposure levels as high as those known to have given rise to inflammatory responses in the lungs of volunteers.

9

- The essential nature of zinc, together with its relatively low toxicity in humans and the limited sources of human exposure, suggests that normal, healthy individuals not exposed to zinc in the workplace are at potentially greater risk from the adverse effects associated with zinc deficiency than from those associated with normal environmental exposure to zinc.

1.10.2 Environment

- Zinc is an essential element in the environment. The possibility exists both for a deficiency and for an excess of this metal. For this reason it is important that regulatory criteria for zinc, while protecting against toxicity, are not set so low as to drive zinc levels into the deficiency area.

- There are differences in the responses of organisms to deficiency and excess.

- Zinc bioavailability is affected by biotic and abiotic factors, for instance: organism age and size, prior history of exposure, water hardness, pH, dissolved organic carbon and temperature.

- The total concentration of an essential element such as zinc, alone, is not a good predictor of its bioavailability or toxicity.

- There is a range of optimum concentrations for essential elements such as zinc.

- The toxicity of zinc will depend on environmental conditions and habitat types, thus any risk assessment of the potential effects of zinc on organisms must take into account local environmental conditions.

2. IDENTITY, PHYSICAL AND CHEMICAL PROPERTIES, AND ANALYTICAL METHODS

Zinc is the twenty-fifth most abundant element. It is widely found in nature and makes up 0.02% by weight of the earth's crust (Budavari, 1989). Zinc normally appears dull grey owing to coating with an oxide or basic carbonate. It is extremely rare to find zinc metal free in nature (Beliles, 1994).

Some zinc compounds, synonyms and formulae are given in Table 1.

2.1 Identity

Pure zinc is bluish-white and lustrous when polished. It has the atomic number of 30 and the relative atomic mass of 65.38, and belongs to group 2b and the fourth period of the periodic table. The configuration of the outermost electrons is $3d^{10}4s^2$. Thus, its valence in chemical compounds is +2. In nature, zinc is a mixture of five stable isotopes: ^{64}Zn (49%), ^{66}Zn (28%), ^{68}Zn (19%), ^{67}Zn (4.1%), and ^{70}Zn (0.62%) (Budavari, 1989). A further 19 radioactive isotopes ($^{57}Zn-^{63}Zn$, ^{65}Zn, $^{68}Zn-^{80}Zn$) are known; ^{65}Zn is the most stable with a half-life of 243.8 days, but most have very short half-lives (Lide, 1991).

2.2 Physical and chemical properties

2.2.1 Zinc metal

Zinc possesses a low to intermediate hardness (Mohs' hardness 2.5) and crystallizes in a distorted hexagonal close-packed structure. Because of its density of 7.13 g/cm^3, it is called a heavy metal. It has an electrical conductivity of 28.3% of the international annealed copper standard (Kirk & Othmer, 1982). At ordinary temperatures the metal is too brittle to roll, but it becomes malleable and ductile when heated to 100–150 °C. At temperatures of > 210 °C, zinc becomes brittle and pulverizable, and, at higher temperatures, again soft and malleable (Budavari, 1989; Beliles, 1994). Since zinc is very reactive, it reacts strongly with other elements, such as oxygen,

Table 1. Chemical names, synonyms and formulae of elemental zinc and zinc compounds

Chemical name	CAS registry number	Formula	Synonyms
Zinc	7440-66-6	Zn	-
Zinc acetate	557-34-6	$Zn(C_2H_3O_2)_2$	-
Zinc arsenite	10326-24-6	$Zn(AsO_2)_2$	zinc meta-arsenite, ZMA
Zinc bromide	7699-45-8	$ZnBr_2$	-
Zinc carbonate	3486-35-9	$ZnCO_3$	-
Zinc chloride	7646-85-7	$ZnCl_2$	butter of zinc
Zinc cyanide	557-21-1	$Zn(CN)_2$	-
Zinc diethyldithiocarbamate	14324-55-1	$Zn[SC(S)N(C_2H_5)_2]_2$	-
Zinc fluoride	7783-49-5	ZnF_2	-
Zinc hexafluorosilicate	16871-71-9	$ZnSiF_6.6H_2O$	zinc silicofluoride; zinc fluosilicate
Zinc iodide	10139-47-6	ZnI_2	-
Zinc laurate	-	$Zn(C_{12}H_{33}O_2)_2$	-
Zinc nitrate	7779-88-6	$Zn(NO_3)_2$	-
Zinc oleate	557-07-3	$Zn(C_{17}H_{33}COO)_2$	-

Table 1 (contd.)

Zinc oxide	1314-13-2	ZnO	Chinese white; zinc white; flowers of zinc; philosopher's wool
Zinc permanganate	23414-72-4	Zn(MnO4)$_2$.6H$_2$O	
Zinc peroxide	1314-22-3	ZnO$_2$	zinc dioxide; zinc superoxide; ZPO
Zinc-1,4-phenolsulfonate	127-82-2	Zn (SO$_3$C$_6$H$_4$OH)$_2$.8H$_2$O	p-hydroxybenzenesulfonic acid zinc salt; zinc sulfophenate; zinc sulfocarbolate
Zinc phosphate	7779-90-0	Zn$_3$(PO$_4$)$_2$	zinc orthophosphate; zinc phosphate, tribasic
Zinc phosphide	1314-84-7	Zn$_3$P$_2$	-
Zinc silicate	13597-65-4	Zn$_2$SiO$_4$	zinc orthosilicate
Zinc sulfate	7733-02-0	ZnSO$_4$.7H$_2$O	white vitriol; white copperas; zinc vitriol
Zinc sulfide	1314-98-3	ZnS	wurtzite; sphalerite; zinc blende
Zinc telluride	1315-11-3	ZnTe	-
Zinc thiocyanate	557-42-6	Zn(SCN)$_2$	zinc thodanide; zinc sulfocyanate
Zinc dimethyldithiocarbamate	137-30-4	Zn(SCSNCH$_3$CH$_3$)$_2$	Ziram
Zinc ethylene-bis(dithiocarbamate)	12122-67-7	Zn(CS$_2$NHCH$_2$)$_2$	Zineb

chlorine and sulfur, at elevated temperatures (Melin & Michaelis, 1983). Zinc has reducing and also several transitional properties (see below).

The metal burns in air with a bluish-green flame. It is stable in dry air, but on exposure to moist air it becomes covered with an adherent film of zinc oxide or basic carbonate ($2ZnCO_3 \cdot 3Zn(OH)_2$), so isolating the underlying metal and retarding further corrosion.

Zinc is amphoteric and dissolves in strong alkalis and mineral acids with evolution of hydrogen and soluble zinc salts. Oxidizing agents or metal ions, e.g., Cu^{2+}, Ni^{2+} and Co^{2+}, accelerate the dissolution of zinc. Zinc is capable of reducing most metals except aluminium and magnesium (E °(aq) Zn/Zn^{2+}, 0.763 eV; Budavari, 1989).

In solution, four to six ligands can be coordinated with the zinc ion. Complexes are formed with polar ligands, e.g., ammonia, amines, cyanide and halogen ions. Zinc is a reactive amphoteric metal. The hydroxide is precipitated in alkaline solution, but with excess base, it redissolves to form "zincates", ZnO_2^{2-}, which are hydroxo complexes such as $Me^+[Zn(OH)_3]^-$, $Me_2^+[Zn(OH)_4]^{2-}$ and $Me_2^+[Zn(OH)_4(H_2O)_2]^{2-}$ (Budavari, 1989).

2.2.2 Zinc compounds

Zinc has a strong tendency to react with acidic, alkaline and inorganic compounds. Because of its amphoteric properties, zinc forms a variety of salts, which are all nonconducting, nonmagnetic and white or colourless, with the exception of those with a chromophore group, such as chromate. Some physical and chemical data for zinc and selected zinc compounds are given in Table 2.

Zinc oxide is a coarse white or greyish powder, odourless and with a bitter taste. It absorbs carbon dioxide from the air and is soluble in acids and alkalis but insoluble in water and alcohol. The compound is used as a pigment in paints and as an ultraviolet (UV) absorber in several products. It has the greatest UV absorption of all commercial pigments (Lide, 1991). Its major use (see section 3.2.2) is as a vulcanizing agent in the production of rubber products (Melin & Michaelis, 1983).

Table 2. Physical and chemical properties of zinc and some of its compounds[a]

Chemical name	Relative atomic/molecular mass	Melting point (°C)	Boiling point (°C)	Relative density (g/cm³) (°C)	Crystalline form	Solubility
Zinc	65.38	419.58	907	7.14 (25)	distorted hexagonal close packed	soluble acid, alkali; insoluble H_2O, organic solvents
Zinc acetate	183.47	237	200[b]	1.735	monoclinic	soluble H_2O, alcohol
Zinc bromide	225.19	394	690	4.201 (25)	rhombic	soluble H_2O, alcohol, ether
Zinc carbonate	125.39	300[b]	ND	4.398	rhombohedral	soluble acid, alkali; slightly soluble H_2O
Zinc chloride	136.29	283	732	2.907 (25)	hexagonal, deliquescent	soluble H_2O, acid, acetone, alcohol
Zinc fluoride	103.38	872	ca. 1500	4.95 (25)	monoclinic or triclinic	soluble HCl, HNO_3, NH_4OH; slightly soluble H_2O, aqueous HF
Zinc hexafluoro-silicate	207.46	ND[b]	ND	2.104	crystalline powder	soluble H_2O
Zinc hydroxide	99.39	125[b]	ND	3.053	rhombic	soluble acid, alkali; very slightly soluble H_2O
Zinc iodide	319.19	446	624[b]	4.736 (25)	hexagonal	soluble H_2O, alcohol, ether

15

Table 2 (contd.)

Chemical name	Relative atomic/mole-cular mass	Melting point (°C)	Boiling point (°C)	Relative density (g/cm³) (°C)	Crystalline form	Solubility
Zinc nitrate, hexahydrate	297.48	36.4	105–131 (-H$_2$O)	2.065 (14)	tetragonal	soluble H$_2$O, alcohol
Zinc oxide	81.38	1975	ND	5.606	hexagonal	soluble dilute acetic acid, alkali; insoluble H$_2$O, alcohol
Zinc phosphate	386.08	900	ND	3.998 (15)	rhombic	soluble acid, NH$_4$OH; insoluble H$_2$O, alcohol
Zinc phosphide	258.09	> 420	1100 (sublimes in H$_2$)	4.55 (13)	tetragonal	soluble benzene, CS$_2$; insoluble H$_2$O, alcohol
Zinc sulfate	161.44	600[b]	ND	3.54 (25)	rhombic	soluble H$_2$O, MeOH, glycerol
α-Zinc sulfide	97.44	1700 ± 20	ND	3.98	hexagonal	very soluble alcohol; insoluble acetic acid
β-Zinc sulfide	97.44	ND[b]	ND	4.102 (25)	cubic	very soluble acid

[a] From: Lide (1991); ND = not determined.
[b] Decomposition.

16

Zinc chloride, chlorate, sulfate and nitrate are readily soluble in water, whereas the oxide, carbonate, phosphates, silicates, sulfides and organic complexes are practically insoluble in water, with the exception of zinc diethyldithiocarbamate (Budavari, 1989).

Zinc halogenides are hygroscopic. Zinc chloride forms hydrates with 1.33–4 mol H_2O and exerts a water-extracting and condensing action on many organic compounds. Owing to the high polarizing effect, zinc protolyses part of the water envelope and forms hydroxo complexes. Thus, concentrated zinc chloride solutions react like strong acids because of the formation of the acids $H[ZnCl_2OH]$ and $H_2[ZnCl_2(OH)_2]$ (Giesler et al., 1983). Zinc chloride and fluoride have catalytic properties and are used in organic synthesis and also in wood preservation and for antiseptic purposes (Budavari, 1989).

Zinc carbonate occurs naturally as zinc spar. When heated to 150 °C, the compound decomposes into zinc oxide and carbon dioxide. Basic zinc carbonate, zinc carbonate hydroxide, is known in variable composition and is usually characterized as $3Zn(OH)_2$ $2ZnCO_3$. It occurs as the mineral hydrozincite, a weathering product of zinc spar.

Zinc sulfide is a white powder that appears in two different modifications: the hexagonal close packed α-modification (wurtzite), the form preferred by the pigment industry (n ≈ 2.37); and the cubic β-modification (sphalerite), which is substantially converted to wurtzite when heated to 725 °C in the absence of air. Because of its semiconducting and luminescent properties, zinc sulfide is used industrially as a pigment and as phosphors in X-ray and television screens (Neumueller, 1983; Budavari, 1989).

Some organo-zinc compounds (diethyl zinc, diphenyl zinc) are sensitive to air and water. The lower alkyl compounds are autoflammable when exposed to air.

Other organo-zinc compounds, such as zineb (zinc ethylene-bis(dithiocarbamate)) and ziram (zinc dimethyl-dithiocarbamate), are used as agricultural fungicides (Neumueller, 1983).

2.3 Analytical methods

2.3.1 Introduction

Because zinc is ubiquitous in the environment, special care is required during sampling, sample preparation and analysis to avoid sample contamination. Precautions must be taken to avoid contamination arising from such sources as sampling apparatus, filtration equipment, and atmospheric exposure during collection and analysis. Clean room conditions and sample handling using apparatus rigorously cleaned with acid by operators wearing polyethylene gloves and appropriate lint-free clothing are desirable (Batley, 1989a). The necessary detection limits for trace analysis are often affected by problems related to inadequate reagent purity or contamination introduced during the course of the sampling and analytical manipulations. With adequate care, however, zinc concentrations as low as 0.006 µg/litre in water and 0.1 mg/kg in solid samples are detectable, using modern instrumental analysis techniques.

For many environmental samples, zinc concentrations are sufficiently high to obviate the need for the precautions described above. Nevertheless, appropriate quality assurance during both sampling and analysis is necessary to ensure confidence in the methods of analysis used and the subsequent data that they generate.

2.3.2 Sampling and sample preparation

The background concentrations of dissolved zinc in many natural water samples are frequently below 1 µg/litre. However, contamination leading to levels as high as 20 µg/litre is quite possible during sampling and filtration of waters. Containers must be carefully selected and precleaned before use. Teflon containers are preferable; polyethylene is acceptable and superior to Pyrex glass, but soda glass should be avoided (Batley, 1989a). Precleaning is best carried out by prolonged soaking in 2 mol/litre nitric or hydrochloric acids, although hot nitric acid has been used (Mart, 1979). The containers should be rinsed with distilled water and thoroughly rinsed with sample before collection. The need for rigorous care with water sampling has been elegantly demonstrated by Ahlers et al. (1990).

18

Water sample preservation is achieved by acidification to < pH 2, generally after filtration if dissolved metals are being sought. For zinc speciation analysis, acidification is unacceptable, and storage at 4 °C minimizes species transformations or losses. Similar constraints apply to biological fluids.

For ultratrace analysis, the use of a clean laboratory or at least a laminar flow work station is highly recommended to avoid contamination from airborne particulates. Typical unfiltered urban room air may contain zinc at concentrations as high as 1 μg/m^3 (Henkin, 1979). In general laboratory operations, care should be taken to avoid galvanized laboratory fittings (especially retort stands and clamps), rubber materials and powdered gloves, all of which contain zinc.

Contamination of soil and sediment samples, in which zinc concentrations may vary in the range 10–2000 mg/kg, is less of a problem. Where sediment samplers are likely to contaminate the sample, the outer sample layers should be discarded and only those portions not in contact with contaminating surfaces should be subsampled. Coring is usually carried out with PVC or Perspex tubes; where metal corers are used, it is usual for them to have polyethylene or polycarbonate liners. Where sieving of samples is undertaken, stainless steel or nylon sieves are unlikely to cause sample contamination.

If the measurement of zinc present in soils or sediments in specific mineral phases is required, the sample should be frozen as soon as possible after collection and air excluded to avoid oxidation of metal sulfides and transformation of chemical forms. When selective extractions are to be undertaken, the sample is thawed and homogenized by mixing. An aliquot of the moist sample is then taken for analysis, with moisture content being determined in replicate aliquots (Batley, 1989a).

Sampling of plant material from the field requires procedures that take into account a number of abiotic and biotic factors (Quevauviller & Maier, 1994; Ernst, 1995). The former include climate, i.e., sampling before or after rain and, in the case of roots, soil type. Biotic factors include age of material and the presence of parasites (e.g., mildew) or mycorrhizal fungi.

For total zinc analysis, sample preparation involves drying at 110 °C followed by acid digestion. Total mineralization requires a mixture of concentrated acids, e.g., nitric, hydrochloric and hydro-fluoric acids, and the digestion is performed on a hot plate in a heated block assembly or microwave oven. Microwave digestion is being increasingly used to minimize sample contamination. The detection of acid-soluble metals, as stipulated by US EPA method 200.8, uses only nitric and hydrochloric acids (Long & Martin, 1991).

Biological samples comprise aquatic and terrestrial organisms and may include human tissue, hair, sweat, blood, urine and faeces. Again, care is required in the handling of samples to avoid contamination (Batley, 1989a), avoiding metal surfaces and using appropriately cleaned plastic containers. The method of sample preparation depends to a large extent on sample type. Animal and human tissue samples are usually analysed without drying, and wet weight concentrations are reported. In some instances freeze-drying has been employed. Plant tissue samples have been dried at 110 °C, freeze-dried and, in some instances, ashed at 500 °C to facilitate dissolution. In recent years, however, it has been realized that temperature can have a significant effect on the quality of plant material during drying and mineralization prior to analysis. Owing to burning of carbohydrates, drying at 110 °C will diminish the real dry mass, leading to overestimation of the zinc concentration. Ashing at 500 °C should be avoided as it causes loss of zinc as volatile compounds. Plant samples are therefore now usually oven-dried at 80 °C for 48 h (Ernst, 1995; Rengel & Graham, 1995). Freeze-drying remains an option, especially in zinc compartmentation studies.

Dissolution is usually undertaken by wet ashing with nitric acid, either on a hot plate or by microwave-assisted digestion (White, 1988). The use of perchloric acid is generally avoided nowadays, and complete decomposition of organic compounds is not required for most spectroscopic analysis techniques. For marine organisms, hydrogen peroxide is usually added during the dissolution process. Tissue solubilizers such as tetramethylammonium hydroxide or potassium hydroxide have been used for effective dissolution of biological tissue samples (Martin et al., 1991).

Care should be taken in the acid dissolution of blood and urine samples, as frothing of natural surfactants in the sample during digestion can lead to losses. Allowing the sample to stand overnight after the addition of acid can often obviate this problem.

It should be noted that in all of the above analyses, care must be paid to the quality of acids and other reagents used. For analysis of zinc at low concentrations, reagents of an appropriately high purity are essential.

For air sampling with high-volume samplers, low-ash filters are required. Glass fibre filters are sources of zinc contamination and membrane filters made of cellulose acetate or Teflon are preferred (Batley, 1989a). Samples are analysed after dissolution of particulates in nitric acid, although ashing has also been used (NIOSH, 1984).

2.3.3 Separation and concentration

Given the low detection limits of modern analytical techniques, separation techniques, such as ion exchange or solvent extraction, that preconcentrate zinc from solution, are less frequently used nowadays, although they are required for ultratrace detection. Any additional sample manipulation, however, increases the opportunity for sample contamination. A range of preconcentration techniques has been applied, but only those currently in common use are discussed here.

Most appropriate is the use of the complexing agents ammonium pyrrolidine dithiocarbamate (APDC) or diethyldithiocabamate (DDC) to extract zinc, using trichloroethane or chloroform as the solvent. Apte & Gunn (1987) have described a micro solvent extraction procedure with analysis by graphite furnace atomic-absorption spectrometry (GF-AAS); detection of zinc concentrations as low as 20 ng/litre in seawater and other natural waters is possible.

Chelating resins have also been widely used for preconcentration. Chelex-100 or equivalent iminodiacetate resins in the sodium or calcium forms effectively remove zinc from seawater or fresh waters at pH values greater than 6. It should be noted that zinc associated with colloids will not be satisfactorily removed. The

21

use of immobilized 8-hydroxyquinoline, dithiocarbamates or other zinc-binding ligands has also been reported. The former is incorporated in at least one *in situ* water sampler (Willie et al., 1983; Batley, 1989b).

In natural water systems, measurements typically involve either total zinc, dissolved zinc or some form of zinc speciation analysis. Water quality criteria are frequently based on total analyses. Acidification of the sample, with heating, is therefore used as a pretreatment option. Filtration through 0.45-μm membrane filters provides the accepted means of separating particulate species, and a separate analysis can then be performed on each phase. For speciation, the principal concern is for bioavailable species, and a range of procedures has been applied, including ultrafiltration, dialysis, ligand exchange, chelating resin separations and measurement techniques, such as anodic and cathodic stripping voltammetry (ASV and CSV) that discriminate between labile and non-labile zinc. These have been comprehensively reviewed elsewhere (Florence & Batley, 1980; Batley, 1989b; Apte & Batley, 1995).

2.3.4 *Detection and measurement*

For environmental and biological samples, the required detection limits necessitate the use of modern instrumental methods of analysis. Traditional titrimetric and gravimetric methods are not sufficiently sensitive. Spectrophotometric methods offer greater sensitivity, but are tedious and subject to numerous interferences (Cherian & Gupta, 1992). A summary of analytical methods for zinc in various environmental media is given in Table 3.

To achieve the necessary detection limits, spectrophotometric methods will usually require some form of sample preconcentration. The achievable detection limit is frequently limited in practice by the purity of the reagents used.

Instrumental techniques offer element-specific detection at low concentrations. The most common are atomic absorption or emission spectrometry (AAS and AES), X-ray fluorescence (XRF), electroanalytical techniques, such as polarography or stripping voltammetry, and neutron activation analysis.

Table 3. Analytical methods for zinc

Sample	Preparation[a]	Analytical method[b]	Limit of detection	Reference
Atmospheric particulates	collection on membrane filter, ashing with HNO_3	F-AAS	2.6 pg/litre	Ottley & Harrison (1993)
Atmospheric particulates	polystyrene filter collection, pressed into pellets	NAA	not given	Zoller et al. (1974)
Atmospheric particulates	cellulose filter collection	NAA	0.4 pg/litre	Amundson et al. (1992)
Water	filtration, acidification	FAAS	50 µg/litre	Greenberg et al. (1992)
Water	APDC/MIBK extraction	FAAS	not given	Greenberg et al. (1992)
Water	filtration, acidification	GF-AAS	0.1 µg/litre	Greenberg et al. (1992)
Water	filtration, acidification	ICP-AES	2 µg/litre	Greenberg et al. (1992)
Water	filtration, acidification US EPA Method 200.8	ICP-MS	1.8 µg/litre	Long & Martin (1991)
Water/seawater	APDC/trichloroethane extraction	GF-AAS	0.02 µg/litre	Apte et al. (1998)
Water/seawater	acidification, ultraviolet irradiation	DP-ASV	0.05 µg/litre	Batley & Farrar (1978)
Seawater	APDC chelation	CSV	0.006 µg/litre	Van den Berg (1986)

23

Table 3 (contd.)

Sample	Preparation[a]	Analytical method[b]	Limit of detection	Reference
Water/seawater	chelating resin preconcentration	ICP-MS	0.05 µg/litre	Sturgeon et al. (1981)
Water, leachates	acidification	XRF	5 mg/litre	Cornjeo et al. (1994)
Soil, sediments	US EPA Method 200.8	ICP-MS	0.7 mg/kg	Long & Martin (1991)
Soil, sediments	HCl/HNO$_3$/HF microwave digestion	ICP-MS	0.7 mg/kg[c]	Dale (unpublished data)
Biota (fish, oysters, mussels, etc.)	HNO$_3$/H$_2$O$_2$ microwave digestion	ICPA-ES	0.2 mg/kg[c]	Martin et al. (1991)
Biota (fish, oysters, mussels, etc.)	tetramethylammonium hydroxide dissolution, US EPA Method 200.11	ICP-AES	0.2 mg/kg[c]	Martin et al. (1991)
Biota (fish, oysters, mussels, etc.)	homogenization, freeze-drying, HNO$_3$/H$_2$O$_2$ dissolution	IDMS	1.5 ng absolute	Waidmann et al. (1994)
Biological samples	solid	XRF	0.1 mg/kg	Heckel (1995)
Plant material	homogenization, digestion in HNO$_3$/HCl in Teflon bomb	AAS/F-AAS	not given	Harmens et al. (1993)
Food	dry ashing, HNO$_3$/H$_2$O$_2$ digestion	ICP-MS (isotope dilution)	not given	Veillon & Patterson (1995)

Table 3 (contd.)

Food	homogenization, freeze-drying, acid microwave digestion	ICP-AES	2 mg/kg[c]	Copa-Rodriguez & Basadre-Pampin (1994)
Blood serum	dilution with HNO_3/HCl	ICP-AES	10–50 µg/litre	Que Hee & Boyle (1988)
Biological tissues, whole blood, faeces	heating with HNO_3, Parr bomb digestion, addition of $HClO_4$	ICP-AES	10–50 µg/litre	Que Hee & Boyle (1988)
Blood, plasma	dilution with water	GF-AAS	6 µg/litre	Schmitt et al. (1993)
Human milk	ultrafiltration	GF-AAS	1.6 µg/litre	Arnaud & Favier (1992)
Faeces	drying, digestion with H_2SO_4/$HClO_4$	F-AAS	not given	Dastych (1990)
Saliva	–	GF-AAS	0.4 µg/litre	Henkin et al. (1975)

[a] APDC = ammonium pyrrolidine dithiocarbamate; MIBK = methyl isobutyl ketone; US EPA = United States Environmental Protection Agency.

[b] CSV = cathodic stripping voltammetry; DP-ASV = differential pulse anodic stripping voltammetry; F-AAS = flame atomic-absorption spectrometry; GF-AAS = graphite furnace atomic-absorption spectrometry; ICP-AES = inductively-coupled plasma atomic emission spectrometry; ICP-MS = inductively-coupled plasma mass spectrometry; IDMS = isotope dilution studies; NAA = neutron activation analysis; XRF = X-ray fluorescence.

[c] Dependent upon the mass of sample taken and the dilution.

25

XRF and other focused particle beam methods require solid samples. The detection limit for zinc by direct microprobe analysis is only around 240 mg/kg (Kersten & Forstner, 1989). For liquid samples, preconcentration by adsorption or complexation onto solid phases has been used. A relatively new XRF procedure based on polarized X-rays has a detection limit for zinc of 0.1 mg/kg in biological materials (Heckel, 1995).

Flame atomic absorption spectrometry (F-AAS) has for many years been the basis of the standard method for determining zinc in waters (Hunt & Wilson, 1986). The method is very sensitive: for direct F-AAS analysis, the instrumental detection limit is 5 μg/litre, although the optimal concentration range is 50–2000 μg/litre. This can be further enhanced with preconcentration by complexation/solvent extraction or using solid-phase adsorbents. GF-AAS offers improved detection limits for direct analysis, but is subject to matrix interferences, particularly in saline waters (Slavin, 1984).

Inductively-coupled plasma atomic emission spectrometry (ICP-AES) is considerably more sensitive than F-AAS, and detection of 2 μg/litre is possible by direct analysis (Greenberg et al., 1992), although with the latest axial plasma instruments with ultrasonic nebulization, the limit is as low as 0.2 μg/litre. Calibration by standard additions is essential. This technique offers adequate sensitivity for zinc in contaminated waters or for acid digests of soil, sediment and biological samples. The multi-element capability offered by ICP-AES is a considerable advantage over AAS methods.

ICP mass spectrometry (ICP-MS) offers excellent sensitivity. The instrumental detection limit for zinc in fresh waters is 20 ng/litre using conventional nebulization systems. With aerosol desolvation devices, the detection limit is about one order of magnitude better. However, these detection limits are not achievable unless stringent procedures to avoid zinc contamination are implemented, including the use of ultrapure reagents. A content of solids in excess of 0.1%, as in seawater samples, creates problems during nebulization. These are best overcome by complexation and extraction of zinc as described earlier. The technique is ideally suited to digests of soils, sediments and biological samples; the greater sensitivity means that any difficulties due to a high content of solids are overcome by dilution. In addition, because of its mass resolution, ICP-MS enables isotopic ratio analysis (^{67}Zn/^{68}Zn/^{70}Zn) or isotope dilution studies using ^{65}Zn (Ward, 1987). Isotope tracers have been used to study zinc absorption following administration of the isotope in food (Johnson, 1982; Watson et al., 1987).

Neutron activation analysis (NAA) is a useful technique for the non-destructive analysis of solid samples, and requires a minimum of sample preparation (Fredrickson, 1989; Heydorn, 1995). Its main advantage is its multi-element capability; the great disadvantage is its limited availability, and long analysis time. It has largely been superseded by ICP-MS, which offers a similar capability and is more widely available. For zinc, the sensitivity of NAA is poor.

Of the electroanalytical techniques, polarography is rarely employed except for samples containing high zinc concentrations (> 10 µg/litre), such as digests of ores. For ambient water concentrations, stripping voltammetric techniques are essential. Differential-pulse ASV (DP-ASV) offers detection limits in natural waters in the ng/litre range (Florence, 1989). An advantage of ASV is the *in situ* preconcentration achieved during the accumulation step, which avoids the contamination problems associated with the greater sample manipulation of other preconcentration techniques. A disadvantage is the potential interference from high concentrations of natural organic compounds in some samples, which may adsorb to the mercury electrode and limit zinc deposition. Although this is not a problem for most natural water samples, complete digestion of biological samples or highly contaminated waters, to decompose interfering surface-active organic compounds, is essential.

CSV has also been successfully applied to the detection of baseline concentrations of zinc in seawater (Van den Berg, 1986). It requires the formation of a zinc complex with APDC, which can be accumulated at a mercury electrode and stripped using a cathodic scan. CSV is best used with pristine samples, where interference due to other metals or adsorbing ligands is less likely.

It should be noted that voltammetric techniques applied to water samples will only measure an operationally-defined labile fraction unless the sample is pretreated by UV irradiation to destroy non-labile zinc complexes, and acidification to dissociate zinc bound to natural colloids. This property can be an advantage in speciation studies, where the ASV-labile concentration has been related to the zinc fraction that is bioavailable (Florence & Batley, 1980; Florence, 1992).

New zinc-specific fluorophores have been developed to measure and visualize intracellular zinc. One of these, Zinquin, has been successfully used in lymphoid, myeloid and hepatic cells to detect labile intracellular zinc (Zalewski et al., 1993; Coyle et al., 1994), although the interaction between Zinquin and the zinc-binding protein, metallothionein (see section 6.5.1.4) needs further study (Coyle et al., 1994).

In all analyses, the use of appropriate quality assurance procedures is required. In particular, standard reference materials are essential. These are currently available for waters, sediments and soils, as well as for plant and other biological materials.

3. SOURCES OF HUMAN AND ENVIRONMENTAL EXPOSURE

3.1 Natural occurrence

Zinc is a chalcophilic element like copper and lead, and a trace constituent in most rocks. Zinc rarely occurs naturally in its metallic state, but many minerals contain zinc as a major component from which the metal may be economically recovered (Table 4). The mean zinc levels in soils and rocks usually increase in the order: sand (10–30 mg/kg), granitic rock (50 mg/kg), clay (95 mg/kg) and basalt (100 mg/kg) (Adriano, 1986; Malle, 1992). Sphalerite (ZnS) is the most important ore mineral and the principal source for zinc production. Smithsonite ($ZnCO_3$) and hemimorphite ($Zn_4(Si_2O_7)$ $(OH)_2 \cdot XH2O$) were mined extensively before the development of the froth-flotation process (Melin & Michaelis, 1983; Jolly, 1989). The main impurities in zinc ores are iron (1–14%), cadmium (0.1–0.6%), and lead (0.1–2%), depending on the location of the deposit (ATSDR, 1994).

Natural levels of zinc in the soil environment can vary by three or four orders of magnitude. When ore-rich areas are included in the analysis this variation is even greater (GSC, 1995). National Geochemical Reconnaissance data of Canada have reported a mean value of 80 mg/kg for stream sediments with 10th and 90th percentile values of 40 mg/kg and 245 mg/kg, respectively (GSC, 1995). The 99th percentile value for lake sediments was 1280 mg/kg with a maximum of > 20 000 mg/kg. Similar variations were noted in zinc levels in agricultural soils and lake sediments.

As a result of weathering, soluble compounds of zinc are formed and may be released to water. US EPA (1980) estimated the input of zinc to waters in the USA resulting from erosion of soil particles containing natural traces of zinc to be 45 400 tonnes/year. The global flux of zinc to water through erosion has been estimated at 915 000 tonnes/year (GSC, 1995). Zinc flux to the oceans from high temperature hydrothermal fluids in mid-ocean ridges has been estimated to be of the order of 681 000 tonnes/year.

Table 4. CAS chemical names and registry numbers, synonyms, trade names and molecular formula of zinc ores[a]

Chemical name	CAS registry number	Synonyms and trade names	Composition	Formula
Zinc oxide	1314-13-2	zincite	80.34% Zn, 19.66% O	ZnO
Zinc phosphate	7779-90-0	hopeite	50.80% Zn, 33.16% O, 16.04% P	$Zn_3(PO_4)_2.4H_2O$
Zinc silicate	13597-65-4	willemite	58.68% Zn, 28.72% O, 12.60% Si	Zn_2SiO_4
Zinc sulfide	1314-98-3	sphalerite, wurtzite	67.09% Zn, 32.91% S, up to 25% Fe	ZnS
Zinc carbonate	3486-35-9	smithsonite, zincspar	52.14% Zn, 38.28% O, 9.58% C	$ZnCO_3$
Hemimorphite	-	-	58.28% Zn	$Zn_4(Si_2O_7)(OH)_2.XH_2O$
Franklinite	-	-	15–25% ZnO, 10–16% MnO	$(Zn, Fe, Mn),(FeMn)_2O_4$
Hydrozincite	-	zinc bloom	-	$Zn_5(OH)_6(CO_3)_2$
Tetrahedrite	-	-	8–9% Zn	$(Cu,Zn)_{12}Sb_4S_{14}$

[a] Adapted from Neumueller (1983) and Melin & Michaelis (1983).

Global emissions to air are mainly due to windborne soil particles, igneous emissions and forest fires, and are estimated to be 19 000 tonnes/year, 9600 tonnes/year and 7600 tonnes/year, respectively. Further natural sources of zinc in air are biogenic emissions and seasalt sprays, with annual amounts calculated at 8100 tonnes and 440 tonnes, respectively (Nriagu, 1989).

Lantzy & Mackenzie (1979) calculated the natural continental and volcanic dust flux to be about 35 800 tonnes annually, based on the average zinc concentration in soils and andesites. Thus, total annual emissions of zinc to air from natural sources are estimated at about 45 000 tonnes/year (Nriagu, 1989). Such estimates of zinc may be low (Rasmussen, 1996), particularly those for zinc transferred by biogenic emissions and from volcanic activity. Long-range dust flux has been estimated at 61–366 million tonnes/year (Pye, 1987). Given an average zinc crustal abundance of 70 mg/kg (70 ppm), this yields up to 25 600 tonnes/year. However, short-range, low-level dust transport can also be included and would increase the windblown dust estimate to 5000×10^6 tonnes/year (Pye, 1987), corresponding to a zinc input of 350 000 tonnes/year. Given such uncertainties in the database, it is very difficult to estimate a ratio of natural to anthropogenic emissions for zinc.

3.2 Anthropogenic sources

3.2.1 *Production levels and processes*

3.2.1.1 *Production levels*

Zinc ore (smithsonite) has been used for the production of brass since 1400. In Europe, the production of elemental zinc started in 1743 (Melin & Michaelis, 1983).

World mine production of zinc was 7 140 000 tonnes in 1992 and 7 089 000 tonnes in 1994 (US Bureau of Mines, 1994; ILZSG, 1995). Global zinc production and consumption are summarized in Table 5.

Secondary zinc production constitutes about 20–30% of current total zinc production (1.9 million tonnes in 1994). Taking the historical consumption and produce life cycles of recovered zinc

31

Table 5. Total zinc production and consumption in 1994 (thousand tonnes)[a]

Geographical area	Mine production	Zinc production	Zinc consumption
Europe	1012	2510	2350
Canada	1008	693	148
Australia	971	323	161
China	755	975	577
Peru	682	158	69
USA	597	356	1191
Mexico	369	212	108
Other countries	1271	1862	2291
World total	6665	7089	6895

[a]   From: ILZSG (1995). Total figures for 1995 were: mine production, 6939×10^3 tonnes; zinc production, 724×10^3 tonnes; and zinc consumption, 7354×10^3 tonnes (ILZSG, 1996).

products into account, recovery rates have been estimated to be as high as 80% from zinc sheet and coated steels (EZI, 1996).

3.2.1.2   Production processes

Zinc ore is mined from underground and open pit mines (approximately 62% underground, 14% open pit, 15% a combination, 9% unspecified) (MG, 1994). The mined ores usually contain zinc at levels of 4–8% and are concentrated at the mine sites to levels of 40–60%. Unwanted impurities (gangue) and other impurities, such as iron, cadmium and lead, which substitute for zinc in the mineral crystal structure, are removed by flotation (Jolly, 1989).

The resulting fine-grained sphalerite concentrates contain 40–60% zinc, 30% sulfur and a number of other metals, in varying quantities, that are of economic significance as extractable by-products. All the world's cadmium (excluding recycles) and a large proportion of the germanium and gallium are extracted as by-products of zinc production. Large quantities of sulfuric acid are also produced (UN ECE, 1979; Melin & Michaelis, 1983). Concentrates are the raw materials for zinc smelting.

Zinc metallurgy can be divided into two basic processes: electrolytic refining, which comprised 83% of primary production in 1993; and pyrometallurgical smelting (ILZSG, 1994).

In the conventional electrolytic process zinc concentrates are roasted to remove sulfur, as sulfur dioxide, which is made into sulfonic acid. The resulting calcine (zinc oxide) is leached with spent electrolyte, the solution is then purified and zinc is recovered by electrowinning. The process produces iron residues, such as goethite and jarosite, and gypsum. In an alternative electrolytic process, pressure leaching, concentrates are treated directly with spent electrolyte under pressure to remove sulfur, iron and other impurities. The zinc dissolves in the spent electrolyte and the solution is purified prior to recovery of zinc by electrowinning. This process also produces iron residues and gypsum, and elemental sulfur as a marketable by-product.

In the pyrometallurgical process, concentrates are roasted to produce sinter, as a solid lump feed for the blast furnace, and sulfur dioxide, which is made into sulfuric acid. Sinter and coke are charged to the imperial smelting blast furnace, which produces metallic zinc and lead, and an iron-rich slag. The zinc is refined by distillation in reflux columns.

Trade in zinc intermediate products (ash, drosses, skimmings and residues) represents an important source of material for secondary zinc production. These products contribute up to 42% of the sources of zinc for recycling purposes in western countries (Henstock, 1996). Recycling provides some 28% of the zinc metal produced.

3.2.2 Uses

Zinc is the fourth most widely used metal in the world after iron, aluminium and copper. Table 6 shows the applications of zinc in western Europe. An overview of the uses of zinc compounds is given in Table 7.

Table 6. Applications of zinc in western Europe (ILZSG, 1995)

Application	Consumption (%)
Galvanizing	43
Brass	23
Alloys, other than brass	13
Wrought zinc	12
Pigments/chemicals	8
Others[a]	1

[a] Including use of zinc in veterinary and human medicines, as a feed additive, and in cosmetics (Bruère et al., 1990; EU, 1996).

Zinc is mainly used as a protective coating of other metals, such as iron and steel. Because the metal lacks strength, it is often alloyed with other metals, e.g., aluminium, copper, titanium and magnesium, to impart a variety of properties. If zinc is the primary constituent of the alloy, it is called a zinc-base alloy, mainly used for casting and for wrought applications. The zinc-copper-titanium alloy has become the dominant wrought-zinc alloy because of its greater strength and dent resistance than other metals of the same thickness (Beliles, 1994). Further important applications are in dye-casting, the construction industry, and other alloys (brass, bronze). Zinc dust is a widely used catalyst; it is also used as a reducing and precipitating agent in organic and analytical chemistry. Inorganic zinc compounds have various applications, e.g., for automotive equipment, storage and dry-cell batteries and organ pipes. Zinc chloride, sulfide and sulfate have dental, medical and household applications. Zinc oxide is frequently used in ointments, powders and other medical formulations. Zinc salts are used as solubilizing agents in pharmaceuticals (e.g., injectable insulin) (Budavari, 1989). Organo-zinc compounds are used as fungicides, topical antibiotics and lubricants (Shamberger, 1979; Sax & Lewis, 1987). Zinc soaps (zinc palmitate, stearate and oleate) are used as drying lubricants and dusting agents for rubber, and as waterproofing agents for textiles, paper and concrete (Budavari, 1989). Zinc phosphide is highly poisonous owing to liberation of phosphine gas; it is used in rat and mouse poisons (Bertholf, 1988).

Table 7. Some uses of zinc compounds[a]

Zinc compound	Uses
Zinc acetate	medicine (astringent), timber preservative, textile dyeing
Zinc antimonide	thermoelectric devices
Zinc arsenate	insecticide, timber preservative
Zinc arsenite	insecticide, timber preservative
Zinc bacitracin	antibacterial agent in ointments, suppositories
Zinc bromide	photographic emulsions, rayon manufacture
Zinc caprylate	fungicide
Zinc carbonate	ceramics, fire-proofing agents, cosmetics, pharmaceuticals (ointments, dusting powder), medicine (topical antiseptic)
Zinc chloride	organic synthesis (catalyst and dehydrating agents), fireproofing, soldering fluxes, electroplating, antiseptic preparations, textiles (mordants, mercerizing agents), adhesives, dental cements, medicine (astringent)
Zinc dibenzyldithiocarbamate	accelerator for latex dispersions and cements
Zinc dichromate	pigment
Zinc fluoride	phosphors, ceramic glazes, timber preservation, electroplating
Zinc fluorosilicate	concrete hardener, laundry sour, preservative, mothproofing agents
Zinc iodide	medicine (topical antiseptic), analytical reagent
Zinc laurate	paints, varnishes, rubber compound manufacture
Zinc linoleate	paint drier, especially with cobalt and manganese soaps
Zinc oxide	accelerator, rubber (reinforcing agent), ointments, paints (pigment, mould-growth inhibitor), plastics (ultraviolet absorber), feed additive, cosmetics, photoconductor, piezoelectric devices
Zinc-1,4-phenolsulfonate	insecticide, medicine (antiseptic)
Zinc phosphate	dental cements, phosphors, conversion coating of steel
Zinc phosphide	rodenticide
Zinc propionate	fungicide on adhesive tape

Table 7 (contd.)

Zinc compound	Uses
Zinc salicylate	medicine (antiseptic)
Zinc stearate	cosmetics, lacquers, ointments, lubricant, mould-release agent, medicine (for dermatitis), dietary supplement
Zinc sulfate	rayon manufacture, dietary supplement, mordant, timber preservative, production of plastics
Zinc sulfide	pigment, glass, ingredient of lithopone, phosphor in X-ray and television screens, luminous paints, fungicide
Zineb	insecticide, fungicide
Ziram	fungicide, rubber accelerator

[a] Adapted from: Sax & Lewis (1987) and Budavari (1989).

Zinc dialkyldithiocarbamates are used as accelerators for the vulcanization of rubber.

In agriculture, zinc-carrying fertilizers are by far the largest source of zinc. About 22 000 tonnes of zinc are used annually as fertilizers in the USA (Adriano, 1986).

3.2.3 Emissions during production and use

Zinc emissions can be classified as follows: *controlled emissions* (e.g., point source emissions) from industrial processes; *fugitive emissions* resulting from mining, handling or transport operations or from leakages from buildings and insufficient ventilation; and *diffuse emissions* from the use of zinc-containing products (OSPARCOM, 1994; Van Assche, 1995). Zinc is released to the atmosphere as dust and fumes from mining, zinc production facilities, processing of zinc-bearing raw materials, brass works, coal and fuel combustion, waste incineration, and iron and steel production. However, refuse incineration, coal combustion, smelter operations, and some metal-working industries constitute the major sources of zinc in air (ATSDR, 1994). More efficient emission control technology and changes in zinc refining methods have resulted in decreases of emissions of 73% to air and 83% to water during the period 1985–1995 (Royal Belgian Federation of Non-Ferrous Metals,

1995). These data are confirmed by the results of the US Toxics Release Inventory during the period 1988–1993.

Additionally, the use of zinc-containing chemical fertilizers and pesticides in agriculture, the application of sewage sludge and manure to fields, and the disposal of zinc-bearing waste may increase zinc concentrations in soil (US EPA, 1980; Cleven et al., 1993).

3.2.3.1 Emissions to atmosphere

During mining, atmospheric zinc loss is estimated to be 100 g per tonne of zinc mined, mostly from handling ores and concentrates and from wind erosion of tailing piles (Lloyd & Showak, 1984). From stationary sources, average emissions of zinc to the atmosphere of 151 000 tonnes/year are reported for 1969–1971 (Fishbein, 1981). Based on emission studies in western Europe, USA, Canada and the former Soviet Union, total worldwide zinc emissions to air were estimated to be in the range 70 250–193 500 tonnes in 1983. Emissions from the non-ferrous metal industry account for the largest fraction of zinc emitted (50–70%). Cement production accounted for 1780–17 800 tonnes/year and the use of phosphate fertilizers was stated to contribute 1370–6850 tonnes/year. Additionally, 1724–4783 tonnes/year were attributed to emissions from miscellaneous sources (Nriagu & Pacyna, 1988).

The above estimates are not generally descriptive of emissions from modern zinc production techniques. Emission factors for industrial point sources have decreased significantly since the 1970s. For pyrometallurgical zinc production, Nriagu & Pacyna (1988) used an emission factor of 100–180 kg of zinc for each tonne of metal produced. Currently, emission factors for releases to air from pyrometallurgical processes do not exceed 0.7 kg per tonne of metal produced in western Europe (EZI, 1996). In addition, industrial production patterns were erroneously estimated at the time these estimates were made. Nriagu & Pacyna (1988) estimated that total world pyrometallurgical zinc production was 4.6×10^6 tonnes in 1983-1984. Given a total of 6.25×10^6 tonnes of global zinc production, including about 1×10^6 tonnes of zinc from new scrap recycling for that period (ILZSG, 1995), 74% of zinc production was considered to be by electrolytic processes. However, 80% of western world zinc production in 1984 was by electrolytic refining and only 20% by non-electrolytic processes (ILZSG, 1994). Even if it is assumed that all zinc production elsewhere (1.5×10^6 tonnes) was by pyrometallurgical technology, the proportion of zinc produced by this technology could not have exceeded 2.4×10^6 tonnes, or 39%.

By 1993, 83% of zinc was being produced by electrolytic techniques (ILZSG, 1994). The emissions to air from hydrometal-lurgical zinc production processes are currently 4–400 g of zinc per tonne produced (EZI, 1996). Taking present-day emission factors and production methods into account, total zinc emissions to air from zinc production are likely to be about 2000 tonnes/year. This contrasts with the estimates of 70 250–193 500 tonnes (for 1983–1984) made by Nriagu & Pacyna (1988), but is in good agreement with the data on controlled emissions in Europe (OSPARCOM, 1994).

Controlled emissions from point sources to air from the German non-ferrous industry were 16.2 tonnes/year in 1993-1994. Zinc emissions of 47.7 tonnes were reported for the French zinc and lead industry in 1991. For the Netherlands, annual emissions of zinc oxide from zinc production in 1990 amounted to 24 tonnes/year. From zinc production, zinc emissions of 58 tonnes were reported for the United Kingdom in 1990 and 6 tonnes for Spain in 1992. For a combined zinc-copper-lead plant in Sweden, zinc emissions to air amounted to 33 tonnes in 1990 (OSPARCOM, 1994). In the USA, industry data for stack/point source emissions indicated a release of 387 tonnes in 1994 with fugitive emissions of 377 tonnes (TRI, 1995). Emissions of zinc from all industrial sources in Canada in 1983 were 1410 tonnes compared to 151 000 tonnes in the period 1969–1971 (NPRI, 1994).

The reduction of atmospheric zinc emissions for European countries near the North Sea over the time period 1985–1995 is summarized in Table 8.

3.2.3.2   Emissions to aquatic environment

Anthropogenic inputs of zinc from mining and manufacturing processes (production of zinc, iron, chemicals, pulp and paper, and petroleum products) into aquatic ecosystems are given as 33 000–178 000 tonnes/year. A further 15 000–81 000 tonnes/year

Table 8. Reduction of zinc emissions to air and surface waters in European
countries in the period 1985–1995
after the NORTHSEA Conference (OSPARCOM, 1994) [a]

Country	Reduction of zinc emissions (%)	
	To air	To water
Belgium	25[b]	5 [b]
Germany	70	no data
Netherlands	25	25
Norway	75	37
Sweden	50	70
Switzerland	5	35
United Kingdom	5	no data

[a] All data are official country data. It must be emphasized that in some countries strong reductions in emissions took place before the reference period 1985–1995. For example, in Belgium the reduction of zinc emissions to water from the non-ferrous metal industry during the period 1980–1985 was 65% (Royal Belgian Federation of Non-Ferrous Metals, 1995).

[b] 1995 figures for Belgium: to air, reduction of 18%; to water, reduction of 32% (VMM, 1996).

originate from domestic waste water, 21 000–58 000 tonnes/year from atmospheric fallout, and 2600–31 000 tonnes/year from the dumping of sewage sludge. Total worldwide input was estimated to be 77 000–375 000 tonnes/year (Nriagu & Pacyna, 1988). In this study, the emission factors for non-ferrous metals smelting and refining were 300–3000 g of zinc per tonne of metal produced. In current zinc production, emission factors are 0.1–50 g of zinc per tonne of metal produced (EZI, 1996). US EPA (1980) calculated that urban runoff accounts for approximately 5200 tonnes/year, and drainage from inactive mines for 4060 tonnes/year. The German chemical industry and the Rotterdam harbour agreed to reduce the annual zinc input into the river Rhine and its tributaries from 270 tonnes in 1995 to 100 tonnes in 2000 (VDI-Nachrichten, 1995).

For the North Sea, a total input of 28 000 tonnes of zinc was estimated for 1987 (Kersten et al., 1988) compared to 15 190 tonnes in 1990 (riverine input, 6900 tonnes; dredgings and direct discharge, 5000 tonnes; atmospheric input, ≥ 2700 tonnes; industrial waste, 440 tonnes; and sewage sludge, 150 tonnes) (UBA, 1994). For the Baltic Sea, the following inputs were estimated for 1987: municipal input, 460 tonnes; rivers, 6709 tonnes; industrial, 1765 tonnes; and atmosphere, 3200 tonnes (UBA, 1992). For German rivers, an input of around 18 tonnes to the Baltic Sea was estimated for 1990 (UBA, 1994). A marked reduction of industrial point source emissions has been observed in western Europe during the last decade, resulting in a substantial decrease of the concentrations in surface water (Van Assche, 1995). Significant reductions are also evident from the US TRI database; discharges to surface waters were 386 tonnes in 1988 and only 30 tonnes in 1993 (TRI, 1993).

The recent general tendency for reduction of zinc emissions to the water, is illustrated by data reported by European countries bordering or close to the North Sea (see Table 8). This general reduction is also reflected in the decrease of zinc deposited in Greenland snow samples after the 1960s (Boutron et al., 1995).

3.2.3.3 Emissions to soil

On an annual basis, an estimated 1–3 million tonnes of zinc from mining and smelter operations are discharged on land worldwide. An additional $689–2054 \times 10^3$ tonnes/year are released to soil from anthropogenic activities: 260–1100 tonnes/year originate from the use of fertilizers and 49 000–135 000 tonnes/year from atmospheric fallout. However, a further significant source of zinc emissions to soil is represented by zinc-containing wastes, such as agricultural and animal wastes, manure, sewage sludge and fly ash, which contribute $640–1914 \times 10^3$ tonnes/year (Nriagu & Pacyna, 1988). On the basis of an average zinc concentration of 60–470 mg/kg in chemical phosphate fertilizers and < 5 mg/kg in non-phosphate fertilizers, and the consumption of commercial fertilizers, the total zinc input into soil from these fertilizers was 745 tonnes in Germany in 1989. The zinc content of manure is given as 12.6–39 mg/kg (UBA, 1992). In Australia, annual consumption of zinc in fertilizers ranged between 900 and 1700 tonnes (Mortvedt & Gilkes, 1993). In the USA, zinc in fertilizer increased from 13 100 tonnes in 1967-1968 to 37 300 tonnes in 1984 (Mortvedt & Gilkes, 1993).

Nriagu & Pacyna (1988) estimated that zinc is discharged on land worldwide during mining and smelting operations at a rate of approximately $310–620 \times 10^3$ tonnes/year in smelter in slags and wastes, and $194–620 \times 10^3$ tonnes/year in mine tailings. The vast majority of such discharges are non-dispersive and occur within the mine or smelter site. Its physical and chemical properties, and the lack of availability of the zinc make it difficult to envisage a large global impact of this material on the environment.

3.2.4 Emissions during combustion of coal and oil, and refuse incineration

Zinc concentrations in oil and coal average 0.25 mg/kg and 50 mg/kg, respectively (Bertine & Goldberg, 1971). On the basis of these data, the global emissions from oil and coal combustion to air were calculated by Lantzy & Mackenzie (1979) to average 140 000 tonnes/year. For 1983, the releases of zinc to atmosphere due to coal and oil combustion were calculated to be 2570–19 630 tonnes/year and 532–3786 tonnes/year, respectively. Estimated emissions from refuse incineration are in the range 2950–8850 tonnes/year (Nriagu & Pacyna, 1988).

3.2.5 Zinc releases from diffuse sources

Several categories of diffuse emissions can be relevant in terms of total environmental input of zinc: zinc wash-off from metallic zinc surfaces exposed to atmospheric conditions (sacrificial zinc corrosion), household emissions, emissions from agricultural practice (see section 3.2.3.3) and traffic, and atmospheric emissions (see section 3.2.3.1).

3.2.5.1 Releases from atmospheric zinc corrosion

In air, acidifying factors, such as sulfur dioxide, nitric oxides and chlorides attack the zinc hydroxide-carbonate layer on the surface of metallic zinc yielding soluble zinc compounds. Sulfur dioxide levels in ambient air are particularly important in this respect. Chloride levels are significant but only at distances smaller than 1.5 km from the seaside (Porter, 1995). Zinc is washed off slowly and forms a diffuse source of zinc release to the environment. Corrosion is increased at pH levels of rain of < 4, corrosion at pH 4–7 amounts to less than 1 μm/year but increases six-fold at pH 3. It has been demonstrated that atmospheric corrosion is strongly and linearly related to the sulfur dioxide levels in ambient air (Knotknova & Porter, 1994). An empirical formula for the reduction of the thickness of zinc layers is:

Rate of zinc corrosion (μm/year) = 0.29 + 0.039 × [SO₂] ([SO₂] in μg/m³ in air)

Since the 1970s, ambient air sulfur dioxide levels have markedly decreased (Iversen et al., 1991). As a consequence, corresponding zinc corrosion rates have also decreased. In Stockholm, for example, ambient air sulfur dioxide levels and experimental zinc corrosion rates have decreased concomitantly by 94% and 73%, respectively (Knotknova & Porter, 1994).

The annual removal of zinc from exposed metal is estimated to be 3.6 μm in rural air, 3.8 μm in urban air, 4.3 μm in industrial air, and 4.5 μm in sea air (Boettcher, 1995). For European countries, annual corrosion rates are estimated to be < 8 g/m², 8–16 g/m², and 16–28 g/m² for rural, urban and industrialized areas, respectively (Van Assche, 1995). A study by Knotkova et al. (1995) indicates that corrosion rates in Europe are now about 1.1 μm/year, corresponding to a potential zinc wash-off of about 8 g/year per m² of exposed zinc surface. The highest corrosion rate reported by Knotkova et al. (1995) was 2.2 μm/year in an industrial site (< 16 g/m² of zinc surface). Similar corrosion rates have been observed in North America (Spence & McHenry, 1994).

3.2.5.2 Releases from sacrificial zinc anodes

In order to protect steel structures from corrosion in the marine environment and in soils, sacrificial zinc anodes are used, resulting in a slow release of zinc to the environment. Current releases to the marine environment from European countries bordering or close to the North Sea are estimated at 1900 tonnes/year (OSPARCOM, 1994).

3.2.5.3 Household zinc emissions

Some household zinc emissions are of natural origin, e.g., background levels in tap water and foodstuffs. Others are of anthropogenic origin: from galvanized water pipes, cosmetics, pharmaceuticals, etc. In the Netherlands, the zinc load from households was estimated to be 8.1 g per person per year, of which 53% originated from food consumption (estimated from faeces), 25% from drinking-water, and 22% from "consumer products" (Coppoolse et al., 1993).

4. ENVIRONMENTAL TRANSPORT, DISTRIBUTION AND TRANSFORMATION

4.1 Transport and distribution between media

4.1.1 Air

Zinc in the atmosphere is primarily in the oxidized form in aerosols (Nriagu & Davidson, 1980). Zinc is found on a particles of various sizes, the size being determined by the source of zinc emission. Waste incinerators release small zinc-containing particles to the atmosphere, whereas wear of vehicle tyres produces large particles (Sohn et al., 1989).

Zinc in urban and industrial areas, including metallurgical plants and brass/zinc production facilities, was present on particles with diameters of up to 5 μm (Nriagu & Davidson, 1980). Dorn et al. (1976) reported that 73% of the atmospheric zinc sampled from a farm near a lead smelter was in the form of particles smaller than 4.7 μm (the upper limit of respirable particles), compared to 54% on a farm not affected by the smelter. Zinc has been reported to be adsorbed to even larger particles from windblown soil and road dust. Zinc bound to soil particulates may be transported to the atmosphere as wind-blown dust (Perwak et al., 1980).

Anderson et al. (1988) examined atmospheric aerosol particles collected from Arizona. The aerosols originated from the nearby urban area, the surrounding desert and several major copper smelters, which were 120 km from the sampling area. The particles containing zinc were divided into five groups: zinc sulfide, ferrous zinc, zinc phosphide, zinc chloride and metallic zinc. The authors suggested that the zinc sulfide particles originated from the copper smelters, and that the zinc phosphide particles may have been emitted during spray-painting of primer on steel, possibly from a construction site.

The proportion of zinc on atmospheric particulate matter collected from a rural area that was in water-soluble form ranged from 12% to 48%, with a mean value of 26% (Lum et al., 1987). The proportion of zinc in the dissolved fraction of rainwater collected from Rhode Island, USA ranged from 52% to 100% (Heaton et al., 1990). Colin et al. (1990) reported zinc in rainwater collected in France. The geometric mean was 78 µg/litre (1.20 µM) for total zinc and 3.25 µg/litre (0.05 µM) for insoluble zinc.

Zinc particles in the atmosphere are transported to soil and water by wet and dry deposition. These processes are dependent upon particle size. Pacyna et al. (1989) derived a model which demonstrated that zinc adsorbed on particles of low diameter and density can be transported through the atmosphere to regions in Norway distant from their source in Central Europe. The dry deposition velocity for zinc was calculated to be 0.5 cm/s. Analysis of Greenland snow samples shows a significant decrease in the atmospheric zinc deposition over time. Between 1967 and 1989, the level decreased by a factor 2.5 (Boutron, 1991). More extensive studies have shown a five-fold increase of zinc deposition in Greenland snow layers in the period after the industrial revolution (from 1800 onwards), with a maximum during the 1960s followed by a significant decrease of 40% between 1960 and 1990 (Boutron et al., 1995).

The deposition rate of airborne zinc downwind of an abandoned metalliferous mine complex was reported to range from 3.10 ± 1.30 µg/cm^2 per month at a site 10 m from the edge of the spoil tip, to 0.61 ± 0.14 µg/cm^2 per month at a site 1000 m from the edge of the spoil tip (Roberts & Johnson, 1978).

Teraoka (1989) reported zinc concentrations in dry atmospheric fallout sampled in Japan to range from 290 to 790 mg/kg of ashed sample. Concentrations in bulk precipitations were 25–67 µg/litre.

Dasch & Wolff (1989) reported zinc concentrations in rain from Massachusetts, USA. The mean concentration was calculated to be 3.7 ± 0.8 µg/litre. The enrichment factor (the degree of enrichment of an element in the atmosphere compared to the relative abundance of that element in crustal material) was calculated to be 110 ± 78. Enrichment factors have to be calculated and interpreted with care. The use of simple enrichment ratios in a sample relative to average crustal abundance does not take into account the fact that organic and inorganic enrichment processes cause trace metal levels to shift by orders of magnitude (GSC, 1995). The source of zinc is therefore not entirely due to crustal material. Similar conclusions were derived following analysis of atmospheric particles collected from near sea level in the North Atlantic, with reported concentrations of 0.3–27 ng/m^3 (Duce et al., 1975).

4.1.2 *Water and sediment*

Zinc in water can be divided into seven classes (Florence, 1980):

- particulate matter (diameter > 450 nm)
- simple hydrated metal ion, e.g., $Zn(H_2O)_6^{2+}$ (diameter 0.8 nm)
- simple inorganic complexes, e.g., $Zn(H_2O)_5Cl^+$, $Zn(H_2O)_5OH^+$ (diameter 1 nm)
- simple organic complexes, e.g., Zn-citrate, Zn-glycinate (diameter 1–2 nm)
- stable inorganic complexes, e.g., ZnS, $ZnCO_3$, Zn_2SiO_4 (diameter 1–2 nm)
- stable organic complexes, e.g., Zn-humate, Zn-cysteinate (diameter 2–4 nm)
- adsorbed on inorganic colloids, e.g., $Zn^{2+}Fe_2O_3$, $Zn^{2+}SiO_2$ (diameter 100–500 nm)
- adsorbed on organic colloids, e.g., Zn^{2+}-humic acid, Zn^{2+}-organic detritus (diameter 100–500 nm).

Zinc compounds hydrolyse in solution to produce hydrated zinc ions, zinc hydroxide and hydrated zinc oxides, which may precipitate. These reactions decrease the pH of the water, although the natural buffering capacity of the water usually prevents any significant change (US DHHS, 1994).

Zinc is adsorbed strongly by ferric hydroxide in alkaline waters (Gadde & Laitinen, 1974). Zinc has also been reported to be adsorbed on sulfides (Hem, 1972), silica (Huang et al., 1977), alumina (Huang et al., 1977), manganese dioxide (Doshi et al., 1973), and humic acid (Guy & Chakrabarti, 1976).

The stability constant (log_k) for zinc-fulvic acid complexes in lake water was reported to be 5.14 (Mantoura & Riley, 1975).

Farrah & Pickering (1976) determined the adsorption of zinc to three clay minerals, kaolinite, illite and montmorillonite. The adsorption capacity of the clays increased between pH 3.5 and pH 6.5. Under alkaline conditions, zinc precipitated as hydroxy species, which adsorbed to the clay. At pH > 10.5 zinc returned to solution as the zincate, although such a high pH is unlikely to exist in the environment. The attachment of the hydroxy species was reported to be the controlling process for kaolinite and illite. The dominant controlling mechanism in montmorillonite was ion exchange at the negative lattice sites.

Zinc sulfide is the most dominant form of zinc in anoxic sediments (Casas & Crecelius, 1994). Only the uppermost sediments are oxic, and here zinc will primarily be associated with hydrous oxides of iron and manganese as components of the clay fraction or as coatings on the surface of other minerals (US National Academy of Sciences, 1977). In waters, zinc forms complexes with a variety of organic and inorganic ligands (Callahan et al., 1979; US EPA, 1984). Up to 50% of the total zinc in acidic fresh waters is in a non-colloidal inorganic form, such as zinc carbonate, zinc hydroxy carbonate or zinc silicate. In alkaline fresh waters, most bound zinc is adsorbed to organic and inorganic colloidal particles.

Hydroxides and hydrous ions of iron and manganese are components of the clay fraction of sediments and they also exist as coatings on the surface of other minerals (US National Academy of Sciences, 1977). When these hydrous oxides are oxidized they may co-precipitate with zinc. As the precipitates form, they trap various ions in their crystal lattice (Callahan et al., 1979).

Zinc is not directly affected by changes in redox potential (E_h), although the valencies and reactivities of the ligands that react with zinc are (Callahan et al., 1979).

4.1.2.1   Fresh water

The pH of most fresh waters is in the range that is critical for the adsorption of heavy metals on particulates. A change in pH of 0.5 can mean the difference between the majority of zinc being in an adsorbed or desorbed form. Florence (1977) reported that zinc in several fresh waters at pH 6.0–6.1 was distributed between labile ionic species and a stable inorganic form. The amount of zinc bound to organic colloids was minor.

Elevated zinc concentrations were reported in water sampled from areas rich in ores (White & Driscoll, 1987). Organic material has an important role in the binding of zinc in fresh water, particularly at high pH values (> 6.5). Spatial and temporal variations in the zinc concentration were reported to be minor. Peak concentrations were reported during snowmelt, but were limited to meltwater in streams and at the lake surface. Zinc did not appear to be retained in the lake. Transport of particulate-bound zinc to sediment represented a minor flux. The authors suggested that long- and short-term variations in retention of zinc in the lake due to surface water acidification may complicate quantitative interpretation of zinc deposition in sediments.

4.1.2.2   Seawater

The chemical pathway of zinc is mainly determined by interactions with dissolved organic complexing agents (Van den Berg et al., 1987). The dissolved zinc concentration throughout the Scheldt estuary (Netherlands) was reported to vary according to the dissolved organic concentration. The proportion of dissolved zinc determined to be in a labile form was 34–69%, owing to the low solubility of iron and competition for dissolved copper and zinc with organic complexing ligands. The concentration of these ligands was calculated to be in the range 1.43–14.3 µg/litre (22–220 nM). The conditional stability constants (log_k values) of the zinc complexing ligands were calculated to be 8.6–10.6. The average product of ligand concentrations and conditional stability constants (α coefficient) was 6×10^2.

Increases in the dissolved and suspended fractions of zinc in estuarine water were reported in the mixing zone between fresh and brackish waters. The increases were attributed to the increased residence time of zinc in the estuary compared to that in the fresh water. There was a five-fold increase in the amounts of leachable zinc in sediments sampled from brackish waters compared with those in sediments from fresh waters. The ratio of suspended zinc to leachable zinc was increased from 20% in fresh waters to 86% in brackish waters (Grieve & Fletcher, 1977).

4.1.2.3   Wastewater

Patterson et al. (1977) demonstrated that zinc hydroxide precipitates at a faster rate from industrial wastewaters than zinc carbonate. The minimum soluble zinc concentration reported was 0.25 mg/litre at pH 9.5. Treatment of the effluent with carbonate increased only the amount of dissolved solids. Optimization of the zinc carbonate system with the use of denser sludges or better filtration methods provided no advantage over the zinc hydroxide system.

Rudd et al. (1988) studied the forms of zinc in sewage sludges during chemical extraction and progressive acidification treatment stages (pH values 4.0, 2.0 and 0.5). Fractionation profiles of samples from sequential extraction demonstrated that the majority of zinc was associated with the tetrasodium pyrophosphate ($Na_4P_2O_7$) fraction, comprising 18–52% of the total zinc content. This fraction corresponds to organic and some insoluble inorganic forms. The remaining zinc was evenly distributed between the ethylene diamine tetra-acetic acid (EDTA) and nitric acid fractions. The potassium fluoride (KF) fraction accounted for 2–13% of the total zinc, with less in the potassium nitrate (KNO_3) fraction. The threshold for mobilization of zinc was reported to approach pH 6.0. The majority of mobilizable metal was extracted at pH 2.0, with only slight increases in the amount released at pH 0.5. Zinc was more easily extracted from raw sludges than from dried forms of activated and digested sludges. The threshold for mobilization from liquid sludge samples was pH 4.0. Acidification of the sludge increased the proportion of zinc in an easily extractable form, e.g., from the predominant $Na_4P_2O_7$ fraction to KNO_3-extractable and KF-extractable forms at pH 0.5.

4.1.2.4   Groundwater

Zinc solubility in groundwater increases with redox potential (E_h) value (Hermann & Neumann-Mahlkau, 1985; Pedroli et al., 1990). The solubility also increases with decreasing pH (Pedroli et al., 1990).

4.1.2.5 Sediment

There are two forms of sediment: suspended sediment and bed sediment. Zinc and other heavy metals are highly partitioned to suspended sediment in the water column. Trefry & Presley (1971) calculated that 90% of the zinc was carried in the particulate phase in a clean stretch of the Mississippi River; 40% was reported for a contaminated river (Kopp & Kroner, 1968). Golimowski et al. (1990) reported ranges in distribution constant K_d (the ratio of chemical concentration in the solid phase to the concentration in the liquid phase) for three rivers in the Netherlands: 10 000–145 000 (Rhine), 10 000–190 000 (Waal) and 75 000–230 000 (Meuse).

Phosphates and iron hydroxides play an important role in the transfer of heavy metals from river water to sediments (Houba et al., 1983). Deposition to the bottom of a water body occurs concurrently with a change in the microenvironment. Organic matter reaching the bedded sediment is oxidized. Because oxygen and nitrate are limited, sulfate is the most prevalent terminal electron receptor. Thus sediments tend to be sulfide-rich. Sulfide reacts with transition metals such as zinc to form metal sulfide compounds of low solubility (Allen et al., 1993).

As bedded sediments change from a reduced to an oxidized state, greater amounts of zinc are mobilized and released in soluble forms (US EPA, 1987). The pH controls the interaction of zinc with dissolved organic carbon, a process which determines the bioavailability of zinc (Bourg & Darmendrail, 1992). Compared with other physical processes, diffusive transport of zinc to and from the sediment pore water is negligible.

Sprenger et al. (1987) recorded the zinc concentrations in water and sediments sampled from six acidic lakes in New Jersey, USA. Increased zinc concentrations were reported in the most acidic lakes. The active growth of macrophytes in one of the lakes resulted in sediment with a high organic matter content, with the subsequent retention of zinc.

In estuaries, desorption of zinc from sediments occurs with increasing salinity (Helz et al., 1975) owing to the displacement of adsorbed zinc ions by alkali and alkaline earth cations, which are abundant in brackish and saline waters (Callahan et al., 1979).

4.1.3 Soil

The major sources of zinc in soils are the zinc sulfide minerals, such as sphalerite and wurtzite, and to a lesser extent minerals such as smithsonites ($ZnCO_3$), willemite (Zn_2SiO_4), zincite (ZnO), zinkosite ($ZnSO_4$) franklinite ($ZnFe_2O_4$) and hopeite ($Zn_3(PO_4)_2 \cdot 4H_2O$).

Zinc in soil is distributed between the following fractions (Viets, 1962):

- dissolved in soil water
- exchangeably bound to soil particles
- bound to organic ligands
- occluded in secondary clay minerals and metal oxides/hydroxides
- present in primary minerals.

Only those fractions of zinc that are soluble or may be solubilized are available to plants (Brümmer, 1986). Zinc undergoes reactions involving precipitation/dissolution, complexation/dissociation and adsorption/desorption. These reactions and the resulting bioavailability of zinc will be controlled by the pH and redox potential of the soil, the concentration of zinc ions and other ions in the soil solution, the nature and number of adsorption sites associated with the solid phase of the soil, and the concentration of ligands capable of forming organo-zinc complexes (Kiekens, 1995).

Under most conditions, the amount of zinc present in adsorbed soil fractions is much higher than the soluble fraction that remains in the pore waters or soil solution. A change in any of the above factors will result in a change in the overall equilibrium of the soil, with zinc transformed to different forms until a new equilibrium is reached. Such equilibrium displacements may occur as a result of plant uptake, losses by leaching, zinc input, changes in soil moisture content, changes in pH, mineralization of organic matter, and changes in the redox status of the soil. The proportion of zinc in soil solution increases with decreasing pH. In high pH soils (> 6.5), the chemistry of zinc is dominated by interactions with organic ligands.

Zinc forms complexes with chloride, phosphate, nitrate and sulfate. The complexes with sulfate and phosphate are the most important with regard to total zinc in solution. Under neutral or alkaline conditions, $ZnHPO_4$ contributes to zinc in solution, although this depends on phosphate activity (Kiekens, 1995). The formation of carbonates is also possible (Misra & Tiwari, 1966), and is probably an important factor in explaining some of the retention of zinc at high pH values. Slow diffusion of zinc into soil reduces the mobility and bioavailability of zinc (Brümmer, 1986).

Humic and fulvic acids are important for the speciation of zinc in soil and aquatic systems. For example, 60–75% of zinc in soil solution has been reported to be bound by fulvates (Hodgson et al., 1966; Geering & Hodgson, 1969). These acids are defined by solubility. Because fulvic acid is soluble, its chelates are mobile in the soil. Stability constants for zinc fulvates and humates have been reported by a number of investigators (Courpron, 1967; Schnitzer & Skinner, 1967; Stevenson, 1991); they are dependent on pH. Because the acids are mixtures, not pure chemicals, the stability constants are averages representing the extent of metal or proton binding over the limited range of concentrations for the titration. Adequate descriptions of the metal binding characteristics of these heterogeneous organic substances can be achieved using models incorporating a number of discrete binding sites (Tipping, 1993) or a continuum of binding sites of varying pK (Perdue & Lytle, 1983).

The selective adsorption of zinc and the occurrence of an adsorption/desorption hysteresis effect is controlled by the following parameters (Kiekens, 1995):

- number of pH-dependent adsorption sites
- interactions with amorphous hydroxides
- affinity for the formation of organomineral complexes, and their stability
- formation of hydroxy complexes
- steric factors
- properties of zinc including: ionic radius, polarizability, thickness of the hydration sheet, equivalent conductance, hydration enthalpy and entropy.

The observed hysteresis effect may have important practical consequences and applications. Addition of soil additives such as lime (calcite), zeolite, hydroxyapatite, vermiculite, bentonite, beringite (a modified clay) and other clay minerals, and other products, such as selective cation exchangers (e.g., polystyrene resins), steel shots, Thomas basic slags and hydrous manganese oxide, can reduce the mobility of zinc and uptake of plants cultivated in a contaminated soil (Van Assche & Jansen, 1978; Kiekens, 1986; Vangronsveld et al., 1990, 1995a,b; Vangronsveld & Clijsters, 1992; Mench et al., 1994).

Soils high in clay or organic matter have higher zinc adsorption capacities than sandy soils with a low organic content (Shuman, 1975). A further reduction in zinc adsorption capacity of sandy soils, compared to soils with a high colloidal-size material content, was reported at low pH. Zinc accumulated in the organic horizon (organic matter layer) of sandy soil, with low concentrations in the mineral horizons (mineral layers) (Pedroli et al., 1990).

The mobility of zinc in soil increases at low soil pH under oxidizing conditions and at a lower cation exchange capacity of soil (Tyler & McBride, 1982: Hermann & Neumann-Mahlkau, 1985). The dominant species under anaerobic conditions is zinc sulfide, which is insoluble and so the mobility of zinc in anaerobic soils is low (Kalbasi et al., 1978; Perwak et al., 1980).

Zinc can be readily displaced by calcium, which can be abundant in the soil solution (Van Bladel et al., 1988). There is greater potential for leaching of zinc in light acidic soils, compared to soils with a high organic matter or calcium carbonate content. MacLean (1974) studied the factors that determined the extractability of zinc with diethylenetriamine-penta-acetic acid (DTPA), magnesium chloride or calcium chloride in soils incubated with zinc solutions. The amount of zinc extracted increased with increasing rates of added zinc and increasing amounts of added phosphorus. Extractable zinc was negatively correlated to the soil organic matter content. Liming reduced the amount of extractable zinc in an acid soil. Pretreatment of the soil with phosphate fertilizer also increased the amount of zinc extracted.

The distribution constant for zinc between soil and water (K_d) has been reported to vary from 0.1 to 8000 litres/kg (Baes & Sharp,

1983). Baes et al. (1984) reported an average K_d value of 40 litres/kg. Anderson & Christensen (1988) reported a range of K_d values for zinc of 1–3540 litres/kg. The K_d value was strongly related to pH, although the presence of extractable manganese oxides and hydroxides and the magnitude of the anion exchange capacity were also important. Gerritse et al. (1982) reported K_d values in a variety of soil types and sewage sludges: 70–100 litre/g for sandy loam soil; 2.1 and 3.2 litre/g for organic soil; 0.2–4 litre/g for sandy soils; 60–90 litre/g for sewage sludge; and 3–4 litre/g for sewage sludge after aeration. Bunzl & Schimmack (1989) calculated the K_d value for zinc in the organic and mineral horizon of podzol forest soil. The median values were 14 and 41 litre/kg , respectively. The organic-horizon K_d values were not significantly correlated with pH.

Zinc supplementation of soils is achieved using sewage sludge or chemical fertilizers. Sanders & Adams (1987) added sewage sludge to a clay loam and two sandy loam soils. The concentration of extractable zinc increased rapidly at pH values below a threshold of 6.2–7.0, with less being extracted from the clay soil than from the sandy soils. Sanders & El Kherbawy (1987) determined zinc adsorption equilibria in United Kingdom soils that had similar textures and zinc concentrations but different pH values. Zinc was added to the soils in the form of zinc nitrate or sewage sludge. There were no differences in the results obtained with the different zinc treatments.

Mehrotra et al. (1989) studied the speciation of zinc in primary, secondary, digested and zinc-spiked sewage sludge. They found that 50% of zinc was organically bound and there were no differences in the zinc speciation or zinc loading of the different sludge types. Zinc added to the sludge is redistributed in a similar fashion to existing zinc. It was concluded that the distribution pattern remains more or less the same whether zinc is added during or after digestion. However, other studies have reported that zinc added to sludges after digestion is more readily bioavailable than zinc added prior to digestion. Bloomfield & McGrath (1982) determined the levels of extractable zinc in sludges to which zinc sulfate had been added either prior to or following anaerobic digestion. All three extractants used (NH_4OAc, HOAc and EDTA) removed zinc adsorbed on pre-digested sludge more readily than those incorporated during the digestion process. Davis & Carlton-Smith (1981) reported increased extractability of zinc in sewage sludges amended with soluble zinc salts compared to those to which insoluble zinc sulfide was added. Speciation of heavy metals in sewage sludge and sludge-amended soils has been reviewed by Lake et al. (1984).

Williams et al. (1984) determined the fate of zinc in soil amended annually for 6 years with sewage sludges, one of which contained industrial waste. Zinc moved up to 10 cm below the area of sludge incorporation. The ratios of DTPA-extracted zinc to nitric acid-extracted zinc were similar over the last 4 years and at all soil depths. The availability of zinc was highest in the soils with the lowest pH.

4.2 Bioavailability

The bioavailable fraction (a physicochemical term) is the maximum fraction of the total zinc concentration that can potentially be taken up by organisms, essentially over and above very stable forms of zinc. Uptake (a biological-physiological term) refers to the fraction that is actually taken up by organisms. The term "bioavailability" is used to describe the interaction in nature of physicochemical properties and physiological factors. For instance, zinc in the aquatic environment interacts with binding agents in the aqueous phase and similarly with biological receptors.

Knowledge of the bioavailable fraction is a critical requirement for risk assessment. Total concentrations in the aquatic and soil environments alone, including food, are not useful for estimating bioavailability.

4.2.1 Factors affecting bioavailability

The most important physicochemical factors affecting bioavailability are: pH, dissolved organic carbon (DOC), water hardness, competing ions, soluble ligands, and binding sites on solid phases (e.g., metal oxides in suspended matter, sulfides in sediments and anaerobic soils) (Florence & Batley, 1980).

The most important physiological factors affecting bioavailability are: adsorption sites at the cell wall (type and quantity), exudation of organic substances, protons, and gaseous substances (e.g., oxygen, carbon dioxide) (Cakmak & Marschner, 1988; Bergman & Dorward-King, 1996).

The above physicochemical and physiological factors apply both to aquatic and to terrestrial ecosystems. Bioavailability is determined on the basis of a combination of these factors as well as the kinetics of the chemical and biological processes concerned.

4.2.2 Techniques for estimation

There are currently two different approaches for estimating bioavailability: correlative (e.g., extractable metals in the terrestrial environment, free metal ion in the aquatic environment; Campbell, 1995); and predictive, which models bioavailability, for instance at gill surfaces (Bergman & Dorward-King, 1996). These approaches need to be further developed and validated against bioassays. In particular, improved analytical techniques are required to measure zinc speciation in environmental and biological compartments (the latter related to human health). Extraction techniques are better developed in some areas than in others, for instance they are particularly well developed to measure zinc deficiencies in agricultural soils (relative to crop production) (Brennan et al., 1993). Leach tests, although they may provide a prediction of potential environmental risk, do not accurately measure the bioavailable fraction.

4.3 Biotransformation

4.3.1 Biodegradation

Zinc is an element and therefore cannot be biodegraded, in contrast to zinc compounds. Some studies have examined microbial or abiotic transformations of zinc compounds which can result in a change in zinc speciation (Touvinen, 1988).

Biomethylation of zinc has not been observed.

Biological degradation of zinc complexes is necessary for the normal functioning of ecosystems to enable the recycling of zinc from litter, faeces and dead organisms. In certain environments, bacteria as well as fungi are able to oxidize zinc sulfide in ores, producing zinc sulfate which can be leached into solution (Ilyaletdinov et al., 1977; Tuovinen, 1988).

4.3.2 Bioaccumulation

The concept of bioaccumulation was originally designed to determine the accumulation of a substance/element in biota in comparison to its occurrence in an environmental compartment, i.e., water, soil or sediment. The ratio between the concentration of a substance/element in biota and that in an environmental compartment was defined as the bioconcentration factor (BCF). For example, for uptake from water the BCF is a unitless value calculated by dividing the "steady state" wet tissue concentration of a particular substance by its "steady state" water concentration. Bioaccumulation factors (BAFs) differ from BCFs in that they assume uptake from water and accumulation from the diet.

In the case of zinc, the BCF is not useful for relating uptake to adverse effects, because it does not consider physiological parameters (Canada/EU, 1996; Chapman et al., 1996). The fact that zinc, as an essential metal, is naturally concentrated by living organisms means that the BCF for zinc bears no relationship to toxicity. Bioaccumulation does not differentiate between zinc adsorbed to the outer surface of organisms, and the zinc within organisms. Rapid bio-inactivation of zinc, for instance compartmentation into vacuoles, may result in elevated BCFs with no difference in the health of the organism (Mathys, 1977). Further, the fact that many organisms are capable of regulating internal zinc concentrations within certain limits means that these organisms can stabilize internal concentrations against perturbations or high concentrations in the external environment. Thus, zinc tissue concentrations do not necessarily reflect ambient concentrations and, in contrast to lipophilic organic compounds, zinc BCFs cannot be considered to be constant ratios between tissue concentrations and external water concentrations. Finally, an inverse relationship has been observed in many biological organisms between the BCF and external water concentrations.

Accumulation of zinc to meet physiological requirements can be mistaken for trophic transfer. However, zinc is not biomagnified (Beyer, 1986; Suedel et al., 1994).

57

4.3.2.1 *Aquatic organisms*

In aquatic environments, organisms tend to have a high surface:volume ratio, which is necessary for exchange processes (oxygen, carbon dioxide, nutrients). Exchange processes are enhanced by the enlargement of the receptive tissues (e.g., gills in fish and some benthic organisms, and soft body surfaces in some benthic organisms) and/or by enhancing water passage through the organism and/or its tissues. The effect on the health of an organism of adsorbed zinc is different to that of incorporated intracellular zinc; however, in most experiments and sampling procedures the impact of adsorption is not considered. Thus, for instance, gelatinous algae such as *Chlamydomonas* spp. and *Gloeococcus* spp. are more zinc-insensitive than other species (Foster, 1982). The presence of other organisms may diminish the adsorption of an element by changes in its chemical speciation (Nakatsu & Hutchinson, 1988). The amount of zinc taken up by an organism will strongly depend on the speciation of the metal in the environment. Within the organism the metal can be compartmented in various ways, either being moved to sites of demand (sinks) or partly bio-activated by storage in vacuoles in plants or, in the case of animals, excreted.

As a general rule in ecology, organisms, except cultivated ones, have had sufficient time to adapt to the concentration of bioavailable elements in their ecosystem. However, interference by humans, causing a rapid change in the concentration in the environment, can break down this adaptation. Diversity in niches is a general ecological rule; active excretion of substance to modify bioavailability is a rising issue in modern ecophysiology. In special situations, the life cycle of an organism may be adapted to seasonal changes in element availability.

Aquatic organisms accumulate zinc from food and water. The relative importance of these sources varies between species (Hare et al., 1991; Timmermans et al., 1992; Weeks & Rainbow, 1993).

The bioavailability of zinc in water is influenced by physicochemical and physiological factors (section 4.2).

In general, animals regulate their internal zinc concentrations. However, in some, such as barnacles, the internal zinc concentration is a consequence of zinc storage in granules (Rainbow, 1987; Powell & White, 1990). The concentration at which zinc is homoeostatically regulated is species-specific (Larson & Hyland, 1987) and the external zinc concentration at which regulation breaks down depends on both intrinsic (e.g., species) and extrinsic (e.g., temperature) factors (Nugegoda & Rainbow, 1987).

Examples of the ranges of zinc concentrations that can be found in aquatic organisms are provided in Table 9. These ranges are not all-inclusive, are provided solely for information purposes, and do not necessarily bear any relationship to toxicity, which is discussed in Chapter 9.

4.3.2.2 *Terrestrial organisms*

Zinc taken up by plant roots is mainly in the form of Zn^{2+}, although absorption of hydrated zinc, zinc complexes and zinc organic chelates has also been reported (Kabata-Pendias & Pendias, 1984). Many factors affect the bioavailability of zinc in soils, including total zinc content, pH, organic matter, adsorption site, microbial activity and moisture content. Bioavailability is also determined by climatic conditions and interactions between zinc and other macro- and micronutrients in soil and plants (Kiekens, 1995). Determining factors can be summarized as follows (Kiekens, 1995):

- Highly leached acid soils may have low zinc levels.
- With increasing pH levels there is an increase in the adsorption of zinc by negatively charged colloidal soil particles, with a subsequent decrease in the solubility of zinc.
- In soils with a low organic matter content, the availability of zinc is directly affected by the content of organic complexing or chelating agents originating from decaying organic matter or root exudates.
- Low temperatures and light intensities restrict root development and therefore zinc uptake.
- Reduced zinc uptake has been reported in soils with high phosphorus levels.
- Interactions with other minerals, such as iron, copper, nitrogen and calcium, also reduce zinc uptake.

Table 9. Zinc concentrations in representative organisms during exposure to waterborne zinc

Species	Duration of exposure	Exposure concentration	Zinc concentration in organism (dry weight)	Experimental conditions	Reference
Gammarus pulex	3 days 15 days	2020 µg/litre 410 µg/litre	555 µg/g 213 µg/g	pH 7.7, temperature 11 °C, hardness 109 mg/litre CaCO$_3$	Xu & Pascoe (1993)
Gammarus pulex	15 days	65 µg/litre 319 µg/litre	1502 µg/g 2159 µg/g	pH 7.1, temperature 10–12 °C, hardness 108 mg/litre CaCO$_3$	Xu & Pascoe (1994)
Daphnia magna	40 days	250 µg/litre	420 µg/g	pH 7.8, temperature 19–22 °C, total hardness 2.2 mmol/litre	Memmert (1987)
Chironomus riparius	28 days	900 µg/litre	880 µg/g	temperature 20 °C	Timmermans et al. (1992)
Brachydnanio rerio	35 days	250 µg/litre	390 µg/g	pH 7.8, temperature 19–22 °C, total hardness = 2.2 mmol/litre	Memmert (1987)

Table 9 (contd.)

Littorina littorea	42 days	150.10^{-10} mol/litre	605 µg/g	full strength seawater	Mason (1988)
Orchestia gammarellus	21 days	32 µg/litre 1000 µg/litre	193 µg/g 412 µg/g	temperature 10 °C, salinity 33%	Weeks & Rainbow (1991)
Orchestia mediterranae	21 days	32 µg/litre 1000 µg/litre	202 µg/g 324 µg/g	temperature 10 °C, salinity 33%	Weeks & Rainbow (1991)
Carcinus maenas	21 days	2–316 µg/litre	82 µg/g	temperature 10 °C, salinity 33%	Rainbow (1985)
Palaemon elegans	21 days	2.5–100 µg/litre	76 µg/g (5 °C) 90 µg/g (20 °C)	temperature 5–20 °C, salinity 32%	Nugegoda & Rainbow (1987)
Fundulus heteroclitus	56 days	210 µg/litre 7880 µg/litre	198 µg/g 355 µg/g	temperature 20–24 °C, salinity 25%	Sauer & Watabe (1984)

The factors listed above primarily affect the fraction of zinc in soil that has been immobilized by readily reversible processes. Long-term bioavailability of zinc in soil is influenced by mineralization processes, such as lattice penetration, which result in irreversible binding of zinc (Kiekens, 1995).

The absorption of zinc by the lichen *Usnea florida* was found to follow the classical Langmuir adsorption isotherm and was therefore reversible (Wainwright & Beckett, 1975). The log stability constant was calculated to be 4.46, suggesting a stable association between the zinc ion and the binding site. Zinc binding was dependent upon pH owing to competition between hydrogen ions and zinc for binding sites.

Falahi-Ardakani et al. (1987) reported uptake of zinc by broccoli, cabbage, lettuce, egg plant, pepper and tomato grown in a medium enriched with composted sewage sludge. The uptake of zinc was calculated to be 4–10 mg per week.

Henry & Harrison (1992) studied the uptake of metals by turfgrass, tomatoes, lettuce and carrots grown in different soils (control soil, soil amended with NPK fertilizer, compost, and a 1:1 soil-compost mixture). The loading rates of zinc in the control soil, compost mixture and compost were 232, 239 and 245 kg/ha, respectively. The order of uptake by plants was in the order lettuce > grass > carrots > tomatoes. Uptake slopes for lettuce, grass and carrots grown in compost were higher for than those for plants grown in soil. Zinc concentrations were higher in lettuce, carrots and grass grown in compost and the compost-soil mixture than in plants grown in either the control or fertilized soils. The zinc concentrations in the tomatoes showed no variation.

Singh & Låg (1976) grew barley (*Hordeum vulgare*) in soil sampled from an area near a zinc-smelting plant. Initial zinc concentrations were 545–710 mg/kg. Zinc sulfate was added to the soils at concentrations of 0, 150, 300, 450 and 600 mg/kg. The zinc concentration in the barley increased with increasing total zinc concentration in the soil. The proportion of soil zinc that was bioavailable to plants was reported to be independent of the zinc application.

Increased uptake of zinc was reported for ryegrass (*Lolium perenne*) grown in sludge-amended soil: the zinc concentration in grass exposed to the sludge was 7.5 times that in controls (Dudka & Chlopecka, 1990).

MacLean (1974) studied the uptake of zinc by maize (*Zea mays*), lettuce (*Lactuca sativa)* and alfalfa ((*Medicago sativa*) grown successively (for 6 weeks, 5 weeks and 16 weeks, respectively) in pots with soil of varying zinc concentrations. The concentration of zinc in the plants increased with increasing soil zinc concentration. Maize and lettuce grown in a soil pre-treated with phosphorus tended to have lower zinc levels than those grown in soils without any pre-treatment. Increased zinc levels were also reported in plants grown in soils with higher organic matter contents.

Jones (1983) grew lettuce (*Lactuca sativa*) and radish (*Raphanus sativus*) in soil collected from plots 10 m and 90 m from a rusty galvanized steel electrical transmission (hydro) tower. The plants were harvested after 45 days. The zinc content was higher in soil sampled nearer the hydro tower. The plants grown in this soil had the highest zinc concentrations; lettuce roots had significantly higher zinc levels than lettuce tops, while radish tops had significantly higher zinc levels than radish roots. No differences between zinc levels in tops and roots were reported for plants grown in the soil sampled 90 m from the tower.

Gintenreiter et al. (1993a) studied the bioaccumulation of zinc by gypsy moth (*Lymantria dispar*) larvae following dietary exposure of first instar larvae to 100 or 500 mg/kg. Zinc concentrations in the first instar larvae were similar at both exposure levels. The subsequent uptake of zinc was dependant upon exposure concentration. An increase in larval zinc concentration was reported at the 500 mg/kg dose. At the 100 mg/kg dose and in the control larvae, a dose-related decrease in larval zinc concentrations was reported. The highest zinc concentration factor was reported to be 3.5. The zinc concentration in larval faeces was reported to be inversely related to the zinc concentration in the larvae. Zinc levels in the exuviae decreased with successive larval stages, whereas constant zinc concentrations were reported for the head capsules in all groups. The total amount of zinc in the larvae increased at every stage with the highest amount detected in the pupae. However, the adult life stages tended to have less zinc.

A positive relationship between the zinc concentrations in the terrestrial amphipod (*Arcitalitrus dorrieni*) and zinc concentration in its food was reported (Weeks, 1992). The mean zinc accumulation rate was calculated to be 2.21 µg/g per day, which was calculated to be equivalent to 1.11% of the total body zinc per day.

Hames & Hopkin (1991) determined the assimilation of zinc in two species of woodlouse, *Oniscus asellus* and *Porcellio scaber* fed for 115 h on leaves treated with zinc chloride. The mean zinc assimilation rate during the exposure period was 29.4% in *O. asellus* and 36.7% in *P. scaber*, with a significant ($P < 0.001$) inter-species difference.

Hopkin & Martin (1985) studied the uptake of zinc by the spider *Dysdera crocata* exposed to zinc through its diet of woodlice (*P. scaber*). Spiders fed on woodlice collected from the same site as themselves consumed 34.5% of the total zinc in the woodlice. Spiders fed on woodlice from an area contaminated with heavy metals consumed 42.4% of the total zinc in the woodlice. There were no differences in zinc content of the spiders fed on woodlice from their own site, and those spiders starved throughout the experiment. The spiders were therefore able to excrete any excess absorbed zinc and did not assimilate it.

Lindqvist & Block, (1994) studied the excretion of zinc by the grasshopper *Omocestus viridulus* during moulting. The grasshopper nymphs were fed grass leaves containing a known amount of radiolabelled zinc, and the zinc contents of the grasshopper faeces, exuviae and carcasses were determined. The exuviae accounted for only a minor part of excreted zinc. After rearing for 15 days, approximately 50% of the ingested zinc remained in the grasshoppers.

Recio et al. (1988) studied the cellular distribution of zinc in slugs (*Arion ater*) exposed to dietary zinc. The highest zinc concentration was reported in the lipofuscin material of the excretory cells. Zinc was also detected in the perinuclear cytoplasm and the spherules of the calcium cells (low zinc exposures and short exposure times only). The authors suggested that slugs could be used to indicate high levels of zinc in the environment.

Simkiss & Watkins (1990) determined the factors that affect the uptake of zinc by the garden snail, *Helix aspersa*. The snails were divided into four treatment groups, receiving an artificial diet (controls), antibiotics, zinc nitrate at a concentration of 1.5 μmol/g in the artificial diet, or a diet supplemented with antibiotics and zinc nitrate. Food consumption in the two groups fed a diet containing zinc was reported to be reduced to about 38% of normal. However, the dry weights of snails from each group did not differ after exposure for either 4 or 8 weeks. A direct linear relationship between soft body zinc content and dietary zinc consumed was reported for the snails that were not fed antibiotics. The dietary intake was also correlated with the zinc concentration of the digestive gland. The same was also evident in the snails fed antibiotics, although the relationship was significantly different. In a second experiment it was reported that snails fed a bacterially contaminated diet absorbed more zinc than snails fed a sterile diet.

5. ENVIRONMENTAL LEVELS AND HUMAN EXPOSURE

5.1 Environmental levels

In nature, zinc occurs only rarely in its metallic state and the vast majority of environmental samples contain the element only in the form of zinc compounds. In the following text, therefore, zinc content relates to those compounds.

5.1.1 Air

Zinc concentrations in air are summarized in Table 10. The proportion of zinc derived from anthropogenic sources remains uncertain (see section 3.2). In air, zinc is primarily adsorbed to particulate matter, which is expected to be short-lived in the atmosphere (Perwak et al., 1980). The mass median diameter for zinc-containing particles in airborne dust is 1.5 μm for rural and urban sites (Lioy et al., 1978).

In general, zinc levels in urban and industrial areas are higher than in rural areas. Natural atmospheric zinc levels due to weathering of soil are almost always less than 1000 ng/m^3. Levels of 10–300 ng/m^3 are given for background concentrations and up to 1000 ng/m^3 for urban industrial areas. Zinc concentrations of 0.3–27 ng/m^3 were found over the Atlantic Ocean, < 0.4–300 ng/m^3 in European rural areas, and 10–2400 ng/m^3 in urban areas (see Table 10). For indoor air in an urban setting, zinc concentrations were in the range 0.1–1.0 μg/m^3 (Henkin, 1979).

Air zinc levels in Belgium have shown a decreasing trend. In 1989–1990, levels were 150–380 ng/m^3 in rural areas and small towns, 140–210 ng/m^3 in large cities, and 500–1270 ng/m^3 in industrial areas (IHE, 1991). By 1992–1993 levels had fallen to 70–100, 100–170 and 390–1020 ng/m^3 for the same locations (VMM, 1994).

In the Netherlands, yearly average zinc levels in the air at four sampling sites varied between 60 ng/m^3 and 80 ng/m^3 in 1990

Table 10. Zinc concentrations in atmospheric particulate matter

Area	Year	Particle size (µm)	Zinc concentration (ng/m³)	Reference
North Atlantic Ocean (various sites)	1970–1972	≥0.1	0.3–27	Duce et al. (1975)
North Sea (various sites; 91 samples)	1988–1989	>0.5	74.6[a] (nd–611)	Ottley & Harrison (1993)
North Sea, Helgoland	1985–1986	not given	32.8[b] (4.7–185)	Kersten et al. (1988)
North Sea (various sites; 98 samples)	1988–1989	not given	41[a] (0.7–250)	Chester & Bradshaw (1991)
North Sea (65 samples)	1980–1985	not given	67.4[b] (5.0–1460)	Baeyens & Dedeurwaerder (1991)
Baltic Sea (various sites; 17 samples)	1985	≥0.1	26.6[c] (7.0–54.5)	Haesaenen et al. (1990)
South Norway, Birkeness (160 samples)	1985–1986	not given	11[d], 15[a] (<0.4–114)	Amundsen et al. (1992)
Germany/Belgium, rural area		not given	100–300	Cleven et al. (1993)
United Kingdom, rural area		not given	41[a] (1.4–237)	Yaaqub et al. (1991)
Netherlands, rural area	1984–1985	not given	30	van Daalen (1991)
Netherlands			65	Cleven et al. (1993)
Netherlands (4 sites)	1990		60–80	CCRX (1991)
	1992		38–57	CCRX (1994)
USA, urban air	1973	MMD 0.58-1.79	100–1700	Lee & Von Lehmden (1973)

Table 10 (contd.)

Area	Year	Particle size (μm)	Zinc concentration (ng/m³)	Reference
USA, San Francisco Bay Area (9 sites)	1970	not given	27–500	John et al. (1973)
USA, New York City (1 site)	1972–1975	MMD: 1.5	330[c] (293–379)	Lioy et al. (1978)
USA, cities (87 samples)			<10–840	Schroeder et al. (1970)
USA, 19 cities (86 samples)	not given	not given	10–2400	Cole et al. (1984)
USA, Idaho (site near lead smelter)	1972	not given	4620[c] (270–15 700)	Ragaini et al. (1977)
Germany/Belgium, urban area			400–1000	Cleven et al. (1993)
Germany, industrial area (48 samples)	1983	not given	140–810	Lahmann (1987)
Germany, industrial area (35 samples)	1984		170–730	Lahmann (1987)
Belgium, Angleur	1986		9300	Cleven et al. (1993)
South-Holland, industrial area	1984–1985	not given	70	van Daalen (1991)

MMD = mass median diameter; nd = not detected

[a] Arithmetic mean.
[b] Mean.
[c] Average.
[d] Median.

(CCRX, 1991). By 1992, the averages had decreased to 43–57 ng/m³ (CCRX, 1994). The 98th percentile of the daily values in 1992 was < 210 ng/m³ (CCRX, 1994).

5.1.2 Precipitation

Deposition of airborne zinc is strongly dependent on particle size and meteorological factors, primarily wind speed and humidity. Wet deposition predominates with estimated values for zinc removal from air of 60–90% (Ohnesorge & Wilhelm, 1991). In rainwater (North Sea, 21 samples), Baeyens et al. (1990) measured average concentrations of dissolved zinc of 500 ± 500 g/litre. In rain sampled on a gas platform in the North Sea, Peirson et al. (1973) found an average zinc level of 2000 µg/litre (includes dry deposition). The annual wet deposition to the North Sea has been estimated to be in the range 14–53 µg/cm² (Peirson et al., 1973; Dedeurwaerder et al., 1982; Baeyens et al., 1990).

The mean annual wet deposition of zinc in the Netherlands was determined to be 1.2 µg/cm² in 1992 (CCRX, 1994).

In Germany, zinc levels in rain were 7–26 µg/litre in rural areas, 23 µg/litre in urban areas, and up to 90 µg/litre in urban industrial areas (Malle, 1992). Peirson et al. (1973) reported an annual average of 85 µg/litre zinc in rain collected in the United Kingdom in 1971. In Nigeria (1976), rain was found to contain 130 µg/litre and annual total zinc deposition was calculated at 10 µg/cm² (Beavington & Cawse, 1979).

Samples from Greenland snows analysed for their heavy metal contents showed a decrease in anthropogenic zinc by a factor of about 2.5 during the period 1967–1989, which is stated to be a consequence of the abatement policies for industrial emissions in the European Union and North America (Boutron et al., 1991). The same authors have more recently reported a five-fold increase of zinc deposition in Greenland ice in the period since the industrial revolution (from 1800 onwards) with a maximum during the 1960s followed by a significant decrease (40%) between 1960 and 1990 (Boutron et al., 1995). However, not all studies have been successful in detecting anthropogenic inputs of zinc distant from point source emissions. For example, studies of the metal content of lake sediments in the Arctic have failed to detect any anthropogenic inputs (Gubala et al., 1995).

5.1.3 Water

When interpreting the data on water concentration of zinc, it is necessary to be aware that the higher values reported in early studies may be due to contamination of the samples.

Zinc concentrations in fresh waters depend significantly on local geological influences and anthropogenic input. As a result of chemical weathering of minerals, soluble zinc compounds, such as zinc sulfate, are formed which may be transferred to surface waters especially at low pH levels (Perwak et al., 1980). Urban runoff, mine drainage and municipal and industrial effluents can also make a considerable contribution to the zinc load of surface water (US EPA, 1980).

In water, zinc is present primarily in the ionic form, but it has a strong tendency to adsorb to suspended organic matter and clay minerals or to precipitate with iron or manganese oxides, resulting in zinc removal from the water column and enrichment of sediments (Perwak et al., 1980). Dissolution of zinc increases at low pH, low hardness and high temperature (Malle, 1992).

5.1.3.1 Fresh water

An overview of zinc concentrations in fresh water is given in Tables 11 and 12. In natural fresh water concentrations rarely exceed 40 µg/litre (Spear, 1981); Bowen (1979) reported a medium background value of 15 µg/litre, with a range of < 1–100 µg/litre. For various rivers worldwide, Holland (1978) reported average values of 5–45 µg/litre. Higher average levels are associated with zinc-enriched ore deposits and anthropogenic sources of pollution. The average zinc concentration of the river Rhine at Lobith was reduced from 57 µg/litre in 1984 to 22 µg/litre in 1993 (see Table 11). Using an erosion model, the natural background level has been estimated at 4 µg/litre (Van der Weijden & Middelburg, 1989), whereas Van Tilborg & Van Assche (1995) have suggested that the natural concentration of zinc in the Rhine is about 10 µg/litre. The estimated zinc load of the Rhine decreased by 42% from

Table 11. Zinc concentrations in fresh water

Area	Year	pH	Particle size (μm)	Zinc concentration (μg/litre)	Reference
Various rivers, worldwide				5–45[a]	Holland (1978)
Canada, unpolluted rivers and lakes				≤ 40	Spear (1981)
USA, nationwide				0.5–10	US EPA (1987)
USA, ambient surface water stations				20[b]	Eckel & Jacob (1988)
Orinoco (9 samples, various sites)	1982	4.3–7.6	< 0.4	0.131	Shiller & Boyle (1985)
9 Rivers in the Orinoco Basin (10 samples)	1982	4.3–7.6	< 0.4	0.02–1.77	Shiller & Boyle (1985)
Yangtze (mouth at low flow)	1981		< 0.4	0.077	Shiller & Boyle (1987)
Amazon (mouth at high flow)	1976	6.7	< 0.4	0.249	Shiller & Boyle (1987)
Amazon (mouth at low flow)	1982	7.5	< 0.4	0.02	Shiller & Boyle (1987)
15 Rivers in the Amazon Basin (26 samples)	1976	5.4–7.1	< 0.4	0.043–1.24	Shiller & Boyle (1985)
India, freshwater lake				200	Prahalad & Seenayya (1989)
Huanghe (10 sites)	1986	8.21	< 0.45	65–327	Zhang & Huang (1993)
Rhine, at Lobith	1984	not given		57[a]	RIWA (1993)
Rhine, at Lobith (25 samples)	1993	7.1–7.9		22[a, b] (<15–38)	RIWA (1993)

Table 11 (contd.)

Area	Year	pH	Particle size (μm)	Zinc concentration (μg/litre)	Reference
Ohio (9 samples; various sites)	1984	7.3–7.5	< 0.4	0.288–3.2	Shiller & Boyle (1985)
16 Ohio tributaries (29 samples)	1984	6.9–8.1	< 0.4	0.072–4.32	Shiller & Boyle (1985)
Mississippi (7 samples; 2 sites in Louisiana)	1982–1984	7.6–8.2	< 0.4	0.194[a] (0.11–0.27)	Shiller & Boyle (1987)
St. Lawrence (10 m depth)	1975–1976	7.63	< 0.4	8.6[a]	Yeats & Bewers (1982)
Twelvemile Creek	1982			< 10	LaPerriere et al. (1985)
Potomac River (3 sites)	1988			14–310	Hall et al. (1989)
Streams with current mining				29–882	LaPerriere et al. (1985)
United Kingdom, River Ystwyth	1973–1975			170–880	Grimshaw et al. (1976)
United Kingdom, Willow Brook, polluted (7 sites; 42 sampling occasions)	1969–1971	7–8.5	< 0.45	320–1150[c]	Solbé (1973)
Spain, surface water near mine	1984				Gonzáles et al. (1985)
– mine (5 samples):				12–3925	
– marshes near mine (16 samples):				7.5–895	
– stabilized sands (6 samples):				9.6–23.8	

[a] Average
[b] Median

Table 12. Dissolved and total zinc concentrations in fresh water

Location	Value (µg/litre)	Measurement	Reference
World	40	dissolved	Spear (1981)
World	15 (0.2–100)	total	Bowen (1979)
World	5–45	total	Holland (1978)
World	12–35	total	Zuurdeeg (1992)
World	0.6–17	total	Zuurdeeg (1992)
Europe	5–43	total	Zuurdeeg (1992)
Rhine	22	total	RIWA (1993)
Rhine	4	total	Van der Weijden & Middleburg (1989)
Rhine	10	total	Van Tilborg & Van Assche (1994)
Belgium	50	total	Goethals (1991)
Alaska	10	total	LaPerriere (1985)
Lake Pontchartrain	< 1	total	Francis & Harrison (1988)
Ohio	0.065–0.65	dissolved	Shiller & Boyle (1985)
Great Lakes	0.09–0.3	dissolved	Nriagu et al. (1996)

3600 tonnes/year in 1985 to 2 100 tonnes/year in 1990. In the Scheldt river at the Belgian-Netherlands border, total zinc concentration declined from about 120 µg/litre in 1975 to approximately 50 µg/litre in 1989 (Goethals, 1991). The annual average zinc concentration in United Kingdom rivers decreased from 42 µg/litre in 1978 to 22 µg/litre in 1992 (UK, 1994).

Drainage from active and inactive mining areas may be a significant source of zinc in water. Waters in acidic mine tailing ponds in Canada were found to contain an average zinc level of 900 µg/litre with a maximum value of 3300 µg/litre (Mann et al., 1989). Gonzáles et al. (1985) reported zinc levels of up to 4 mg/litre in surface water collected near a mine. Elevated zinc levels of up to

73

175 µg/litre were found in Birch Creek, a heavily mined river, compared to <10 µg/litre in an unmined stream (LaPerriere et al., 1985).

The natural background range (total zinc) based on a data set of 8000 analyses of clean streams for Northern European lowland rivers is 5–43 µg/litre. Two data subsets are given by Zuurdeeg (1992) for clean rivers of the world (outside Europe): the first gives a range of 12–35 µg/litre, and the second a range of 0.6–17 µg/litre (total zinc). The latter data set indicates the existence of areas low in natural zinc.

Dissolved average zinc concentrations of the Great Lakes Superior, Erie and Ontario determined by ultraclean techniques were 0.09–0.3 µg/litre (Nriagu et al., 1996). A depletion of zinc in surface waters and an increase in concentration with depth were observed. Similarly, Francis & Harrison (1988) reported zinc concentrations for lake Pontchartrain of < 1 µg/litre (total zinc). In relatively undisturbed rivers of the Ohio valley, dissolved zinc concentrations of 0.06–0.6 µg/litre were measured by Shiller & Boyle (1985).

The inadequacy of many zinc data for fresh waters is well illustrated in the study by Windom et al. (1991). Their results for dissolved zinc obtained using clean sampling and analysis techniques were lower by 1–2 orders of magnitude than those obtained in a national monitoring programme.

5.1.3.2   Seawater

Zinc concentrations in seawater are summarized in Table 13. Baseline levels in seawater are typically in the range 0.0005–0.026 µg/litre (Sprague, 1986), with 0.002–0.1 µg/litre (US EPA, 1987; Yeats, 1988) in open ocean waters. Concentrations are lower at the ocean surface than in deeper water (Bruland et al., 1978). Zinc correlates well with dissolved silicate concentrations (Yeats, 1988). In estuarine waters, anthropogenic inputs result in seafood concentrations with typical values of 1–15 µg/litre (Van den Berg et al., 1987). Concentrations decrease in an offshore direction (Duinker & Nolting, 1982). It should be noted that many of the data, particularly those reported prior to 1975, may be unreliable, if inadequate care was taken to avoid contamination during sample collection and analysis. For example, older data for zinc in open

Table 13. Zinc concentrations in seawater

Area	Year	Sampling depth (m)	Zinc concentration (µg/litre)	Reference
Ocean, surface water			0.002–0.1	US EPA (1987)
North-east Pacific Ocean, California coast	1977	0.2 100 1020 2500	0.0084 0.08 0.43 0.61	Bruland et al. (1978)
North-east Pacific Ocean	1981	100 1000 2500 3500	0.23 0.54 0.60 0.40	Yeats (1988)
North Atlantic Ocean, Sargasso Sea	1984	20 165 1000 1915 3715	0.065 0.026 0.092 0.137 0.124	Yeats (1988)

Table 13 (contd.)

Area	Year	Sampling depth (m)	Zinc concentration (µg/litre)	Reference
North Sea, Southern Bight (128 samples; < 0.45 µm	1975		≥ 0.3	Duinker & Nolting (1982)
Skagerak			0.27–0.81	Kersten et al. (1988)
Western Mediterranean, coastal			0.001–0.002	Sprague (1986)
Western Mediterranean, estuary			max. 0.01	Sprague (1986)
Western Mediterranean, near shore			0.0036	Sprague (1986)
Australia (polluted)			0.134	Sprague (1986)
USA, San Diego coastal			0.0005	Sprague (1986)
USA, San Diego harbour			0.0026	Sprague (1986)
United Kingdom, heavily polluted			0.026	Sprague (1986)
United Kingdom, polluted			0.007–0.012	Sprague (1986)
Pacific Ocean, Australia, New South Wales coast	1995	surface	< 0.04	Batley (1995)

ocean waters can be up to three orders of magnitude higher than current values (Preston et al., 1973).

5.1.4   Soil

Zinc concentrations in igneous and sedimentary rocks were reported to be 48–240 mg/kg for basaltic igneous rock, 5–140 mg/kg for granitic igneous rock, 18–180 mg/kg for shales and clays, 34–1500 mg/kg for black shales and 2–41 mg/kg for sandstones (Thornton, 1996). Zinc levels in geochemically anomalous parent materials in the United Kingdom were found to be 1% and more (Thornton, 1996).

Zinc levels in soils are given in Table 14. Zinc levels and speciation in soil may vary with the soil profile, especially in natural ecosystems (see section 3.1). The mobility of zinc in soils is dependent on its speciation, the soil pH, and content of organic matter. For non-contaminated soils worldwide, Adriano (1986) reported average zinc concentrations of 40–90 mg/kg, with a minimum of 1 mg/kg and a maximum of 2000 mg/kg. Low levels are found in sandy soils (10–30 mg/kg), while high contents are found in clays (95 mg/kg). Wet and dry deposition, the use of zinc compounds as fertilizers and the application of municipal sludges and manure to cropland are considerable sources of zinc in soils (Chang et al., 1987).

Zinc concentrations of up to 118 000 mg/kg are associated with industrial contamination (Eisler, 1993). Significant relationships were found between the distance from smelters or roads and the levels of easily extractable zinc in soil, and the zinc content in herbage (Hogan & Wotton, 1984; Beyer et al., 1985; Reif et al., 1989).

Old geological formations that have been extensively leached may have background concentrations of natural zinc in water that are one order of magnitude below those of the mineral-rich European alluvial flow system.

In soil samples from gardens next to a zinc and lead ore mine in Wales (United Kingdom), average levels of 2923 mg/kg were found, compared to 94 mg/kg in uncontaminated soils (Davies & Roberts, 1975).

Table 14. Zinc concentrations in soils

Area	Zinc concentration (mg/kg dry weight)	Remarks	Reference
Worldwide	10–300	> 2000 soils	Swaine (1955)
Worldwide	90 (1–900)		Bowen (1979)
Worldwide	50		Vinogradov (1959)
Worldwide	40		Berrow & Reaves (1984)
Worldwide	59.8^a (1.5–2000)	7402 soils	Ure & Berrow (1982)
Canada	74 (10–200)		McKeague & Wolynetz (1980)
Canada, Ontario	47.6^b (5–162)	296 soils; 0–15 cm	Frank et al. (1976)
USA	54	3045 soils	McKeague & Wolynetz (1980)
USA	5–264 (53, 50th percentile)	307 soils series	Holmgren et al. (1993)

Table 14 (contd.)

Germany, rural area	85[b]	73 samples	UBA (1994)
United Kingdom	77 (5–816)	748 samples	Adriano (1986)
United Kingdom, Scotland	58[a] (<0.7–987)	725 samples from 83 soil profiles	Berrow & Reaves (1984)
Ukraine, Poles'ye	14–95[c]		Golovina et al. (1980)
USSR, Eastern European Plain	25–120		Vinogradov (1959)
China	100 (9–790)		Liu et al. (1983)
Germany, Rhine-Main plain	3–30		Kauder (1987)
Germany, urban industrialized area	311[b]	371 samples	UBA (1994)
Germany, Harz, polluted area	up to ca. 10 000		Aurand & Hoffmeister (1980)
USA, Ohio, soil treated with sewage sludge for 10 years	107		Levine et al. (1989)

Table 14 (contd.)

Area	Zinc concentration (mg/kg dry weight)	Remarks	Reference
Electrical transmission tower (galvanized, corroded):		4 samples; 0–5 cm	Jones & Burgess (1984)
near tower	11 480[a]		
1 m distance	10 431[a]		
5 m distance	362[a]		
10 m distance	160[a]		
50 m distance	54[a]		
United Kingdom, Scotland, vicinity of chemical waste disposal facility	69–236[a]		Eduljee et al. (1986)
United Kingdom, Wales, zinc–lead ore mine	2923		Davies & Roberts (1975)
USA, Palmerton, 2 zinc smelters:			Beyer et al. (1985)
2 km downwind from smelter	24 000	O₂ horizon	
10 km upwind of smelter	960		

Table 14 (contd.)

Peru, zinc smelter:			Reif et al. (1989)
1 km distance	575		
13 km distance	183		
27 km distance	154		
35–55 km distance	16–29		
Zinc smelter for 50 years:		10–15 cm	Hogan & Wotton (1984)
6 km distance	260		
35 km distance	80		
USA, Idaho, lead smelter	200–29 000		Ragaini et al. (1977)
United Kingdom, revegetated mine tailings dam	1915–2160	1–8 cm	Andrews et al. (1989)
USA, Idaho, reclaimed phosphate mine waste dumps	443–1112[c]	10-15 sites; 0–25 cm	Hutchison & Wai (1979)

[a] Mean
[b] Median
[c] Average

Soil and vegetation in Palmerton, Philadelphia, USA were found to be highly contaminated with zinc by fumes escaping from two zinc smelters. Zinc concentrations in the organic horizon (decomposed leaf litter) increased by regular gradations from a minimum of 67 mg/kg dw at a site 105 km west of the smelters to a maximum of 35 000 mg/kg dw 1.2 km east of the smelters. At a depth of 0–15 cm, concentrations of about 10 000 mg/kg were measured (Beyer et al., 1984, 1985; Beyer, 1988). Soil sampled at a distance of 1–40 km from the smelters showed that approximately 90% of the metal deposited on the soil surface was retained in the top 15 cm of the soil profile (Buchauer, 1973). Soils (0–15 cm) from 40 gardens in rural areas at different distances from the smelters contained average zinc concentrations of 5830 mg/kg dw; background levels of 346 mg/kg dw and 311 mg/kg dw were reported (Chaney et al.,1988; see Table 14).

In the Netherlands, reference values for agricultural soils have been related to soil types on the basis of the percentage by weight of clay (C) and organic matter (H) according to the following equation: Zn (mg/kg dw) = 50 + 1.5 (2C + H)

This relationship was based on a large data set of widely varying uncontaminated Netherlands soils. The derived values are regarded as ambient levels from which no detrimental effects are expected. For a standard soil (C = 25% and H = 10%) the reference value is 140 mg/kg (Cleven et al., 1993).

5.1.5 Sediments and sewage sludge

The concentration of zinc and other metals in sediment spans at least three orders of magnitude. The concentration is related to particle size, mineralogy and input sources. The biological effects of metals in sediment are not related to total concentrations (Allen, 1996). The binding of zinc by sulfide, organic matter and metal oxides should be taken into account (Allen, 1996). Standards for metals should not be based on total concentration (Allen, 1996).

As a consequence of adsorption to organic substances and other inorganic minerals, zinc is precipitated from waters and thus enriched in sediments (Malle, 1992).

Zinc concentration in sediments from aquatic systems in northern Greece was 40 mg/kg dw (Sawidis et al., 1995). In the Elbe estuary, a median zinc concentration of 1400 mg/kg (range 440–2920 mg/kg; 60 samples, < 20-µm fraction) was observed in 1992; the background level was reported to be 95 mg/kg (Mueller & Furrer, 1994). The zinc concentration in the sediments of the Rhine harbour in Rotterdam amounted to 1900 mg/kg dw in 1980 (Cleven et al., 1993). A time series in the same area showed a decrease in zinc content in this sediment of more than 50% between 1979 and 1986 (Malle, 1992). Zinc levels in the Scheldt estuary at the Belgian-Netherlands border declined from 520 mg/kg dw in 1951 to 350 mg/kg dw and 229 mg/kg dw, respectively in 1971 and 1974 (Van Alsenoy et al., 1990). In sediments from the highly polluted river Vesdre (Belgium), mean zinc concentrations were 2920 mg/kg dw (1629–4806 mg/kg dw; 26 samples) at a site located 6 km downstream from a zinc factory; 2 km upstream from the factory, the concentration was 1317 mg/kg dw (823–1666 mg/kg dw; 5 samples) (Houba et al., 1983). For the <20-µm fraction of sediments from the Wadden Sea, baseline levels of 100 mg/kg dw were given (UBA, 1994). For the zinc concentrations in sediments in the upper layer (0–10 cm) from the Baltic Sea, average values of 180 mg/kg and 337 mg/kg, with maximum levels of up to 2290 mg/kg, were reported. Zinc levels in this layer were 1.5–10 times higher than in the next layer (10–20 cm) (UBA, 1994).

Chaney et al. (1984) stated that the levels of zinc in composted sewage sludge vary between 101 and 49 000 mg/kg dw with a mean of 1700 mg/kg dw. A compilation of more than 100 values from the literature gives a mean of 2420 mg/kg, while an analysis of 80 samples from sewage treatment plants (USA) gives 6380 mg/kg dw (Dean & Smith, 1973; Ohnesorge & Wilhelm, 1991). Zinc contents of 2300 mg/kg dw and 2000–3000 mg/kg were reported for sewage sludge from Switzerland and Germany, respectively (Ohnesorge & Wilhelm, 1991). Additionally, average zinc concentrations of 1480 mg/kg dw were determined in Germany (Schweiger, 1984). Sewage sludge applied to arable land in Ohio, USA was found to contain zinc at a concentration of 866 mg/kg dw (Levine et al., 1989). In the European Union, the maximum zinc concentration permitted in biosolids (sludge) for application to agricultural soils cannot raise the zinc level above 300 mg/kg of dry soil (Berrow & Reaves, 1984).

5.1.6 *Aquatic and terrestrial organisms*

Zinc is present in all tissues of all organisms, as it is essential for growth. Concentrations are higher in organisms near anthropogenic point sources of zinc pollution. Interspecies variations in zinc content are considerable; intraspecies levels also vary, for instance with life stage, sex and season. In general, zinc-specific sites of accumulation in animals are bone, liver and kidney (Spear, 1981).

5.1.6.1 *Aquatic plants and animals*

For rooted aquatic plants and algae, zinc concentrations are generally in the range 20–120 mg/kg dw (Spear, 1981). Zinc concentrations of 38 and 90 mg/kg dw were reported for marine phytoplankton and seaweeds, respectively (Young et al., 1980). In eelgrass (*Zostera marina*), zinc levels were found to increase with age of leaf but were independent of the zinc load at the sampling site (Brix & Lyngby, 1982). Background levels in aquatic moss (*Fontinalis squamosa*) were < 400 mg/kg dw, whereas moss from a contaminated river contained maximum concentrations of 2810 mg/kg dw (Young et al., 1980). Sawidis et al. (1995) analysed aquatic macrophytes from aquatic systems in northern Greece for zinc metal concentrations. Levels of between 10.2 mg/kg and 145 mg/kg dw were reported for *Ceratophyllum demersum*, *Cladophora glomerata*, *Myriophyllum spicatum*, and *Potamogeton nodosus*. Many species of algae in Canadian mine tailing environments were found to have zinc contents of > 1000 mg/kg (Eisler, 1993).

Baseline zinc levels in invertebrates are in the range 50–300 mg/kg dw (Spear, 1981). In molluscs, which are known to be good accumulators of trace metals, concentrations may be elevated. In soft parts from mussels (*Mytilus edulis*), zinc ranged from 28 mg/kg dw (visceral mass) to 3410 mg/kg dw (kidney). In scallops (*Pecten* sp.), zinc levels of 200 mg/kg dw were found in soft parts, but concentrations of up to 32 000 mg/kg dw and 120 000 mg/kg dw were reported in kidney and kidney granules, respectively (Eisler, 1993). Luten et al. (1986) examined the zinc content of mussels from the North Sea, the Wadden Sea, and three estuaries over the period 1979–1983. The median zinc content in mussels was 87–234 mg/kg dw. In the marine bivalve *Macoma balthica* in the Westerschelde

Estuary (Netherlands), zinc concentrations of 377–692 mg/kg dw were recorded; concentrations were higher in winter and lower in summer (Bordin et al., 1992).

Bullfrogs (*Rana catesbeiana*) caught downstream from the source of mine tailings had markedly higher zinc levels in most tissues (Niethammer et al., 1985).

Spear (1981) reported background levels for fish usually ranging from 4 mg/kg to 20 mg/kg fresh weight (fw). In muscle of marine fish, such as the northern anchovy (*Engraulis mordax*) and the Atlantic menhaden (*Brevortia tyrannus*), concentrations of up to 20–25 mg/kg fw were measured (US National Academy of Sciences, 1979). Schmitt & Brumbaugh (1990) reported results of a US national contaminant biomonitoring programme in which the concentration of zinc in freshwater fish was measured. Zinc concentrations were highest in the common carp (*Cyprinus carpio*); maximum zinc concentrations were 168.1 μg/g fw in 1978–1979, 109.2 μg/g fw in 1980–1981 and 118.4 μg/g fw in 1984. Maximum values for other common fish species measured in the 1984 survey were 32.66 μg/g fw for channel catfish (*Ictalurus punctatus*), 24.39 μg/g fw for white sucker (*Catostomus commersoni*), 23.91 μg/g fw for large-scale sucker (*Catostomus macrocheilus*), 19.93 μg/g fw for largemouth bass (*Micropterus salmoides*) and 12.78 μg/g for lake trout (*Salvelinus namaycush*). Lowest zinc levels are found in muscle, highest (5–10 times higher) in eggs, viscera and liver (Eisler, 1993; Stanners & Bourdeau, 1995). In Toronto Harbour, Ontario, Canada, various species of fish contain only slightly elevated zinc levels (36 mg/kg fw) in muscle tissues. In acidic lakes near Sudbury (mining area), Canada, the zinc content of fish liver tissues is generally 1–2 orders of magnitude higher than in muscle tissues.

5.1.6.2 *Terrestrial plants and animals*

Studies have shown that the uptake of zinc by terrestrial plants is significantly increased at a low soil pH, but reduced when there is a high content of organic matter (Jones & Burgess, 1984; Chaney et al., 1987). Normal levels of zinc in most crops and pastures range from 10 mg/kg to 100 mg/kg. Some plant species are zinc accumulators, but the extent of the accumulation in plant tissues varies with soil properties, plant organ and tissue age.

Application of fertilizers (including sewage sludge) to soil increases the zinc concentration in plants (Mortvedt & Gilkes, 1993).

Earthworms (various species) from uncontaminated soils were found to contain 120–650 mg/kg dw, from mining sites 200–950 mg/kg dw, from industrial sites 320–1600 mg/kg dw, and near galvanized towers 340–690 mg/kg dw (Beyer & Cromartie, 1987). For various species of moths, background loads of 140–340 mg/kg dw were reported. Beetles (various species) were found to contain 470 mg/kg dw zinc, and zinc levels in caterpillars (*Porthetria dispar*) averaged 170 mg/kg dw (Beyer et al., 1985)

Zinc concentrations in birds were found to range between 6.4 mg/kg fw in eggs (*Pelicanus occidentalis*) and 150 mg/kg fw in liver (*Pandion haliaetus*), but the highest values, 250 mg/kg, fw were reported in the liver of the Californian condor (*Gymnogyps californianus*) (Wiemeyer et al., 1988).

For white-footed mouse (*Peromyscus leucopus*), short-tailed shrew (*Blarina brevicauda*) and different species of songbirds, background levels of 140, 200 and 120 mg/kg dw were reported, whereas in animals from a polluted area, concentrations were 190, 380 and 140 mg/kg dw, respectively (Beyer, 1988). Andrews et al. (1989) reported tissue levels of 103 µg/kg dw (muscle) to 226 µg/kg dw (pelvic girdle) for the field vole (*Microtus agrestis*) and 126 µg/kg dw (heart) to 547 µg/kg dw (femur) for the common shrew (*Sorex araneus*). The kidney of a white-tailed deer (*Odocoileus virginianus*) collected 4 km from zinc smelters contained 600 mg/kg dw, well above the mean of 145 mg/kg detected in five deer collected at least 100 km from the smelters (Sileo & Beyer, 1985).

5.2 General population exposure

5.2.1 Air

Negligible quantities of zinc are inhaled in ambient air (approximately 0.7 µg/day) (Cleven et al., 1993). For urban room air, zinc concentrations range from 0.1 to 1.0 µg/m^3 (Henkin, 1979). Harrison (1979) analysed dust samples collected in Lancaster, United Kingdom, for their metal content. The total zinc levels detected were

1600 mg/kg in dust from car parks, 534 mg/kg in urban dust, 940 mg/kg in household dust, and 297 mg/kg in dust from rural roads. In Germany, mean zinc concentrations of 496 mg/kg in household dust and 2.9 $\mu g/m^2$ per day in dust deposition were reported (UBA, 1992).

Elevated amounts may be inhaled by people who work in facilities for smelting and refining zinc material or in coal mines, or who live near waste sites and smelters. At zinc levels of 1 $\mu g/m^3$, the general population would inhale about 20 $\mu g/day$.

5.2.2 Food

Zinc is an ubiquitous and essential element. For diets with moderate bioavailability of zinc, Sandstead et al. (1990) proposed the following daily dietary zinc intakes as adequate: 3–5 mg for infants, 5–10 mg for children, 9–18 mg for adults, and 13–25 mg for pregnant or lactating females (Sandstead et al., 1990). Similar values have been published by the US National Academy of Sciences (1989): 12–15 mg/day for adults, and 15–25 mg/day for pregnant or lactating females. The mean dietary zinc intakes of women in industrialized countries are listed in Table 15.

The zinc content of some foods is shown in Table 16. Zinc levels of 10–150 mg/kg of fresh edible portion are found in vegetables, with values as high as 550 mg/kg in mongo beans. In general, meat, eggs and dairy products contain more zinc than plants; liver is a particularly rich zinc source, with average values of 44–84 mg/kg of edible portion. High zinc levels are also found in wheat and rye germ, yeast and oysters; white sugar and pome and citrus fruits provide among the lowest, usually with < 1 mg/kg of fresh edible portion (Adriano, 1986; Scherz et al., 1986).

Information on the concentration, distribution and variation of zinc in the 234 food items comprising the US Food and Drug Administration's Total Diet Survey from 1982 to 1991 has recently been published (Pennington & Young, 1991; Pennington et al., 1995). Major food group sources of zinc (> 10% of daily intake) were identified as meat, mixed dishes and ready-to-eat cereals. Zinc contents in foods were presented as mean and median values per 100 g and per serving portion. Zinc intakes (mg/day) in the period

Table 15. Dietary zinc intake of women in industrial countries

Country	Age group in years (No. n)	Zinc (mg/day)	Method	Reference
Canada	30.0 ± 6.1 ($n = 100$)	10.1 ± 3.3	3-day diet record	Gibson & Scythes (1982)
USA (NHANES III)	20–29 ($n = 838$)	9.7 ± 0.28[a]	single 24-h recall	Alaimo et al. (1994)
USA (NHANES III)	20–29 ($n = 838$)	9.6 ± 0.26[a]	single 24-h recall	Alaimo et al. (1994)
Germany	25–34	8.9[b]	7-day protocol	Van Dokkum (1995)
Germany[b]	35–44	9.2[b]	7-day protocol	Van Dokkum (1995)
United Kingdom	25–34	8.2	7-day weighed record	Gregory et al. (1990)
United Kingdom	35–49	8.7	7-day weighed record	Gregory et al. (1990)
Ireland[b]	25–40 ($n = 122$)	9.4 ± 3.3	7-day dietary history	Van Dokkum (1995)
Netherlands[b]	22–50	9.7	2-day diet record	Van Dokkum (1995)
Sweden	30–79 ($n = 60$)	8.0	24-h duplicate diet composites	Van Dokkum (1995)
New Zealand	25–44	9.0 ± 6.0	single 24-h recall	LIZ (1992)
Australia	18–60+	11.2	semiquantitative food-frequency questionnaire	Baghurst et al. (1991)

[a] SEM = standard error of the mean
[b] Data compiled by Van Dokkum (1995).

Table 16. Zinc concentrations in some foodstuffs[a]

Food	Zinc concentration (mg/kg of edible portion)
Meat	
Beef	31.7 (25.9–42.1)
Mutton	31
Pork	19 (14–62)
Liver	44 (sheep)–84 (calf)
Kidney	3.7 (pig)–28 (sheep)
Poultry	
Chicken	08.5
Chicken liver	32
Turkey	20 (17–23)
Chicken eggs	8–20
Fish and seafood	
Sea fish	5 (haddock)–14 (anchovy)
Freshwater fish	4.8 (trout)–12 (eel)
Oysters	65–1600
Shrimps	23.1
Dairy products	
Butter	2.3
Cow's milk	3.8
Milk powder	21
Cheese	11–106
Fruit	
Apple	1.2 (0.4–2.2)
Banana	2.2
Fig	2.5
Stone fruits	0.2 (peach)–1.5 (cherry)
Berries	0.8 (grape)–2.5 (cranberry)
Exotic fruits	0.8 (mandarin)–9 (guava)
Nuts	5 (coconut)–48 (cashew nut)
Vegetables	
Vegetable fruits	10–30
Leaves, stems, flowers	13 (rhubarb)–140 (onions)
Roots and tubers	2.7 (potato)–170 (taro)
Legumes and oilseeds	124 (chickpea)–550 (mungo bean)
Carrot	6.4 (1.8–21)
Tomato	2.4 (0–2.5)
Lettuce, cabbage	2.2 (1.6–15)
Mushroom	3.9 (2.8–5)

Table 16 (contd.)

Food	Zinc concentration (mg/kg of edible portion)
Cereal products	
Whole grain	13 (rye)–45 (oats)
Flour	7.7 (rye)–34 (wheat wholemeal)
Germs	120 (wheat)–208 (rye)
Rye bread	8.6 (5.9–12)
Wheat bread	5 (2–8)
Wheat wholemeal bread	21
Corn flakes	3
Rolled oats	44 (35–69)
Brewer's yeast	80
Pasta	16 (10–22)

[a] From: Scherz et al. (1986).

1982–1989 ranged from 8.7 to 9.7 mg/day for women aged 60–65 and 25–30 years, respectively; comparable estimates for men were 12.9 and 16.4 mg/day.

In areas highly polluted with zinc, accumulation by plants, especially leafy vegetables, may occur. Machholz & Lewerenz (1989) reported zinc concentrations of 301 mg/kg for contaminated lettuce compared to 77 mg/kg in uncontaminated lettuce. In other crops, baseline zinc levels of 0.4–35 mg/kg were found, but in contaminated samples levels of 4–400 mg/kg were measured (Fiedler & Roesler, 1988).

Food processing can alter the zinc content of food and usually results in a decrease. For example, the zinc content of spinach is reduced by about 20% during freezing and thawing (Kampe, 1986); during the milling of wheat flour, up to 80% of zinc is removed. Increased zinc contents in acidic foods attributed to storage in galvanized zinc containers has been reported (Halsted et al., 1974).

5.2.3 Drinking-water

Zinc concentrations in drinking-water have been reported as follows: Canada, 10–750 µg/litre (Meranger et al., 1981);

Netherlands, 20–400 µg/litre (Zoeteman, 1978); and in other European countries, from 2 µg/litre in Bordeaux, France to 688 µg/litre in Frankfurt, Germany (Zoeteman, 1978). In general, concentrations of 1–2 mg/litre, rarely up to 5 mg/litre, may occur in water after passage of corrosive water through galvanized pipes or after standing in galvanized pipes, especially in combination with elevated chloride and sulfate concentrations (Hoell et al., 1986). On the basis of taste, such water would be considered of extremely poor quality for drinking (WHO, 1996a).

In USA, drinking-water from 35 areas (100–110 samples) was found to contain zinc concentrations of 0.025–1447 µg/litre zinc (Greathouse & Osborne, 1980). Median concentrations in water from galvanized pipes were about 10 times higher than those in water from copper pipes; for homes older than 5 years, reported values were 547 µg/litre and 70 µg/litre, respectively (Sharrett et al., 1982a,b).

5.2.4 Miscellaneous exposures

Intentional consumption of large doses of zinc supplements in excess of dietary intake and chronic treatment of patients with drugs containing zinc salts, e.g., injectable insulin, may result in high-level zinc exposure (Bruni et al., 1986). People with copper deficiency are at particular risk. Zinc exposure may also occur after extensive application of zinc-containing powder or ointments to wounds (Seeger & Neumann, 1985).

5.3 Occupational levels

Occupational exposure to dusts and fumes of metallic zinc and zinc compounds occurs during production of zinc (e.g., mining, smelting) and zinc compounds, and during their use. Many countries regulate workplace levels of zinc oxide fume and dust at levels between 5 and 10 mg/m^3 (ILO, 1991) to prevent adverse respiratory effects and metal-fume fever.

Size distribution and chemical composition of condensation aerosol particles generated in metallurgical plants in Germany during high-temperature processes were studied by Reiter & Poetzl (1985). Zinc concentrations detected during smelting of iron from scrap in induction furnaces averaged 0.190–0.287 mg/m^3 (main particle size 0.3–0.4 μm); during sprays and hot-dip galvanizing of tubes, 0.101 and 0.076 mg/m^3 (main particle size $\geq$ 0.09 μm) or 0.067 and 0.122 mg/m^3 (main particle size 0.3–0.4 μm); and during the electrolytical production of aluminium, 0.6 μg/m^3.

A zinc concentration of 0.540 mg/m^3 (range 0.110–0.800 mg/m^3) was measured in the breathing zone during welding of painted unalloyed steel in large rooms without local exhaust ventilation in Netherlands industries (Van der Wal, 1990). During decorative chrome plating, Lindberg et al. (1985) found zinc levels of 2.2–1.3 mg/m^3 in air. A collective of 35 workers (Germany) was exposed to 2.2 mg/m^3 zinc oxide in the breathing zone during welding of coated and uncoated steel (Zschiesche, 1988). Gun metal founders were exposed to a mean concentration of zinc oxide fumes of 0.680 mg/m^3 (Murata et al., 1987).

Personal breathing zone samples collected throughout the production area of a brass foundry (USA) for 7 employees contained zinc concentrations (time-weighted average) ranging from 4 μg/m^3 to 0.732 mg/m^3 depending on the working area (Clark et al., 1992). In a plant using zinc stearate releasing agent (USA), area samples were collected from lathe operators, a steam autoclave operator, and three areas near these operators. The total zinc concentrations ranged from 2 μg/m^3 to 0.120 mg/m^3 (Letts et al., 1991).

Zinc oxide concentrations were monitored at a non-ferrous foundry for three different job classifications from 1989 to 1990. The respirable fraction of personal air samples showed levels of 1.4 mg/m^3 for assistant caster-top, 1.34 mg/m^3 for assistant caster-pit, 0.896 mg/m^3 for caster, and 1.240 mg/m^3 for all casting positions; maximum values were 3.830 mg/m^3, 6.230 mg/m^3, 5.590 mg/m^3 and 6.230 mg/m^3, respectively. However, only 35% of the total collected zinc oxide was respirable. Total zinc oxide concentrations of up to 20.2 mg/m^3 were measured for caster positions (Cohen & Powers, 1994).

Borroni et al. (1986) measured zinc concentrations in the plating room of an electromechanical factory (Italy) before and after reorganization. Before reorganization, zinc concentrations were 39 μg/m^3 in the anticorrosive treatment plant, 0.388 mg/m^3 in the zinc barrel plating area, and 0.693 mg/m^3 in the zinc rack plating area; after reorganization the levels were 0.021, 0.003 and 0.009 mg/m^3, respectively. Dust exposure levels monitored during catalyst handling (loading and unloading) in the chemical industry (France) ranged from 0.210 to 2.18 mg/m^3 (mean values 0.400–1.29 mg/m^3) (Hery et al., 1991).

Airborne samples in a Philadelphia waste incinerator plant (USA) contained zinc at a level of 1.2 mg/m^3 in the personal breathing zone. With area sampling, the level was 0.0028 mg/m^3 (Bresnitz et al., 1992).

In Sweden, the average 8-h exposure of 12 painters to zinc from various water-based paints was reported to be < 0.001–0.080 mg/m^3 (mean 0.020 mg/m^3) (Wieslander et al., 1994).

5.4 Total human intake from all sources

5.4.1 General population

For humans, the most important route of exposure to zinc is through the ingestion of food. The dietary intakes of zinc in several countries are summarized in Table 17. Daily dietary intake ranges from 4.7 to 18.6 mg/day. Low zinc intakes have been reported for populations in Papua New Guinea, while intakes of zinc from vegetarian diets in India have been reported to be as high as 16 mg/day (WHO, 1996b). The major food sources of dietary zinc for adult women are outlined in Table 18.

The zinc requirement is mainly met by consumption of meat in omnivorous diets, or unrefined cereals, legumes and nuts diet patterns that are mostly vegetarian. High zinc concentrations are also found in seafood, especially oysters, whereas fruits and vegetables contain relatively low zinc concentrations. Absorption of dietary zinc is estimated to range from < 15% to 55%, depending on the composition of the diet; absorption is facilitated by foods containing animal protein.

5.4.2 Bioavailability in mammalian systems

"Bioavailability" is the term used to refer to the proportion of the external dose of a compound (in this case zinc) that is actually

93

Table 17. Estimated mean dietary intakes of zinc

Country	Zinc intake (mg/day)	Reference
Australia	adult males (18–60+ years) 12.8 adult females (18–60+ years) 11.2	
Germany	adults (25–34 years) 8.9 adults (35–44 years) 9.2 adults (45–54 years) 9.2	Van Dokkum (1995)
Germany	children (4–9 years) 5.3	Laryea et al. (1995)
India	8.0 16.1	Pfannhauser (1988) Adriano (1986)
Netherlands	male children (4–10 years) 7.7 female children (4–10) years) 7.1 adult males (22–50 years) 12.1 adult females (22–50 years) 9.7	Van Dokkum (1995)
New Zealand	adult females (15–65+ years) 9 adult males (15–65+ years) 13	LIZ (1992)
Ireland	male children (8–12 years) 10.1 female children (8–12 years) 8.9 adult males (25–40 years) 14.4 adult females (25–40 years) 9.4	Van Dokkum (1995)
United Kingdom	adult males 10.5–11.6 adult females 8.3–8.5 children (1.5–4.5 years) 4.3–4.8	Gregory et al. (1990)
USA	adults (20–80+ years) 8.8–12.4 infants (2–11 months) 6.0 children (3–11 years) 8.0–10.0 children (12–19 years) 12.3–13.0	Pennington et al. (1995); Alaimo et al. (1994)
USA (1994)	infants 5.6–6.3 adult males 12.3–13.3 adult females 8.4–8.9	Sandstead & Smith (1996)

absorbed by living organisms. In pharmacology and nutrition, the proportion utilized is also included. For mammalian systems, nutrient intakes calculated for food composition data or determined by direct chemical analysis represent the external dose. For most studies, the amount actually absorbed and utilized by the body (i.e., the internal dose) is much lower than the administered dose. Although there are certain common aspects of bioavailability, as discussed in section 4.2, there are various other factors that can affect the bioavailability of zinc in mammals (Table 19). These include the chemical form of the nutrient (speciation), the composition of the food ingested (e.g., fibre and phytic acid content), the body stores of the nutrient and related chemicals, the physiological status of the organism, and nutrient and nutrient–dietary interactions (WHO, 1996b).

Bioavailability from foods of plant origin is impaired by inositol phosphate (phytate), and possibly components of dietary fibre, increased levels of calcium in the presence of phytate, and certain metals, if consumed at high levels as dietary supplements (Sandstead, 1981) (see section 6.1.2.1).

Mean daily intake of zinc from drinking-water is estimated to be < 0.01 mg/day (Cleven et al., 1993), but may be higher due to water treatment or zinc leaching from transmission and distribution pipes, especially at low pH. A study in Seattle (USA) revealed zinc levels of 2 mg/litre in standing and 1.2 mg/litre in running water from galvanized pipes and 0.44 and 0.16 mg/litre in standing and running water from copper pipes, respectively (Sharrett et al., 1982b). Where drinking-water is drawn from systems with corroded fittings, galvanized piping or private wells, it can provide up to 10% of the daily zinc intake (0.5–1 mg/day) (WHO, 1996a). However, in general, this source provides only a small part (< 0.1 mg/day) of the total oral intake.

In conclusion, the total intake of zinc from all environmental sources by the general adult population varies between 4.7 and 16 mg/day. In most circumstances, over 95% of this comes from food, with negligible amounts from air, and between <1% and 10% from drinking-water. Comparison of the recommended daily intakes (section 5.2.2) and the lower intakes shown in Table 17 indicates that the risk of zinc deficiency is a worldwide public health concern (see section 8.3.5).

Table 18. Major food sources of dietary zinc for adult women

Study	Bread and cereal products	Meats, eggs, legumes, nuts and seeds	Vegetables and fruit	Milk and dairy products	Fats, sugars, beverages, alcohol	Reference
Canada. Women in 1982 aged 30.0 ± 6.1 years	19.0	43.0	12.5	23.7	1.7	Gibson & Scythes (1982)
USA. Women in 1986 (NFCS, CSF11 Report No. 86-3)	22.5	51.3	8.1	14.2	4.4	Moser-Veillon (1990)
Germany. Women (National Food consumption Survey, 1985–1988)	21.0[b]	18.0[a]	?	24.0	?	Van Dokkum (1995)
United Kingdom. Women aged 16–64 years (n = 2200)	28.0	36.0[a]	?	15.0	?	Gregory et al. (1990)
Ireland Adults >18 years (n = 676) plus children 8–18 years (n = 538)	12–24[b]	35–49[a]	5–9[c]	9–20	?	Van Dokkum (1995)

Table 18 (contd.)

Netherlands Women (National Food Consumption Survey in 1987–1988)	16[b]	28.0[a]	?	28.0	?	Van Dokkum (1995)
New Zealand. Women aged 23–50 years (Market Basket Survey based on 8.4 MJ energy intake)	17.0	45.4[d]	1.5	13.2	< 20.5	Pickston et al. (1985)
Australia. National sample of women aged 18–60+ years (n = 763)	12.3[e]	13.3	?	15.5[f]	?	Baghurst et al. (1991)

[a] Meat only
[b] Bread only
[c] Potatoes only
[d] Meat, fish and eggs
[e] Bread and breakfast cereal
[f] Milk and cheese only.

Table 19. Factors affecting the bioavailability of zinc in the diet for mammals

Factor	Major food
Extrinsic	
Diet	chemical form of element in diet
	presence of competitive antagonism between ions (e.g. Cu–Zn; Cd–Zn; Fe–Zn; Ca–Zn; Mg–Zn)
	antagonistic ligands: decrease in gastrointestinal lumen solubility of zinc (products of Maillard browning; lignin; casein phosphopeptides; ethanol; inositol hexaphosphates, i.e., phytate)
	ligands that enhance zinc absorption (e.g. citrate; histidine; cysteine; picolinic acid)
Intestinal lumen	pH and redox state
Intrinsic	
Genetic influences	inhibitor condition: inborn absorption errors (e.g. acrodermatitis enteropathica)
Age	infants: poor postnatal regulation of zinc absorption
	elderly: reduced efficiency of zinc absorption with age
Metabolic function	enhancer conditions: growth in infancy, childhood; pregnancy and lactation
Homeostatic regulation	enhancers: feedback stimulation of absorption in deficiency
	inhibitors: decreased absorption when excess amounts are available and nutritional status is normal (mechanism unclear: possibly metallothionein)
	adaptation to low zinc intake and existing low zinc status by increasing absorption of exogenous zinc and conserving endogenous zinc
Physiological stress	
Disease	intestinal malabsorption syndromes

5.4.3 Occupational exposure

Levels of zinc in the diverse facilities worldwide that manufacture, utilize or repair zinc and zinc compounds vary widely (see section 5.3) and are largely dependent upon the quality of industrial hygiene practices. It is thus difficult to estimate with certainty the exposure of workers globally. In the non-ferrous metal industries, where levels of zinc can be high (0.8–1.3 mg/m³), workers may inhale an amount of zinc about equivalent to that taken orally by food (see Table 17). However, it is the level of zinc on respirable particulates and absorption from the lung that will determine the amount absorbed (see Chapter 6). In most other industries in developed countries, the intake of zinc by workers will be lower (< 10 µg to 5 mg per shift).

6. KINETICS AND METABOLISM IN MAMMALS

6.1 Absorption

6.1.1 Inhalation

6.1.1.1 Human studies

While quantitative data on the absorption of zinc following inhalation exposures were not determined, the increased zinc levels demonstrated in plasma, blood and urine of occupationally exposed workers indicated that absorption from the pulmonary tract does occur (Hamdi, 1969; Trevisan et al., 1982).

6.1.1.2 Animal studies

Gordon et al. (1992) exposed guinea-pigs, rats and rabbits (nose-only) to zinc oxide aerosols at concentrations of 4.3–11.3 mg/m^3 for 3 h and guinea-pigs for 6 h. Particle mass median diameter was 0.17 μm. Retention values were 19.8% and 11.5% in rabbits, and 4.7% in the lungs of guinea-pigs and rats.

6.1.2 Oral

6.1.2.1 Human studies

In humans, the absorption of zinc in the diet ranges widely. Bioavailability can be affected by abnormalities in the gastro-intestinal tract, in transport ligands or in substances that interfere with zinc absorption. A decreased absorption was noted for elderly subjects. Bioavailability also depends on the amount of zinc ingested or the amount and kind of food eaten (Sandstroem & Cederblad, 1980; Aamodt et al., 1982, 1983; Istfan et al., 1983; Seal & Heaton, 1983; Bunker et al., 1984, 1987; Wada et al., 1985; Hunt et al., 1987, 1991; Sandstroem & Abrahamson, 1989; Sandstroem & Sandberg, 1992).

A significantly reduced absorption of zinc in humans and laboratory animals was observed after oral uptake of phytate (from

100

grain and vegetable components) owing to the formation of insoluble zinc-phytate-complexes in the upper gastrointestinal tract (O'Dell & Savage, 1960; Oberleas et al., 1962; Reinhold et al., 1973, 1976; Davies & Nightingale, 1975; Solomons et al., 1979; Loennerdal et al., 1984; Turnland et al., 1984; Sandstroem et al., 1987; Ferguson et al., 1989; Ruz et al., 1991; Sandstroem & Sandberg, 1992). In a study with human volunteers, the absorption of zinc decreased with increasing gastric pH (Sturniolo et al., 1991). Other components that have been shown to reduce the availability of zinc are binding to casein and its phosphopeptides as a result of tryptic or chymotryptic digestion. Maillard products, dairy products such as milk or cheese, and interactions with calcium in the diet, coffee or orange juice (Pécoud et al., 1975; Walravens & Hambidge, 1976; Spencer et al., 1979; 1992; Harzer & Kauer, 1982; Flanagan et al., 1985; Lykken et al., 1986). The availability of zinc from diets rich in foods prepared from unrefined cereals tends to be poor owing to the content of phytate, fibre and lignin (Prasad et al., 1963a; Reinhold et al., 1973, 1976; Pécoud et al., 1975; Solomons et al., 1979).

The mechanism and control of zinc absorption from the intestine has not yet been fully elucidated, although absorption is known to be regulated homoeostatically, and depends on the pool of zinc in the body and the amount of zinc ingested. In humans and laboratory animals, increased uptake is associated with decreased absorption and increased excretion. Persons with adequate nutritional levels of zinc absorb approximately 20–30% of all ingested zinc, while greater proportions of dietary zinc are absorbed in zinc-deficient subjects if presented in a bioavailable form. Both a passive, unsaturable pathway and an active, saturable carrier-mediated process are involved. At low luminal zinc concentrations the binding of zinc is to specific sites, whereas at higher concentrations a non-specific binding occurs (Smith et al., 1978; Davies, 1980; Smith & Cousins, 1980; Flanagan et al., 1983; Istfan et al., 1983; Menard & Cousins, 1983; Cousins, 1989; Lee et al., 1989b; Oestreicher & Cousins, 1989; Tacnet et al., 1990; Gunshin et al., 1991; Hempe & Cousins, 1991, 1992).

In a study with human volunteers, most of the zinc in a zinc acetate solution (0.1 mmol/litre) administered by intestinal perfusion was absorbed from the jejunum, followed by the duodenum and the ileum (357, 230 or 84 nmol/litre per min per 40 cm respectively).

The absorption showed a linear increase at concentrations of 0.1–1.8 mmol/litre (Lee et al., 1989b).

6.1.2.2 Animal studies

Metallothionein and a low-molecular-mass zinc-binding protein, cysteine-rich intestinal protein (CRIP), which was isolated from intestinal mucosa of rats, play an important role in the gastrointestinal absorption of zinc. Like several other metals, zinc can rapidly induce metallothionein production in intestinal mucosal cells, liver, pancreas, kidney and lungs; in the intestine in particular, binding to metallothionein leads to retention of zinc and may so prevent absorption of excess zinc into the body (Richards & Cousins, 1975, 1976; Hall et al., 1979; Foulkes & McMullen, 1987; Bremner & Beattie, 1990; Rojas et al., 1995). In rats, a direct correlation between dietary zinc intake and the binding of zinc to mucosal metallothionein was observed following administration of 30–900 mg/kg (Hall et al., 1979). In rats fed a low-zinc diet, more zinc was associated with CRIP and lesser amounts were bound to metallothionein (40% compared to 4%), while with a high-zinc diet lesser amounts were associated with CRIP and most zinc was bound to metallothionein (14% compared to 52–59%) (Hempe & Cousins, 1991, 1992).

In the rat, the major site of zinc absorption was shown to be the duodenum followed by the more distal portions of the small intestine; absorption was rapid. Only minimal amounts were absorbed from the stomach, caecum and colon (Methfessel & Spencer, 1973; Davies, 1980).

6.1.3 Dermal

6.1.3.1 Human studies

An increase in serum zinc levels was observed in 8 burn patients treated with adhesive zinc-tape (zinc oxide content 7.5 ± 0.05 g/100 g dw); the maximum value (28.3 μmol/litre) was reached within 3–18 days of treatment (Hallmans, 1977).

The mean release rate of zinc to normal skin to which a zinc oxide (25%) medicated occlusive dressing was applied was 5 μg/cm^2

per hour. After 48 h of treatment, a 6- to 14-fold increase in zinc concentration in the epidermis was noted. The zinc flux was found to increase with decreasing pH (Ågren, 1990). The transport of zinc through intact human skin was enhanced by gum rosin (Ågren, 1991).

6.1.3.2 Animal studies

After topical application of zinc chloride (as ^{65}Zn) at a concentration of 0.005–4.87 mol/litre to guinea-pigs at pH values in the range 1.8–6.1, the loss of radioactivity in most cases was < 1% within 5 h; at a concentration of 0.08 mol/litre at pH 1.8 only, an increased loss of radioactivity (1–3% within 5 h) was observed. An increase of radioactivity in liver, kidney, intestine and faeces was noted (Skog & Wahlberg, 1964).

Two applications of zinc were administered to rabbits (3 per group) as zinc oxide, zinc omadine, zinc sulfate or zinc undecylenate. Each application provided 2.5 mg of zinc containing 5 µCi of ^{65}Zn. The animals were killed 6 or 24 h after the second application. All the zinc compounds were absorbed equally after one or two applications. The ^{65}Zn retention on excised skin blocks ranged from 3% to 65% of the applied dose. ^{65}Zn was located mostly in the keratogenous zone of the hair shaft and in the subcutaneous muscle layer (Kapur et al., 1974).

Keen & Hurley (1977) studied the effect of topical application of zinc chloride in female Sprague-Dawley rats (5–7 per group). Four groups were fed a zinc-deficient diet for 24 h. Half of the animals were treated during this period with a topical application of oil saturated with zinc chloride, for the full 24 h in one group, and for the last 8 h in the other. Plasma zinc levels in rats receiving zinc supplementation for 8 h were similar to those of the control group (114 µg/100 ml), while levels in rats receiving zinc supplementation for 24 h were significantly increased (182 µg/100 ml).

In a comparative study with zinc oxide and zinc sulfate in rats with full-thickness skin excision, the application of zinc oxide resulted in a sustained delivery of zinc ions, while zinc sulfate delivered zinc ions rapidly, resulting in decreasing wound-tissue zinc levels. About 450 µg of zinc (12% of the initial dose) was delivered to each wound from the zinc oxide dressing and about 650 µg of zinc (65% of the initial dose) from the zinc sulfate dressing over a 48-h treatment period (Ågren et al., 1991).

6.2 Distribution

Zinc is one of the most abundant trace metals in humans and is found in all tissues and all body fluids. The total zinc content of the human body (70 kg) is in the range 1.5–3 g. Most of this is found in muscle ($\approx$ 60%), bone ($\approx$ 30%), skin and hair ($\approx$ 8%), liver ($\approx$ 5%) and gastrointestinal tract and pancreas ($\approx$ 3%). In all other organ systems, the zinc content is $\leq$ 1% (Wastney et al., 1986; Aggett, 1994). After ingestion, zinc in humans is initially transported to the liver and then distributed throughout the body (Aamodt et al., 1979). Glucocorticoids have been shown to enhance the uptake of zinc to liver cells *in vitro* (Failla & Cousins, 1978). Interleukin-1 and ACTH also cause increased liver uptake of zinc (Sandstead, 1981; Hambidge et al., 1986). Glutathione may be involved in the release of zinc from intracellular protein ligands and its transfer to the blood by forming complexes in the mucosa, which pass by passive diffusion across basolateral membranes (Foulkes, 1993).

The highest concentrations of zinc in humans were found in liver, kidney, pancreas, prostate and eye (Forsséen, 1972; Yukawa et al., 1980; Hambidge et al., 1986). Zinc is also present in plasma, erythrocytes and leukocytes. In healthy subjects, the normal plasma zinc concentration is $\cong$ 1 mg/litre (Spencer et al., 1965; Juergensen & Behne, 1977; Whitehouse et al., 1982; Ohno et al., 1985). Zinc is mostly bound to albumin (60–80%) and to a lesser extent to α-2-macroglobulin and transferrin (Prasad & Oberleas, 1970; Giroux et al., 1976; Smith & Cousins, 1980; Wastney et al., 1986; Bentley & Grubb, 1991).

After oral uptake in humans, peak levels in plasma are reached within 3 h (Nève et al., 1991; Sturniolo et al., 1991). An excess of dietary zinc in humans and animals resulted in high concentrations accumulated in kidneys, liver, pancreas and bone (Allen et al., 1983; Bentley & Grubb, 1991; Schiffer et al., 1991). As shown by Giugliano & Millward (1984) in a study with zinc-deficient rats, a redistribution of zinc from bone mainly into muscle may occur; the authors described a marked increase in muscle zinc with a similar loss from bone.

Placental transfer of zinc in pregnant ewes has also been demonstrated (James et al., 1966). In an isolated perfused single-cotyledon human term placental model, the normal zinc transfer was shown to be slow. Only up to 3% of maternal zinc reaches the fetal compartment in 2 h (Beer et al., 1992). The uptake of zinc in the placenta seems to involve a potassium-dependent zinc transport mechanism, as shown in studies with microvilli isolated from human term placenta (Aslam & McArdle, 1992). Baseline values for zinc were measured in parenchyma, membrane and cord from placental tissue taken from 23, 24 and 22 healthy pregnant women, respectively. Ranges reported were: parenchyma, 12.8–89.9 µg/g dry tissue; membrane, 21.4–80.3 µg/g dry tissue; and cord, 13.7–97.2 µg/g dry tissue (Centeno et al., 1996).

6.3 Excretion

In humans, most ingested zinc is eliminated in the the faeces (5–10 mg/day), and comprises unabsorbed zinc and endogenous zinc from bile, pancreatic fluid and intestinal mucosa cells. In humans and animals, a considerable amount of zinc is excreted into the small intestine through pancreaticobiliary secretion (Davies, 1980; Matseshe et al., 1980; Johnson & Evans, 1982). In rats, biliary zinc excretion seems to be a glutathione-dependent process; glutathione probably acts as a carrier molecule (Alexander et al., 1981). Human pancreatic secretions contain zinc levels of 0.5–5 µg/ml (Hambidge et al., 1986). As shown by Spencer et al. (1965), up to 18% of an intravenous dose of radiolabelled zinc was excreted into the intestine within 45 days. In a study by Matseshe et al. (1980) in healthy volunteers, zinc was recovered from the duodenum at levels greater than the dose ingested.

Zinc is also reabsorbed from the intestine. In a study with ligated duodenal and ileal loops of rats, approximately 35% of zinc secreted into the gut lumen was reabsorbed (Davies & Nightingale, 1975). In humans, a mean reabsorption of 70% of the administered dose (45 mg of zinc as zinc sulfate) was demonstrated by Nève et al. (1991).

Yokoi et al. (1994) measured the disappearance of zinc from blood after an injection of isotope in women with normal (> 20 μg/litre) and low (< 20 μg/litre) levels of ferritin. The disappearance of zinc from blood in women with low ferritin was accelerated in the range usually found in subjects with zinc deficiency. This confirmed the early studies of Prasad (1963).

Rats fed a diet supplemented with zinc oxide at a rate of ≅ 96–672 mg/kg per day (1200–8400 ppm) for 21 days (Ansari et al., 1976) or ≅ 48 mg/kg per day (600 ppm) for 7–42 days (Ansari et al., 1975) showed a linear increase in excretion in the faeces with an increase in dietary intake.

In healthy humans, only small amounts (≈ 0.5 mg/day) are excreted via urine (Halsted et al., 1974; Elinder et al., 1978). In patients with taste and smell dysfunction given zinc sulfate at a rate of 8–13 mg/day for 290–440 days followed by an additional 100 mg/day over the next 112–440 days, a 188% increase in daily renal excretion with only a 37% increase in plasma zinc was observed; this was possibly due to an increase in filtration and/or decreased reabsorption in the kidney (Babcock et al., 1982). In zinc clearance studies in anaesthetized dogs, proximal secretion and distal reabsorption in the nephron was described (Abu-Hamdan et al., 1981).

Hohnadel et al. (1973) reported that zinc concentrations in cell-free sweat from healthy human subjects averaged 0.50 mg/litre (range 0.13–1.46 mg/litre) in 33 men and 1.25 mg/litre (range 0.53–2.62 mg/litre) in 15 healthy women. Zinc in sweat appeared to be related to the level of dietary zinc in three studies in male volunteers (Milne et al., 1983), with whole body sweat losses averaging 0.49 mg/day when zinc intakes averaged 8.3 mg/day, compared with losses of 0.29 mg/day (range 0.18–0.38 mg/day) with a zinc intake of 3.6 mg/day, and 0.62 mg/day (0.46 and 0.77 mg/day in two studies) when zinc intake was 33.7 mg/day

6.4 Biological half-life

After oral, intravenous or intraperitoneal application, equilibrium is quickly achieved between plasma zinc and a rapidly changing non-plasma pool, probably located within the liver

(Fairweather-Tait et al., 1993). In humans, the liver is the organ of highest initial zinc uptake, and with a slow turnover (Spencer et al., 1965). A kinetic two-compartment model can be applied for the estimation of the rapid initial flow of zinc out of the plasma to the liver following administration of radiolabelled zinc (Foster et al., 1979; Aamodt et al., 1979). After intravenous injection of ^{65}Zn, the average biological half-life of zinc in the smaller compartment was 12.5 days and the turnover of the larger compartment averaged 322 days (Spencer et al., 1965). In another human isotope study, the estimated half-life was approximately 280 days (Wastney et al., 1986).

In the rat, the biological half-life of ^{65}Zn decreased with an increase in dietary zinc (5 mg/kg, 52 days; 160 mg/kg, 4 days) (Coppen & Davies, 1987).

6.5 Zinc status and metabolic role in humans

6.5.1 *Methods for assessment of zinc status in humans*

Diagnosis of zinc deficiency in humans is hampered by the lack of a single, specific and sensitive biochemical index of zinc status. A large number of indices have been proposed, but many are fraught with problems that affect their use and interpretation. At present, the most reliable method for diagnosing marginal zinc deficiency in humans is a positive response to zinc supplementation. Such an approach is time-consuming; it necessitates good compliance with follow-up visits, making it impractical for community studies. Consequently, dietary and/or static and functional biochemical and physiological functional indices are frequently used to evaluate zinc status. However, some of these indices are affected by biological and technical factors other than depleted body stores of zinc, which may confound the interpretation of the result. The potential impact of these confounding factors should be taken into account by standardizing sample collection and analytical procedures.

6.5.1.1 *Dietary methods to predict the proportion of the population at risk of inadequate intakes of dietary zinc*

A quantitative dietary assessment method designed to measure the quantity of foods consumed by an individual over more than one day must be used to calculate the proportion of the population at risk of inadequate intakes of dietary zinc. The number of days required depends on the day-to-day variation in zinc intakes for the population group under study. Suitable methods include recalls and records and, in some circumstances, semi-quantitative food frequency questionnaires. A detailed description of these methods can be found in Gibson (1990). Energy, nutrient and antinutrient intakes can be calculated from the quantitative food consumption data using food composition tables or a nutrient data bank. Alternatively, chemical analyses of representative samples of staple foods collected from the study area can be performed.

The adequacy of the zinc intakes can then be evaluated by comparison with an appropriate set of tables of recommended nutrient intakes for the population group under study. Several such tables are available; they are discussed in detail in Gibson (1990). For studies in developing countries, the newly revised requirement estimates for zinc set by WHO (1996b) can be used. Because the adequacy of dietary zinc depends on both its amount and bioavailability in the diet, however, an estimate of the bioavailability of zinc in the diets under study is also required for this evaluation. Direct measurements of the bioavailability of zinc in the plant-based diets consumed in many developing countries are limited; some have been made using metabolic studies or stable isotope techniques (WHO, 1996b). The zinc absorption data have been used by WHO (1996b) to develop a model for classifying diets as having high, moderate and low zinc bioavailability. The model is based on the dietary content of animal and/or fish protein calcium (< 1 or > 1 g/day), and phytate/zinc molar ratios (< 5, 5–15 or > 15) per day.

Once the bioavailability of the zinc in the diets has been estimated, the zinc intakes can be evaluated, preferably using the probability approach, which attempts to assess more reliably the risk of nutrient inadequacy both for an individual and for the population. The method predicts the number of persons within a group with nutrient intakes below their own requirements and provides an estimate of the population at risk, or the prevalence of inadequate intakes. For the individual, the method estimates the relative probability that the zinc intake does not meet his or her actual requirement (Anderson et al., 1982).

Probability estimates for risk of zinc deficiency do not identify actual individuals in the population who are deficient or define the severity of the zinc inadequacy, however. Such information can only be obtained when the dietary intake data are combined with biochemical and functional physiological indices of zinc status. This is especially important in developing countries, where the coexistence of many other multifaceted health problems often confounds the diagnosis of zinc deficiency.

6.5.1.2   *Static tests of zinc status*

Static tests measure the total quantity of zinc in various accessible tissues and body fluids, such as blood or some of its components, urine, hair or nails. Ideally, the tissue or fluid selected should reflect total body content of zinc, or at least the size of the body pool most sensitive to zinc depletion. Unfortunately, the tissues containing the most zinc (i.e., bone and muscle) are not readily accessible for human studies, and their zinc content is not measurably reduced, even in severe zinc deficiency. Consequently, the choice of biopsy materials for static tests is based primarily on their accessibility, convenience and ethical acceptability (Aggett, 1991).

Serum/plasma

Serum/plasma zinc is the most widely used index of zinc status in humans. Only a small proportion ($< 1\%$) of body zinc circulates in plasma. Hence, plasma zinc does not necessarily reflect total body zinc content. Nevertheless, in persons with severe zinc deficiency, serum/plasma zinc concentrations are usually low (Arakawa et al., 1976; Hess et al., 1977; Prasad et al., 1978a; Gordon et al., 1982; Baer & King, 1984). Concentrations return to normal following zinc supplementation.

Serum/plasma zinc concentrations are not useful for detecting mild zinc deficiency states, when values are often within the normal range (Milne et al., 1987; Gibson et al., 1989a; Ruz et al., 1991). Serum/plasma zinc is also not very specific. Concentrations are modified by a number of non-nutritional factors, some of which (e.g., acute infection, inflammation and stress) decrease levels by inducing hepatic uptake of zinc (Beisel et al., 1976). During periods of rapid tissue synthesis, pregnancy and use of oral contraceptive agents, serum zinc levels are also decreased (Swanson et al., 1983; Breskin et al., 1983; King, 1986). Chronic disease states associated with hypoalbuminaemia also induce low serum zinc; zinc circulates in serum bound principally to albumin.

Serum/plasma zinc concentrations are also affected by haemolysis: erythrocytes have a high zinc content. Haemolysis may be particularly important in cases of zinc deficiency, when red cell fragility is increased (Bettger et al., 1978). In addition, blood samples should be taken under carefully controlled conditions standardized with respect to time of day, fed or fasted state, position of the subject during blood collection, refrigeration of blood samples and length of time prior to the separation of serum/plasma; all these factors influence serum/plasma zinc concentrations (Markowitz et al., 1985; Wallock et al., 1993; Tamura et al., 1994).

Blood samples for zinc analysis must also be taken carefully to avoid contamination from sources such as preservatives, evacuated tubes, rubber stoppers and anticoagulants. Certain anticoagulants (e.g., citrate, oxalate and EDTA) efficiently chelate metallic ions; if these agents are used, plasma zinc values will be lower than if zinc-free heparin is used. The cut-off point generally used to assess risk of zinc deficiency for both plasma and serum values is < 10.71 μmol/litre (< 70 μg/dlitre), a value approximately two standard deviations below the adult mean. This value may only be appropriate for morning fasting blood samples. For nonfasting morning, and for afternoon samples, lower cut-off points of < 9.95 μmol/litre (< 65 μg/dlitre) and < 9.18 μmol/litre (< 60 μg/dlitre), respectively, have been recommended.

Erythrocytes

Relatively few investigators have used erythrocytes as a biopsy material for assessing zinc status because the analysis is difficult and the response during experimentally induced zinc depletion–repletion studies has been equivocal (Prasad et al., 1978a; Baer & King, 1984; Ruz et al., 1992). The half-life of erythrocytes is quite long (120 days), so that erythrocyte zinc concentrations will not reflect recent changes in body zinc stores.

Zinc uptake by erythrocytes is influenced by many other factors, including protein intake, stress and endotoxins (Chesters & Will, 1978). Age-related changes in erythrocyte zinc in infants and children (Nishi, 1980) and adolescent females have been reported (Kenney et al., 1984).

Leukocytes

Leukocytes have a shorter half-life than erythrocytes and should therefore reflect changes in zinc status over a shorter time-period. They also contain up to 25 times more zinc than erythrocytes. Concentrations of zinc in mixed leukocytes and specific cellular types (e.g., neutrophils and lymphocytes) have been examined as potential indices of zinc status in humans. Some investigators (Prasad & Cossack, 1982) but not all (Milne et al., 1985; Ruz et al., 1992) have suggested that they are more reliable as indices of zinc status than plasma zinc. Relatively large volumes of blood are required, and isolation of the leukocytes and their subsequent analysis is lengthy and technically difficult, limiting the use of these indices, especially for infants and young children.

Milne et al. (1985) have emphasized that the zinc content of leukocytes is a function of the type of separation used; contamination with zinc from the anticoagulant, reagents, the density gradient system or erythrocytes and platelets may occur. Changes in the relative proportions of leukocyte subsets with physiological state (e.g., pregnancy) and haematological disorders (Aggett, 1991) must also be taken into account in the interpretation of the results. Finally, comparison of results between different studies is difficult because no consensus exists as to how to express zinc concentrations in the cell types (Thompson, 1991).

Urine

Depletion of body zinc stores causes a reduction in urinary zinc excretion (Hess et al., 1977; Ruz et al., 1991), often before any detectable changes in serum/plasma zinc concentrations (Baer & King, 1984). Supplementation with high (100 mg) but not moderate (50 mg) zinc intakes increases urinary zinc excretion (Verus & Samman, 1994). Several factors can affect urinary zinc concen-trations, however, making interpretation of the results difficult. For example, despite the presence of zinc deficiency in sickle-cell anaemia, hyperzincuria occurs (Prasad, 1985). Hyperzincuria is also present in disorders such as cirrhosis of the liver and diabetes mellitus, after injury, burns and acute starvation, in certain renal diseases and infections, and after treatment with chlorothiazide. Hypertensive patients on long-term therapy with chlorothiazide may therefore be vulnerable to zinc deficiency (Prasad, 1983). The measurement of zinc in urine is therefore helpful for diagnosing zinc deficiency only in apparently healthy persons. Zinc levels in the urine usually range from 300 to 600 µg per day. In general, 24-h urine collections are preferred because diurnal variation in urinary zinc excretion occurs.

Hair

The use of hair zinc concentrations as an index of zinc status has been controversial (Hambidge, 1982). Available evidence suggests that low zinc concentrations in hair samples collected during infancy and childhood probably reflect a chronic suboptimal zinc status when the confounding effect of severe protein-energy malnutrition is absent (Hambidge et al., 1972a; Gibson, 1980; Smit Vanderkooy & Gibson, 1987; Gibson et al., 1989b). Hair zinc cannot be used in cases of very severe malnutrition and/or severe zinc deficiency, when the rate of growth of the hair shaft is often diminished. In such cases, hair zinc concentrations may be normal or even high (Erten et al., 1978; Bradfield & Hambidge, 1980).

Low hair zinc concentrations have been reported in infants and children with impaired linear growth (Hambidge et al., 1972; Walravens & Hambidge, 1976; Buzina et al., 1980; Walravens et al., 1983; Smit Vanderkooy & Gibson, 1987; Gibson et al., 1989b; Ferguson et al., 1989) and taste acuity (Hambidge et al., 1972; Gibson et al., 1989a; Cavan et al., 1993), two clinical features of mild zinc deficiency in children. Moreover, in some of these studies, the low hair zinc concentrations have been related to low availability of dietary zinc (MacDonald et al., 1982; Ferguson et al., 1989; Cavan et al., 1993). In some but not all of these cases of suboptimal zinc status, hair zinc concentrations have increased in response to zinc supplementation. The discrepancies may arise from variations in the dose, duration of zinc supplementation, and confounding effects of season on hair zinc concentrations. Periods of 6 weeks or less are probably too short for a response, since hair zinc reflects only chronic changes in zinc status (Greger & Geissler, 1978; Lane et al., 1982). Unfortunately, when studies are made over a longer term, seasonal changes in hair zinc concentrations must also be taken into account when interpreting the results (Hambidge et al., 1979; Gibson et al., 1989a).

Standardized procedures for sampling, washing and analysing hair samples are essential in all studies. Variations in hair zinc concentrations with hair colour, hair beauty treatments, season, sex, age, anatomical site of sampling (scalp or pubic), and rate of hair growth have been described (Hambidge, 1982; Taylor, 1986; Klevay, 1987; Gibson et al., 1989b). The effects of these possible confounding factors must be considered in the interpretation of hair zinc concentrations.

Many investigators have failed to find any positive correlations between the zinc content of hair and serum/plasma zinc concentrations (Klevay, 1970; Lane et al., 1982; Gibson et al., 1989b). These findings are not unexpected. The zinc content of the hair shaft reflects the quantity of zinc available to the hair follicles over an earlier time interval. Positive correlations between hair zinc concentrations and serum zinc are only observed in chronic, severe zinc deficiency in the absence of confounding factors.

Clinical signs of marginal zinc deficiency in childhood, such as impaired growth, poor appetite and reduced taste acuity, are usually associated with hair zinc concentrations of less than 70 µg/g (1.07 µmol/g) (Hambidge et al., 1972; Smit Vanderkooy & Gibson, 1987) Therefore, this value is frequently used as the cut-off point for hair zinc concentrations indicative of suboptimal zinc status in children. The validity of hair zinc as a chronic index of suboptimal zinc status in adults is less certain, and further studies are required.

Saliva

Zinc concentrations in mixed saliva, parotid saliva, salivary sediment and salivary supernatant have all been investigated, but their use as indices of zinc status is equivocal (Greger & Sickles, 1979; Freeland-Graves et al., 1981; Lane et al., 1982; Baer & King, 1984).

6.5.1.3 Functional tests of zinc status

Functional tests measure changes in the activities of certain enzymes or blood components that are dependent on zinc. Alternatively, physiological functions dependent on zinc, such as growth, taste acuity and immune competence, can be assessed. Such tests have greater biological significance than static biochemical tests because they measure the extent of the functional consequences of zinc deficiency. Nevertheless, because functional physiological tests are not very specific, they must always be interpreted in combination with a biochemical test.

Zinc-dependent enzymes

Over 300 zinc metallo-enzymes have been identified. They vary in their response to zinc deficiency, depending on the tissues examined, their affinity to zinc, and rate of turnover of the enzyme (Cousins, 1986). The activity of serum alkaline phosphatase is the most widely used to assess zinc status (Adeniyi & Heaton, 1980), although its response has been inconsistent in humans (Ishizaka et al., 1981; Nanji & Anderson, 1983; Baer et al., 1985; Weismann & Hoyer, 1985; Milne et al., 1987). In general, activity is reduced in severe zinc deficiency states (Kasarskis & Schuna, 1980; Sachdev et al., 1990) but this parameter is probably not sensitive enough for detecting mild zinc deficiency (Walravens & Hambidge, 1976; Walravens et al., 1983, 1989; Ruz et al., 1991). Response of the enzyme during zinc supplementation studies has been inconsistent (Hambidge et al., 1983; Walravens et al., 1983; Weismann & Hoyer, 1985; Gibson et al., 1989b. Measurements of alkaline phosphatase activity in neutrophils (Prasad, 1983, 1985), leukocytes (Baer et al., 1985; Schilirò et al., 1987), erythrocytes (Samman et al., 1996) and red cell membranes (Ruz et al., 1992) have also been investigated as indices of body zinc status in humans. Although some promising results have been obtained, more studies are needed before any definitive conclusions can be reached.

Other zinc metallo-enzymes that have been investigated as indices of zinc status in humans include δ-amino-laevulinic acid dehydratase in erythrocytes (Faraji & Swendseid, 1983; Baer et al., 1985), angiotensin-1-converting enzyme (Reeves & O'Dell, 1985; Ruz et al., 1992), α-D-mannosidase in serum/plasma (Apgar &

Fitzgerald, 1985) and nucleoside phosphorylase in whole, lysed cells (Prasad & Rabbani, 1981; Ballester & Prasad, 1983). To date, there is no universally accepted zinc-dependent enzyme that can be used to assess marginal zinc deficiency in humans.

Taste acuity tests

Diminished taste acuity (hypogeusia) is a feature of marginal zinc deficiency in children (Hambidge et al., 1972: Buzina et al., 1980; Gibson et al., 1989a,b; Cavan et al., 1993) and adults (Wright et al., 1981; Henkin, 1984). Several methods for testing taste acuity have been used (Desor & Maller, 1975; Bartoshuk, 1978). In studies of Canadian and Guatemalan children, significant inverse relationships between the detection threshold for salt and hair zinc concentrations were noted (Gibson et al., 1989a,b; Cavan et al., 1993). These results suggest that impaired taste acuity can be used as a functional test of suboptimal zinc nutriture in children, in conjunction with a biochemical index of zinc status.

Growth

Impairments in ponderal and linear growth are characteristic features of mild zinc deficiency in infancy and childhood. Some of the double-blind studies in infants and children (Ronaghy et al, 1974; Walravens & Hambidge, 1976; Walravens et al., 1983, 1989, 1992; Castillo-Duran et al., 1987; Gibson et al., 1989a) but not all (Ronaghy et al., 1974; Udomkesmalee et al., 1992; Cavan et al., 1993; Bates et al., 1993) have demonstrated significant improve-ments in weight and/or length or height in the zinc supplemented group compared to those receiving a placebo. In some cases, these changes have been observed only in the males. Possible reasons for failure of studies to show an efficacious effect of zinc on growth may include the presence of other limiting deficiencies and binding ligands in diets, which lower bioavailability of the zinc supplement, the form and level of the dose given, and the duration of the study period.

Until recently, a period of 6 months has been said to be the minimum interval for the provision of reliable growth data. For shorter intervals, measurement errors were too large in relation to the mean increments. Recently, however, it has been shown that increments in knee height measured using a device developed by Spender et al. (1989) can be accurately assessed over 60 days, and possibly even over a 28-day period. Using this instrument, future studies may measure knee height to monitor growth over a shorter time interval. The efficacy of this technique has been shown in controlled repletion trials in children (Sandstead et al., 1998).

6.5.1.4 New approaches

Plasma metallothionein concentrations have been suggested as a useful indicator of poor zinc status (King 1990). Metallothionein appears to play a role in zinc absorption, inter-organ zinc transport and tissue detoxification (Grider et al., 1990; King, 1990). Levels fall in zinc depletion and deficiency as a result of impaired synthesis. Specificity is poor: levels are also affected by iron deficiency, diurnal rhythm and acute infection. Metallothionein in erythrocytes appears to be much less responsive to stress and infection than in plasma, and may provide a useful index of zinc status in infancy and childhood.

Serum thymulin has also been assessed as a potential index of zinc status; thymulin is a zinc metallopeptide, the activity of which falls in mild zinc deficiency (King, 1990). Serum insulin-like growth factor (IGF), a peptide of low molecular weight regulated by growth hormone, nutrition and insulin, is also affected by zinc status (Cossack, 1986). Zinc-deficient rats showed a reduced growth rate, which was associated with a significantly lower serum IGF and with growth hormone receptor genes (McNall et al., 1995). Zinc repletion of Vietnamese children was followed by enhanced growth and increased serum concentrations of IGF-1 (Ninh et al., 1996).

Based on the essentiality of zinc for alcohol dehydrogenase, ethanol metabolism has been examined as a functional test of zinc status. Ethanol tolerance was shown to be impaired in women fed a diet marginal in zinc (Milne et al., 1987).

The essentiality of zinc for the activity of retinol reductase was first demonstrated in rats (Huber & Gershoff, 1975). Based on these findings, zinc was shown to be essential for human dark adaptation (Morrison et al., 1978). Retinol reductase is required for the regeneration of rhodopsin from retinol, a reaction essential for normal rod function, which in turn is responsible for dark adaptation.

More recently, Udomkesmalee et al. (1992) successfully used the dark adaptation test for assessing the response to zinc repletion in schoolchildren in north-eastern Thailand. The test is not appropriate for pre-school children who are too young to perform it accurately. Age influences dark adaptation and must be taken into account when interpreting test results.

The essentiality of zinc for brain function was established in laboratory animals (Sandstead, 1985). Studies in men showed that intakes of 1–4 mg of zinc daily for intervals of 35 days impaired neuromotor and cognitive function (Penland, 1991). Observations in children showed that zinc repletion improved neuromotor and cognitive function (Penland et al., 1997).

6.5.2 Metabolic role

Zinc is an essential trace element in all biological systems studied, and health disorders as a result of zinc deficiency have been well documented in humans and animals (Prasad, 1966, 1976, 1988, 1993; Sandstead, 1982c; Hambidge et al., 1986). The metabolic changes underlying human zinc deficiencies are incompletely understood (Hambidge, 1989); it is known, however, that zinc has a fundamental role in the structure and function of numerous proteins, including metalloenzymes, transcription factors and hormone receptors. The widespread role of zinc in metabolism is underscored by the occurrence of zinc in all tissues, organs and fluids of the human body (see section 6.2). Chapters 7 and 8 provide further information on the effects of zinc deficiencies in animals and humans respectively.

6.5.2.1 Zinc metalloenzymes

Many zinc metalloenzymes have been identified in humans and other mammals since the first report in 1940 of a zinc metalloenzyme, carbonic anhydrase, purified from ox erythrocytes (Keilin & Mann, 1940). The number of zinc metalloenzymes identified in all phyla is now reported to exceed 300 (Vallee & Auld, 1990a; Coleman, 1992), and the list encompasses all major enzyme classes (Vallee & Auld, 1990b). A summary of the metabolic role of major classes of zinc-containing enzymes has been given by Walsh et al. (1994). Information on some zinc metalloenzymes that have been widely studied in humans and other mammals has also been summarized by Vallee & Falchuk (1993).

Zinc metalloenzymes contain stoichiometric amounts of zinc, which may serve a functional and/or structural role, depending on the particular enzyme. In its functional role, zinc is considered to be located at the active site in many enzymes, and to participate directly in the catalytic process. Indirect evidence of a role of zinc in catalysis has been provided for many enzymes by the reversible inhibition or abolition of enzyme activity by metal chelating agents *in vitro*. In a structural role, zinc may stabilize protein structure or influence protein folding. In a comparison of the 12 zinc metallo-enzymes for which the structures have been determined by X-ray crystallography, Vallee & Auld (1990a,b) noted that, at each catalytic site, zinc is generally coordinated by three amino acid residues, most commonly histidines, and a water molecule, whereas at structural sites zinc is coordinated by four cysteine residues. The water molecule at catalytic sites has a critical role in the catalytic process (Vallee & Auld, 1990b).

DNA and RNA polymerases

DNA polymerases purified from the bacterium *Escherichia coli* and nuclei of the sea urchin *Strongylocentrotus franciscanus* have been reported to contain about 2 and 4 gram-atoms of zinc per mole of polymerase, respectively (Slater et al., 1971). Activity of DNA polymerase from each source was inhibited by the metal-chelator ortho-phenanthroline (Slater et al., 1971). However, Wu & Wu (1987) suggested that the use of this agent may be misleading, as it can form a complex with DNA that prevents the polymerase activity of DNA polymerase. The DNA polymerase of *E. coli* may retain its polymerase activity in the absence of stoichiometric amounts of zinc, indicating that zinc may have another role in the bacterium.

Eukaryotic RNA polymerases I, II and III are involved in the synthesis of ribosomal, messenger and transfer RNAs, respectively. The DNA-dependent RNA polymerases I (Falchuk et al., 1977), II (Falchuk et al., 1976) and III (Wandzilak & Benson, 1977) of the unicellular eukaryote *Euglena gracilis* have all been shown to be zinc metalloenzymes, each binding about 2 gram-atoms of zinc per mole of protein, with enzyme activity being reversibly inhibited by a variety of metal-chelating agents.

Zinc transcription factors

A comparatively recent development in the study of zinc metabolism has been the elucidation of the potential role of zinc in many protein transcription factors. This development was initiated by the demonstration that transcription factor IIIA (TFIIIA), isolated from *Xenopus laevis* oocytes, is a zinc metalloprotein and requires zinc for specific binding to DNA (Hanas et al., 1983). Examination of the amino acid sequence of *Xenopus* TFIIIA revealed a repeated structural domain, termed the "zinc finger", which has been postulated to bind zinc and interact with DNA. The TFIIIA type of zinc finger is a compact globular structure containing a single zinc atom, coordinated by 2 cysteine and 2 histidine residues (Lee et al., 1989a) The zinc atom maintains the finger structure (Frankel et al., 1987), and the zinc finger binds in the major groove of DNA, wrapping partly around the double helix (Pavletich & Pabo, 1991).

The zinc finger motif first characterized in TFIIIA has subsequently been identified in the cDNA sequences of numerous transcription factors, although in only a few instances has the presence of zinc been confirmed analytically (Vallee & Auld, 1990a). The zinc finger proteins include a substantial number of human proteins (Berg, 1990; South & Summers, 1990).

Steroid hormone receptors have also been identified as a group of transcription factors in which zinc may play an important role in DNA binding. These receptors are located in the cytoplasm or nucleus. Upon binding to the respective hormone, the activated receptor also binds to a DNA element known as the hormone response element, and modulates gene transcription (Tsai & O'Malley, 1994). The DNA-binding domains of the oestrogen receptor (Schwabe et al., 1990) and the glucocorticoid receptor (Freedman et al., 1988) have been shown to contain zinc-binding sites at which two zinc ions are each coordinated by four cysteine residues, and the DNA-binding site is located between and attached to the two zinc complexes. This structural arrangement differs significantly from that of the TFIIIA type of zinc finger (Pavletich & Pabo, 1991), and has been termed the "zinc twist" (Vallee et al., 1991). Zinc is required for the proper folding of the complex into its active structure (Freedman et al., 1988), which binds with DNA (Luisi et al., 1991). The conservation of structural amino acids among the DNA-binding steroid receptors (Evans, 1988) suggests that all members of the steroid receptor superfamily may have a similar zinc-containing structure for DNA recognition (Freedman et al., 1988; Schwabe et al., 1990).

6.5.2.2 Metallothionein

Metallothioneins (MT) are a group of low-molecular-weight (6000-dalton) metalloproteins with many proposed biological functions but no known enzymatic activity (Hamer, 1986). In mammals they occur in four structurally similar isoforms (MT-1,-2, -3 and -4) with several distinct features: a very high cysteine content (30%) of 20 cysteine residues in the total of 61 amino acids (Kissling & Kagi, 1977), and a high zinc and/or cadmium-binding ratio of about 7 gram-atoms of zinc and/or cadmium per mole of protein (Pulido et al., 1966). They bind metals through mercaptide bonds with a tetrathiolate motif by terminal and bridging cysteine ligands, similar to those found in zinc transcription factors (Berg & Shi, 1996). In addition, metallothioneins contains two distinct adamantine-like metal-binding clusters with three and four metal ions, respectively (Furey et al., 1986; Messerle et al., 1990; Robbins et al., 1991). MT-1 and MT-2 are the major isoforms and are found at low basal levels in most adult tissues, especially liver, kidney and pancreas. MT-3 and MT-4 have organ-specific expression; MT-3 is expressed specifically in brain and MT-4 is expressed in stratified squamous epithelium of tongue, cornea, intestine and stomach. High concentrations of MT-1 and MT-2 are found in fetal and neonatal livers, certain proliferating cells and human tumour cells, and they have been localized in the nucleus of these cells (Cherian, 1994). Metallothionein synthesis is induced by various chemicals, including metals, such as zinc, copper, cadmium and mercury, cytokines and stress conditions. The regulation of induced synthesis is at the transcriptional level by both *cis*- and *trans*-acting elements, which involve metal regulatory elements and transcription factors (Palmiter et al., 1993).

The exact physiological role of metallothionein is unclear and suggested functions include detoxification of heavy metals, zinc and copper homeostasis, scavenging of free radicals, zinc storage, and its exchange to other zinc proteins or enzymes (Zeng et al., 1991). Metallothionein is the major zinc- and copper-binding protein in fetal human liver. It is also involved in the altered zinc homeostasis during inflammation and is increased in the liver in response to cytokines and the stress hormone response, leading to hepatic zinc accumulation (Philcox et al., 1995). The induction of metallothionein synthesis can protect animals and cultured cells from some metal toxicity and free radical injury. In addition, recent studies in transgenic mice showed that metallothionein-null mice are very susceptible to cadmium toxicity (Michalska & Choo, 1993; Masters et al., 1994), and embryonic cells from these mice are extremely sensitive to metal toxicity (Lazo et al., 1995). For discussions of the potential roles for metallothionein see Hamer (1986) and Bremner & Beattie (1990).

6.5.2.3   Other metabolic functions of zinc

For example purposes, the role of zinc in two metabolic functions is outlined below.

Hormone storage

Zinc may play a role in the synthesis and storage of insulin. Insulin forms insoluble zinc-insulin crystals (Adams et al., 1969), and is stored in crystalline form in granules of the β-islet cells of the pancreas following synthesis from its soluble precursor, proinsulin (Grant et al., 1972). Grant et al. (1972) showed that at low zinc concentrations *in vitro*, insulin and proinsulin form soluble hexamers, whereas at high concentrations insulin, but not proinsulin, forms a precipitate. Yip (1971) found that pancreatic zinc (Zn^{2+}) minimized the degradation of bovine insulin by a purified pancreatic protease *in vitro*.

Cunningham et al. (1990) proposed that the dimerization of human growth hormone by zinc may prolong the hormone's storage life in the secretory granules of the anterior pituitary. It was shown that zinc (Zn^{2+}) induces the dimerization of human growth hormone, and retards its denaturation by guanidine-HCl *in vitro* (Cunningham et al., 1990).

Neurotransmission

There is some evidence that zinc may influence neurotransmission in the central nervous system, particularly in relation to the inhibitory neurotransmitter γ-aminobutyric acid (GABA). Westbrook & Mayer (1987) demonstrated that zinc (Zn^{2+}) is a potent antagonist of the excitatory neurotransmitter N-methyl-D-aspartate (NMDA) and GABA in cultured mouse hippocampal neurons. The non-competitive antagonism of NMDA suggested that the NMDA receptor channel contains a third binding site, in addition to Mg^{2+} and glycine (Westbrook & Mayer, 1987). Xie & Smart (1991) found that the naturally occurring, spontaneous giant depolarizing potentials (GDPs) in hippocampal neurons in brain slices from young postnatal rats could be inhibited by specific chelation of endogenous zinc, and that GDPs could be induced in adult brain slices by bath application of zinc. It was proposed that GDPs are generated by an inhibitory action of zinc on pre- and postsynaptic $GABA_B$ receptors (Xie & Smart, 1991).

6.5.3 Human studies

6.5.3.1 Copper

Impaired copper nutriture in humans has been noted following the chronic, elevated intake of zinc. These effects are reported in section 8.3.5.

7. EFFECTS ON LABORATORY MAMMALS AND *IN VITRO* TEST SYSTEMS

7.1 Single exposure

7.1.1 Lethality

In a study to determine the acute toxicity of four zinc compounds (acetate, nitrate, chloride and sulfate) administered by the oral or intraperitoneal route to male Swiss mice and male Sprague Dawley rats (Table 20), the majority of deaths occurred during the first 48 h. The clinical and physical signs of toxicity included miosis, conjunctivitis, decreased food and water consumption, haemorrhage and haematosis in the tail. These effects were reported to diminish with time, indicating rapid elimination of zinc from the body. The acute toxicity of zinc varied with the zinc salt used, and ranged from 237 to 623 mg/kg in rats and from 86 to 605 mg/kg in mice after oral administration; the acute toxicity following an intraperitoneal dose ranged from 28 to 73 mg/kg in rats and from 32 to 115 mg/kg in mice (Domingo et al., 1988). LC_{50} values following inhalation exposure to zinc chloride were 11 800 mg/min per m^3 in mice (Schenker et al., 1981) and 2000 mg/m^3 in rats (Karlsson et al., 1991).

7.1.2 Acute studies: summary of key findings

Zinc chloride can produce significant lung damage in rats when instilled directly into the lung; zinc oxide, by contrast, does not produce lung damage even when administered at relatively high concentrations. This is possibly due to the respective solubilities of the two compounds: zinc chloride is readily soluble in water whereas zinc oxide is not. Zinc chloride induces intra-alveolar oedema which closely resembles the effects of inhaled zinc oxide/hexachloroethane smoke in experimental animals. (Zinc oxide/hexachloroethane when burned produces zinc chloride with a residue of zinc oxide.) The oedema correlates with increased levels of protein in the lavage fluid fraction, which represents a plasma exudate. Onset is very rapid, with the greatest effects generally being noted within 3 days when high doses are used. The condition was found to regress between

Table 20. Acute lethal dose toxicity of various zinc compounds in rats and mice: LD_{50} values (mg/kg)[a]

Compound	Route of administration			
	Rats		Mice	
	Oral	Intraperitoneal	Oral	Intraperitoneal
Zinc acetate	794	162	287	108
as zinc:	237	48	86	32
Zinc nitrate	1330	133	926	110
as zinc:	293	39	204	32
Zinc chloride	1100	58	1260	91
as zinc:	528	28	605	44
Zinc sulfate	1710	200	926	316
as zinc:	623	73	337	115

[a] Animals were observed for 14 days.
From: Domingo et al. (1988).

days 3 and 7. Key findings from these studies are summarized in Table 21.

7.2 Short-term exposure

7.2.1 Oral exposure

Reduced growth rates, reduced body weights and anaemia were observed in a number of rat studies and also in a mouse and a sheep study, following high oral or dietary intakes of zinc (Van Reen, 1953; Maita et al., 1981; Allen et al., 1983; Zaporowska & Wasilewski, 1992). Copper deficiency induced by high doses of zinc, was implicated in these effects, as copper supplementation reversed the zinc-induced anaemia (Smith & Larson 1946).

Exposure to high doses of zinc was associated with pancreatic atrophy and histological changes in kidneys, accompanied by changes in kidney function in rats, mice and sheep (Maita et al., 1981; Allen et al., 1983; Llobet et al., 1988). Changes in the liver, including decreased activities of cytochrome P450 and liver catalase,

Table 21. Key findings from acute studies in experimental animals

Species	Exposure	Compound	Effects	Reference
Rats (male)	inhalation, 1 h 0, 11, 580 mg/min/m³	zinc oxide/hexachloro-ethane	11/40 deaths, pulmonary oedema	Brown et al. (1990)
Rats (male)	intratracheal 0, 2.5 mg/kg	zinc chloride	respiratory distress, alveolitis, parenchymal scarring	Brown et al. (1990)
Guinea-pigs and rats	inhalation (nose only), 3 h 2.5, 5 mg/m³	zinc oxide (median diameter 0.06 μm)	inflammatory changes in the lung at both levels of exposure	Gordon et al. (1992)
Guinea-pigs (male, Hartley)	aerosol, 3 h 0, 7.8 mg/m³	zinc oxide (projected area diameter 0.05 μm)	decrease in the lung volume capacity	Lam et al. (1982)
Rats (male, Porton Wistar)	intratracheal 0.3 mg/kg	zinc oxide, zinc chloride	elevated alveolar surface protein levels with zinc chloride exposure only	Richards et al. (1989)

Table 21 (contd.)

Species	Exposure	Compound	Effects	Reference
Rats (male, Porton Wistar)	intratracheal 0, 0.25, 0.5, 1, 2, 4, 5 mg/kg	zinc oxide, zinc chloride	intra-alveolar oedema at doses above 0.5 mg/kg	Richards et al. (1989)
Rats (male, Porton Wistar)	inhalation 2.5 mg/kg	zinc chloride	pulmonary oedema	Richards et al. (1989)
Sheep (weaner)	drench 3 g	zinc	14/100 deaths, oedema of abomasum and duodenum, fibrosing pancreatitis	Allen et al. (1986)
Rabbits (New Zealand)	inhalation 0, 0.6, 0.81 g/m^3	hexachloroethane/zinc	acute inflammation of lungs, pulmonary oedema at both doses	Marrs et al. (1983)

and decreased *de novo* synthesis of high-density lipoprotein, were seen in rats exposed to high levels of zinc (Van Reen, 1953; Woo, 1983; Cho et al., 1989). Minor degenerative changes in the brain, accompanied by elevated neurosecretion and increased activity in the neurohypophysis were seen in rats exposed intragastrically to zinc oxide for 10 days at 100 mg/day (Kozik, 1981).

Key findings from these studies are summarized in Table 22.

7.2.2 Inhalation exposure

Guinea-pigs (3 per group) were given 1, 2 or 3 consecutive, daily, 3-hour, nose-only exposures to freshly generated zinc oxide particles with a projected area diameter of 0.05 μm at concentrations of 0, 2.3, 5.9 or 12.1 mg/m^3. Exposure to zinc oxide at 5.9 or 12.1 mg/m^3 was associated with increased protein and neutrophils and increased activities of β-glucuronidase, acid phosphatase, alkaline phosphatase, lactate dehydrogenase and angiotensin-converting enzyme in lavage fluid. These increases were concentration-dependent, were detected after the second exposure, and generally increased after the third exposure. Significant morphological changes observed at concentrations of 5.9 or 12.1 mg/m^3 consisted of inflammation and type 2 pneumocyte hyperplasia in the proximal alveolar ducts. No evidence of inflammation was present in animals exposed to zinc oxide at 2.3 mg/m^3. It was concluded that exposure of guinea-pigs to ultrafine atmospheric zinc oxide at concentrations of 5.9 or 12.1 mg/m^3 causes significant pulmonary damage. Detection of injury was stated to correlate well with pulmonary lavage fluid changes (Conner et al., 1988).

Male Hartley guinea-pigs exposed to zinc oxide at a concentration of 7 mg/m^3 for 3 h/day for 5 consecutive days showed pulmonary impairment (as measured by lung oedema, lung volume carbon monoxide diffusing capacity and pulmonary mechanics). Exposures at 2.7 mg/m^3 using the same time frame did not cause pulmonary impairment (Lam et al., 1988). A single exposure at 8 mg/m^3 was also without effect (Lam et al., 1982). Guinea-pigs exposed to zinc oxide at a concentration of 5 mg/m^3 for 3 h/day for 6 consecutive days showed significant reductions in vital capacity, functional residual capacity, alveolar volume, and lung volume carbon monoxide diffusing capacity following the last exposure,

128

Table 22. Key findings from short-term exposure studies in experimental animals

Species	Exposure	Compound	Effects	Reference
Mice	0.6 g/kg of diet for 4 weeks	zinc sulfate	no adverse effects on immune responsiveness	Schiffer et al. (1991)
Mice, rats	0, 300, 3000, 30 000 µg/g diet for 13 weeks	zinc sulfate	NOEL for both species was set at 3000 µg/kg; retarded growth, anaemia and pancreatic atrophy at 3000 µg/kg level	Maita et al. (1981)
Rats	0, 0.5, 1% of diet for 15 days	zinc oxide	death at 1% level; reduced body weight, reduced fat content of the liver and impaired bone development at both doses	Van Reen (1953)
Rats	500–700 mg/100 g diet for 4–5 weeks	zinc in zinc carbonate	growth reduction, reduced levels of liver catalase and cyto-chrome oxidase activity, effects reversed by copper supplement	Van Reen (1953)
Rats	100 mg/day intra-gastrically for 10 days	zinc oxide	elevated neurosecretion in hypothalamus, increased release of antidiuretic hormone in neurohypophysis	Kozik (1981)

Table 22 (contd.)

Rats	100 mg/day intra-gastrically for 10 days	zinc oxide	morphological and histoenzymic changes in the brain	Kozik (1981)
Rats	0, 0.12 mg/ml drinking-water for 4 weeks	zinc as zinc chloride	decreased body weight, anaemia and increased lymphocyte count	Zaporowska & Wasilewski (1992)
Rats	0, 160, 320, 640 mg/kg body weight/day for 3 months	zinc acetate	no effect on weight gain or on red blood cells, histological changes in kidneys and increased concentrations of urea and creatinine in plasma at 640 mg/kg body weight per day	Llobet et al. (1988)
Rats	0, 0.7, 1% in diet for 4 weeks	zinc carbonate	subnormal growth, anaemia and reproductive failure at both dose levels, anaemia reversed by copper supplement, growth retardation reversed by liver extract supplement	Smith & Larson (1946)

which had not returned to normal values by 72 h, although increases to flow resistance and decreases in compliance and total lung capacity did return to normal (Lam et al., 1985)

Key findings from these studies are summarized in Table 23.

7.3 Long-term exposure

The long-term studies on the effects of zinc vary in quality and tend to be limited in their usefulness in determining chronic toxicity in animals, as the study design generally does not lend itself to elucidation of dose-related effects. The available studies do, however, provide some information on target organ toxicity resulting from zinc exposure. Key findings from these studies are summarized in Table 24.

7.3.1 Oral exposure

Osborne-Mendel rats (4 per sex per group) were fed diets containing zinc sulfate at 0, 100, 500 and 1000 µg/g for 21 months. While only minimal monitoring of toxic effects was carried out, it was reported that food intake, body weights, haemoglobin values, and erythrocyte, leukocyte and differential counts were unaffected by the treatment. Microcytosis coupled with polychromasia or hyperchromia appeared at 16 months in rats receiving the highest dose and at 17 months in the other zinc-treated groups. However, it was stated that the erythrocyte count returned to normal later in the study (time not specified). Counts of the bone marrow smears taken at autopsy revealed a dose-related decrease in the myeloid:erythrocyte ratio in all of the treated groups. The kidneys of male rats in the 500 and 1000 µg/g groups were enlarged and an increased incidence of nephritis was seen in male, but not female, rats (Hagan et al., 1953).

In a chronic study (Aughey et al., 1977), C3H mice were administered zinc sulfate in the drinking-water at a concentration of 0.5 g/litre for 14 months. Control and zinc-treated mice were removed from the colony in groups of five per sex, usually at monthly intervals, for estimation of plasma zinc, glucose and insulin, tissue zinc, and histological, histochemical and electron microscopy examinations. Plasma zinc increased to a plateau at levels about 1.5–2 times those in controls within the first 30 days. Levels of zinc

Table 23. Key findings from repeated dose inhalation studies in guinea-pigs

Species	Exposure	Compound	Effects	Reference
Guinea-pigs (male Hartley, 3 per group)	0, 2.3, 5.9 or 12.1 mg/m³, 3 h/day for 1, 2 or 3 days	zinc oxide (projected area diameter 0.05 µm)	inflammation and hyperplasia of the lung at 5.9 and 12.1 mg/m³ after the second exposure; NOEL, 2.3 mg/m³	Conner et al. (1988)
Guinea-pigs (male Hartley, 5–8 per group)	0, 2.7 or 7 mg/m³, 3 h/day for 5 days	zinc oxide (median diameter 0.05 µm)	oedema, decrease in total lung capacity and diffusing capacity for CO at 7 mg/m³, oedema; no effects observed at 2.7 mg/m³	Lam et al. (1988)
Guinea-pigs (male Hartley, 18–38 per group)	0 or 5 mg/m³, 3 h/day for 6 days (nose only)	zinc oxide (projected area diameter 0.05 µm)	inflammation, decrease in vital capacity, functional residual capacity, total lung capacity and diffusing capacity for CO	Lam et al. (1985)

Table 24. Key findings from long-term exposure studies in experimental animals

Species	Exposure	Compound	Effects	Reference
Rats	0, 0.1, 0.5 or 1% in diet for up to 39 weeks	zinc carbonate	reduction of growth at 1% and indications of anaemia in the 0.5 and 1% groups	Sutton & Nelson (1937)
Rats	0, 100, 500 or 1000 µg/g in diet for 21 months	zinc sulfate	minimal monitoring; no effect on growth and no anaemia; dose–related decrease in myeloid/erythrocyte ratio in all treated groups; enlarged kidneys at 500 and 1000 µg/g in all male groups; NOEL, <100 µg/g	Hagan et al. (1953)
Mice	0.5 g/litre in drinking-water for 14 months	zinc sulfate	pancreatic hypertrophy, pituitary gland hypertrophy	Aughey et al. (1977)
Dogs	200 mg/kg body weight per day in diet reduced to 50 mg/kg body weight/day by week 35	zinc sulfate	emesis, 1/4 deaths, hypochromic anaemia, hyperplastic bone marrow	Hagan et al. (1953)
Rabbits	5 mg/g in diet for 22 weeks	zinc carbonate	no effects on growth, decrease in haemoglobin and serum copper concentrations	Bentley & Grubb (1991)

Table 24 (contd.)

Mink	0, 500, 1000 or 1500 mg/kg for 144 days	zinc sulfate	no effect on body weights, haematological parameters or survival; no histological lesions in liver, pancreas or kidney; NOEL, 1500 mg/kg	Aulerich et al. (1991)
Ferrets	0, 500, 1500 or 3000 mg/kg for up to 6 months	zinc oxide	body weight loss, reduced food intake and death at 3000 mg/kg on days 9–13 and at 1500 mg/kg on days 7–21; diffuse nephrosis and active haemopoiesis in bone marrow and spleen in the 3000 and 1500 mg/kg groups; pancreatitis in one animal in each group at 3000 and 1500 mg/kg; no toxicity observed at 500 mg/kg except some evidence of effect on red blood cells	Straube et al. (1980)
Mice, rats	0, 1.2, 12 or 120 mg/m^3 air for 1 h/day, 5 days per week for 20 weeks	zinc in smoke produced by ignition of zinc oxide/hexa-chloroethane	no effect on body weight; increase in mortality in mice at 120 mg/m^3; macrophage infiltration of the lung in rats and mice at the highest dose; significant increase in the frequency of alveologenic carcinoma in high dose mice Note: carbon tetrachloride may be present in smoke	Marrs et al. (1988)

in the liver, spleen and skin remained unchanged. The pancreatic islet cells in treated mice were hypertrophied and contained an increased number of secretory granules; however, plasma glucose and insulin levels remained comparable to those in control animals. Hypertrophy of the pituitary gland, suggestive of a pituitary feedback effect, was also observed.

Adult and juvenile mink (3 per sex per group) were fed diets supplemented with zinc as zinc sulfate at 0, 500, 1000 or 1500 mg/kg for 144 days. No adverse effects on food consumption, body weight gains, haematological parameters, fur quality or survival were observed. Zinc concentrations in the liver, kidneys and pancreas increased in direct proportion to the zinc content of the diet. No histological lesions indicative of zinc toxicity were detected in the liver, kidneys or pancreas (Aulerich et al., 1991).

Ferrets (3–5 per group) were fed diets containing zinc administered as zinc oxide at 0, 500, 1500 or 3000 mg/kg for up to 6 months. The three ferrets in the 3000 mg/kg group lost a significant amount of their body weights, had greatly reduced food intakes, and died or were killed *in extremis* between days 9 and 13 of the dosing period. The ferrets exposed to 1500 mg/kg zinc were killed at 7–21 days, by which time they presented with poor condition, weight loss and up to 80% reduction in food intake. Histological examination of the three animals from the 3000 mg/kg group and the four animals from the 1500 mg/kg group revealed diffuse nephrosis and active haematopoiesis in the bone marrow and the extramedullary area of the spleen. One animal from each dose group had acute pancreatitis. Haematograms indicated a severe but responding macrocytic hypochromic anaemia, with high reticulocyte counts in the two highest dose groups. In the liver and kidneys of treated animals, the zinc concentration was significantly increased and the copper concentration was lower than control values. These changes were associated with a high concentration of iron in the liver. Increased incidences of elevated serum urea and blood glucose concentrations and decreased serum ceruloplasmin oxidase activity were observed at the two highest doses, and protein, blood, glucose and bilirubin were present in the urine. None of the ferrets given zinc at 500 mg/kg in the diet developed clinical signs. These animals were killed on days 48, 138 and 191 respectively; they showed signs of extramedullary haematopoiesis in the spleen and slight increases in kidney zinc concentration and decreases in liver copper concentration. Although the number of animals used was small, given the lack of dose–response studies, the threshold of zinc toxicity in ferrets was proposed to be between 500 and 1500 mg/kg, with the kidney identified as the target organ of toxicity in this species (Straube et al., 1980).

The consequences of copper deficiency may be relevant to some of the effects noted in studies using elevated zinc levels. The occurrence of anaemia in animals receiving high doses of zinc is generally attributed to induction of copper deficiency. Some otherwise unexplained effects of high doses of zinc may also be secondary to impaired copper utilization. Relevant studies are described in section 7.8.

7.4 Skin irritation

Zinc chloride, applied daily as a 1% aqueous solution in an open patch test for 5 days, was severely irritant in rabbits, guinea-pigs and mice, inducing epidermal hyperplasia and ulceration. Aqueous zinc acetate (20%) was slightly less irritant. Zinc oxide (20% suspension in dilute Tween 800), zinc sulfate (1% aqueous solution) and zinc pyrithione (20% suspension) were mildly irritant, and induced a marginal epidermal hyperplasia and increased hair growth. Zinc undecylenate (20% suspension) was not irritant. Epidermal irritancy was related to the interaction of zinc with epidermal keratin (Lansdown, 1991).

7.5 Reproductive toxicity, embryotoxicity and teratogenicity

The available studies are limited in their usefulness in determining the reproductive and developmental effects of zinc owing to poor study design and inadequate reporting, although they do provide an indication of the effects of zinc exposure. Key findings from these studies are summarized in Table 25.

In a study in mice (Mulhern et al., 1986), female weanling C57BL/6J mice were fed diets containing zinc as zinc carbonate at a concentration of 2000 mg/kg until they were mated at 6 weeks of

Table 25. Key findings from studies on reproduction, embryotoxicity and teratogenicity in experimental animals

Species	Exposure	Compound	Effects	Reference
Mice	2000 mg/kg until mating at 6 weeks of age, then in various combinations during gestation/lactation and after weaning; second generation killed at 8 weeks	zinc carbonate	in second generation mice exposed throughout gestation/lactation and after weaning, elevated levels of zinc in bones, decreased blood copper levels, signs of anaemia and reduced body weights; alopaecia at 5 weeks, reversed at 8 weeks	Mulhern et al. (1986)
Mice	0, 12.5, 20.5 or 25 mg/kg body weight i.p. on days 8, 9, 10 or 11 of gestation	zinc chloride	increases in skeletal defect incidence, usually ripple ribs; effects were dose-related and seen at all dose levels; no soft tissue anomalies attributed to zinc; greatest effect at 20.5 mg/kg on day 10 of gestation, causing 4/10 deaths	Chang et al. (1977)
Mouse embryos	100 μmol/litre in vitro for 24 h at the 1-, 2-, 4- and 8-cell stage	zinc	40% increase in cell death at 1-cell stage; embryo development affected more in early than late stages; relevance to fetal development uncertain	Vidal & Hidalgo (1993)

136

Table 25 (contd.)

Species	Dose	Compound	Effects	Reference
Rats	0, 0.1%, 0.5% or 1% in diet for 39 weeks and during pregnancy	zinc carbonate	reproduction adversely affected in the 0.5% group: all second litters dead, no further pregnancies thereafter; no pregnancy achieved in the 1% group; anaemia in 0.5% and 1% groups; anaemia and sterility reversed in 0.5% group but sterility remained the 1% group when zinc removed from diet	Sutton & Nelson (1937)
Rats	0, 4 or 500 mg/kg in diet for 8 weeks to weanlings.	zinc chloride	testicular cell development examined only: excess zinc had no effect	Evenson et al. (1993)
Rats	0, 500, 1000 or 2000 µg/g diet during pregnancy	not given	significant decrease in total number of pups born and increased percentage of stillbirths at 2000 µg/g; no increase in the incidence of malformations	O'Dell (1968)
Rats	0 or 150 mg/kg in diet throughout gestation	zinc sulfate	effects on fetus assessed on day 18: incidence of resorption significantly increased in treated animals	Kumar (1976)

Table 25 (contd.)

Species	Exposure	Compound	Effects	Reference
Rats	0, 0.25 or 0.5% in diet during gestation and 14 days of lactation	zinc oxide	maternal body weight, gestation period and viable pups/litter were unaffected at either dose level at birth or on day 14; no malformations observed in any pup; dose-related reduction in pup body weight; some changes in iron and copper distribution in newborn pups at both treatment levels	Ketcheson et al. (1969)
Rats	500 mg/kg in diet during gestation	zinc carbonate	no effect on maternal haematocrits; no effects on litter numbers, viability, implantation sites, fetal length and weight, placental weights or incidence of resorptions; no increase in the incidence of malformations or skeletal ossification	Uriu-Hare et al. (1989)
Rats	4000 mg/kg, 18 days post coitus	zinc sulfate	incidence of conception reduced; no increase in stillbirths or malformations in exposed groups	Pal & Pal (1987)

age. The dams and offspring were distributed into 10 different dietary groups, exposing the second generation to various combinations during gestation, lactation and postweaning development. Second-generation mice were killed at 8 weeks of age. Second-generation mice exposed to high doses of zinc throughout the gestation, lactation and postweaning period had elevated levels of zinc in their bones, decreased blood copper levels, lowered haematocrit values and reduced body weights. Mice in this group began to lose fur at 2–4 weeks of age, with severe alopecia developing at 5 weeks of age, accompanied by thinner than normal skin. The fur grew back by 8 weeks of age, albeit lighter in colour.

The feeding of low (4 mg/kg of diet), normal (12 mg/kg) and high (500 mg/kg) levels of zinc as zinc chloride to weanling Sprague-Dawley rats (10 males per group) for 8 weeks indicated that a diet deficient in zinc is associated with a significant deviation in the ratio of testicular cell types present in the testes, including a reduction in the numbers of cells in S phase and total haploid cells. In rats fed zinc-deficient diets, about 50% of epididymal spermatozoa had a significant decrease in resistance to DNA denaturation *in situ*. Excess zinc in the diet had no effect on rat testicular cell development as defined by sperm resistance to DNA denaturation, distribution of testicular cell types and sperm chromatin structure integrity (Evenson et al., 1993).

A diet supplemented with zinc (source not identified) at 0, 500, 1000 and 2000 μg/g, and with adequate levels of copper (10 μg/g) was administered to pregnant rats (strain and number not given). There was a significant decrease in the total number of pups born and an increase in the percentage of stillbirths at the highest dose of zinc, but no effect on the survival of offspring allowed to nurse for 1 week. The data were reported to indicate that the incidence of hydrocephalus was increased in rat embryos of zinc-treated dams. However, there was no obvious correlation between dose and the incidence of hydrocephalic fetuses associated with the treatment: 0.1% in controls, 0.2% in the 500 μg/g group, 0.7% in the 1000 μg/g group, and 0.1% in the 2000 μg/g group (O'Dell, 1968).

Pregnant rats (10–12 per group, strain not identified) were fed a diet supplemented with zinc sulfate (150 μg/g) throughout the entire gestation period. On day 18 of pregnancy, the incidence of resorptions in pregnant rats increased from 2/101 implantation sites in the 12 control rats (2 females affected) to 11/116 implantations in the supplemented rats (8 females affected) This difference was statistically significant, indicating that even moderately high levels of zinc in the diet of rats may be associated with harmful effects on pregnancy (Kumar, 1976).

Diets high in zinc (0.2 and 0.5%), added as zinc oxide, were fed to pregnant albino rats (10 per group) for the entire period of gestation and for the first 14 days of lactation. The zinc content of the basal diet was 9 mg/kg. Four pups from each litter were killed at birth and the remaining pups were killed and examined on day 14 of lactation. Maternal body weights, food intake, gestation period and the number of viable young per litter were unaffected by the increased zinc levels in the diet, either at birth or on day 14 of lactation. Two dams fed 0.5% zinc had stillborn litters containing oedematous pups. Four stillborn pups were born to dams fed 0.2% zinc (number of dams not given); these pups were not oedematous. Anatomical malformations were not observed in any pup. The body weights of the newborn and 14-day-old pups in the 0.5% group were significantly reduced whereas the size of newborn pups, but not the 14-day-old pups from the 0.2% group were significantly greater than pups from dams fed the basal diet. The dry liver weights of pups at birth were unaffected by the zinc treatment but were significantly reduced in day-14 pups in the 0.5% group. Total zinc in newborn pups and day-14 weanlings was elevated in a dose-related manner in pups from the dams exposed to 0.2% and 0.5% zinc. Bodies (viscera removed) of newborn and day-14 pups from mothers fed the zinc diets contained significantly lower total iron than those from mothers receiving the basal diet: the reduction was dose-dependent. In contrast, the livers of newborns from zinc-treated dams contained significantly elevated total iron than the basal diet pups. These changes in liver iron levels were not observed in day-14 pups. Total copper in the whole animal and body (viscera removed) of the newborn rats was not altered by the treatment. However, liver copper levels were significantly lower only in the newborns of mothers fed 0.5% zinc. After 14 days, total copper was significantly lower in the whole animal, liver and body (viscera removed) of pups from dams fed both zinc diets; this reduction was dose-dependent. Livers of dams fed excess zinc contained elevated zinc and reduced iron and copper levels (Ketcheson et al., 1969). Another study reported no resorption in Sprague-Dawley rats receiving 0.5% zinc as zinc carbonate in the diet (Kinnamon, 1963).

Pregnant Sprague Dawley rats (8 per group) were exposed to basal (24.4 mg/kg of diet) or high levels of zinc (500 mg/kg) in the diet, supplemented as zinc carbonate, for the duration of the gestation period. Ingestion of high zinc levels had no effect on maternal food intake or on body weight throughout the pregnancy. Maternal haematocrits on gestational day 20 were similar in the basal and high-zinc groups. High dietary zinc levels had no effect on litter numbers, litter viability, implantation sites, fetal lengths and weights, placental weights or number of resorptions. There was no significant increase in the incidence of malformations associated with high zinc exposure or in the ossification of the fetal skeleton. Zinc, copper and iron content of the maternal liver, and maternal kidney weights in the basal and high-zinc groups remained comparable. Plasma of dams exposed to the high-zinc diet contained significantly increased zinc levels and significantly decreased iron levels, whereas copper levels remained similar to those found in rats fed the basal diet. The absolute concentrations of zinc bound to albumin and α_2-macroglobulin proteins were significantly increased in the high-zinc group as were maternal liver metallothionein concentrations (Uriu-Hare et al., 1989).

Exposure of Charles Foster rats (12 per group) to diets containing zinc as zinc sulfate at a concentration of 4000 mg/kg reduced the incidence of conception in females treated for 18 days post coitus, indicating that high zinc intake interferes with implantation of fertilized ova. However, exposure to this level of zinc 21–26 days before mating and throughout gestation for 18 days did not affect the incidence of conception. This apparently contradictory finding was interpreted to be due to an adaptation to zinc feeding, which is known to decrease the body burden of zinc. No stillborn or malformed fetuses were observed in zinc-treated animals in either study (Pal & Pal, 1987).

7.6 Mutagenicity and related end-points

Genotoxicity studies conducted in a variety of test systems have failed to provide evidence that zinc is mutagenic. However, there are indications of some weak clastogenic effects following zinc exposure. The findings from genotoxicity studies are detailed below and are summarized in Table 26.

7.6.1 In vitro *studies*

Exposure to zinc does not increase mutation frequencies in the majority of bacterial or mammalian cell culture test systems (Nishioka, 1975; Amacher & Paillet, 1980; Kada et al., 1980; Gocke et al., 1981; Marzin & Vo, 1985; Rossman et al., 1987; Thompson et al., 1989; Karlsson et al., 1991). However, gene mutation effects following exposure to zinc were observed in the $TK^{+/-}$ mouse lymphoma and Chinese hamster ovary cells *in vitro* cytogenetic assays (Thompson et al., 1989), and chromosomal effects were obtained in human lymphocyte cultures (Deknudt & Deminatti, 1978). Zinc chloride did not induce mutations at the thymidine kinase locus in L5178Y mouse lymphoma cells (Amacher & Paillet, 1980) and did not induce mispairing between nucleic acid bases *in vitro* (Murray & Flessel, 1976).

Zinc sulfate inhibited the activity of DNA polymerase-1 activity *in vitro*, but had no effect on the fidelity of DNA synthesis in an assay measuring misincorporation of nucleotides into the new strand of DNA (Sirover & Loeb, 1976; Miyaki et al., 1977). Zinc chloride at concentrations of up to 20 µg/ml did not cause morphological transformation of Syrian hamster embryo cells *in vitro* (Di Paolo & Casto, 1979); however, zinc chloride and zinc sulfate gave equivocal results in an *in vitro* test for the capacity of these metal salts to enhance viral transformation of Syrian hamster embryo cells, producing enhancement in 3/6 and 3/7 trials respectively (Casto et al., 1979). Exposure to zinc had no effect on the induction of unscheduled DNA synthesis in primary cultures of rat hepatocytes (Thompson et al., 1989).

7.6.2 In vivo *studies*

The induction of chromosome aberrations has been studied in bone marrow cells harvested from animals exposed to elevated levels of zinc. Taken as a whole, studies of this end-point yield equivocal and sometimes contradictory results—a likely reflection of inter-study differences in routes, levels and duration of zinc exposure, the nature of lesions scored (gaps compared to more accepted structural alterations) and great variability in the technical rigour of individual studies. Increased aberrations have been reported in rats after inhalation exposure (zinc oxide at 0.5–1.0 mg/m^3 for 5 months; Voroshilin et al., 1978), in rats after oral exposure (zinc chloride in water at 249 mg/litre for 14 days; Kowalska-Wochna et al., 1988) and in mice after multiple intraperitoneal injections of zinc chloride (at 2–5 mg/kg body weight; Gupta et al., 1991). In contrast, other studies have produced negative findings, for example, after intraperitoneal injection of mice (zinc chloride at 15 mg/kg body weight; Vilkina et al., 1978), or have suggested that the induction of aberrations is contingent upon concomitant calcium deficiency. Those studies do not provide compelling evidence for significant clastogenic activity. Negative results have also been reported in the mouse micronucleus test (intraperitoneal injection of zinc sulfate; Gocke et al., 1981). Negative micronucleus test results are consistent with a lack of significant clastogenic activity.

There was no increase in the frequency of dominant lethal mutation in germ cells of mice injected by the intraperitoneal route with zinc chloride at 15 mg/kg body weight (Vilkina et al., 1978).

Zinc sulfate (5 mmol/litre), which is an almost-lethal dose) fed to adult *Drosophila melanogaster* did not increase the incidence of sex-linked recessive lethal mutations when tested in three successive broods (Gocke et al., 1981). In contrast, zinc chloride fed to adult *D. melanogaster* at 0.247 mg/ml significantly increased the incidence of dominant lethal mutations and sex-linked recessive lethal mutations in treated flies (Carpenter & Ray, 1969).

7.7 Carcinogenicity

No adequate experimental evidence has been found to indicate that zinc salts administered orally or parenterally are tumorigenic.

Deficiency and supplements of zinc can have an influence on carcinogenesis, possibly as a result of the influence of zinc on cell growth (Petering et al., 1967; Barr & Harris, 1973; Phillips & Sheridan, 1976; Rath et al., 1991), although zinc has also been reported to inhibit tumour induction (Kasprzak et al., 1988). Zinc has been demonstrated to inhibit the mutagenic action of some genotoxic carcinogens (Francis et al., 1988; Leonard & Gerber, 1989) but has

Table 26. Genotoxicity studies with zinc

Test	Zinc source doses	Concentrations of zinc	Results	Reference
Non-mammalian systems Prokaryotes Gene mutation				
Salmonella typhimurium (TA102)	zinc sulfate	10–1000 nmol/litre per plate	- (no S9)	Marzin & Vo (1985)
S. typhimurium (TA98, TA100, TA1535, TA1537, TA1538)	zinc acetate	50–7200 µg/plate	- (with and without S9)	Thompson et al. (1989)
	zinc 2,4-pentanedione	400 µg/plate	+ (with and without S9)	
S. typhimurium (TA98, TA1538)	zinc sulfate	up to 3600 µg/plate	- (with and without S9)	Gocke et al. (1981)
S. typhimurium (TA98, TA100, TA1535, TA1537, TA1538)	zinc oxide/hexachloro-ethane smoke		- (with and without S9)	Karlsson et al. (1991)
S. typhimurium (TA98, TA100, TA1535, TA1537)	zinc chloride	0.05 mol/litre	-	Nishioka (1975)
	zinc chloride	not given	-	Kada et al. (1980)
Bacillus subtilis H17, M45	zinc chloride	0.4 mmol/litre	-	Rossman et al. (1987)
Escherichia microscreen assay	zinc chloride	0.4 mmol/litre	-	Rossman et al. (1987)
λ prophage induction Trp+ reversion comutagenesis	zinc chloride	0.4 mmol/litre	-	Rossman et al. (1987)

Table 26 (contd.)

Fidelity of DNA synthesis	zinc sulfate	0.2 mmol/litre	-	Miyaki et al. (1977)
DNA polymerase	zinc chloride	0.4 mmol/litre	-	Sirover & Loeb (1976)
Plants				
Chromosomal aberrations				
Vicia faba	zinc sulfate	0.1% solution	+	Herich (1969)
Insects				
Sex-linked recessive lethal test	zinc sulfate	5 mmol/litre	-	Gocke et al. (1981)
Sex-linked recessive lethal test	zinc chloride	0.247 mg/ml of food	+	Carpenter & Ray
Dominant lethal test	zinc chloride	0.247 mg/ml of food	+	(1969)
Mammalian systems				
In vitro animal cells				
Gene mutation				
mouse lymphoma	zinc chloride	0.12–12.13 µg/ml	-	Amacher & Paillet (1980)
mouse lymphoma	zinc acetate	0–13 µg/ml and 4.2–42 µg/ml	+ (with and without S9)	Thompson et al. (1989)
Chromosomal aberration				
Chinese hamster ovary cells	zinc acetate	25–45 µg/ml and 45–80 µg/ml	+ (with and without S9)	Thompson et al. (1989)

Table 26 (contd.)

Test	Zinc source doses	Concentrations of zinc	Results	Reference
Cell transformation				
Syrian hamster embryo cells	zinc chloride	0–20 µg/ml	-	DiPaolo & Casto (1979)
enhancement of cell transformation	zinc sulfate	0.05–0.6 mmol/litre	+/-	Casto et al. (1979)
enhancement of cell transformation	zinc chloride	0.05–0.6 mmol/litre	+/-	
Unscheduled DNA synthesis				
rat hepatocytes	zinc acetate	10–1000 µg/ml	-	Thompson et al. (1989)
rat hepatocytes	zinc 2,4-pentanedione	10–1000 µg/ml	-	
In vitro **human cells**				
Chromosomal aberration				
human lymphocytes	zinc chloride	$3 \times 10^{-4} – 3 \times 10^{-5}$ mol/litre	+	Deknudt & Deminatti (1978)
In vivo **animal**				
Sister chromatid exchange				
sheep bone marrow cell	emission dust	32 g/day	+	Bires et al. (1995)
rat bone marrow	zinc chloride	240 mg/kg	+	Kowalska-Wochna et al. (1988)

146

Table 26 (contd.)

Test	Substance	Dose	Result	Reference
Micronucleus test				
mouse	zinc sulfate	0.1–0.3 mmol/litre per kg	-	Gocke et al. (1981)
mouse	zinc oxide/hexachloro-ethane smoke	0.1 ml smoke condensate	-	Karlsson et al. (1991)
Chromosomal aberration				
rat bone marrow	zinc oxide	$0.5–1\ mg/m^3$	+	Voroshilin et al. (1978)
mouse	zinc chloride	15 g/kg	-	Vilkina et al. (1978)
mouse bone marrow	zinc chloride	0.5% zinc for 1 month	-	Deknudt & Gerber (1979)
mouse bone marrow	zinc chloride	0–15 mg/kg	+	Gupta et al. (1991)
rat bone marrow	zinc chloride	240 mg/kg for 14 days	+	Kowalska-Wochna et al. (1988)
sheep bone marrow cell	emission dust	32 g/day	-	Bires et al. (1995)
Dominant lethal mutation				
mouse	zinc chloride	15 mg/kg	-	Vilkina et al. (1978)
In vivo human				
Chromosomal aberration	zinc smelter dust cadmium plant fumes/dust		+	Bauchinger et al. (1976)
			-	Deknudt & Leonard (1975)

also been shown to be co-carcinogenic in other studies (Wallenius et al., 1979).

7.8 Interactions with other metals

In general, zinc shows a low toxicity to animals, but at high exposure levels it can interact with other trace elements, especially copper, resulting in toxicity, which is usually due to depletion of these elements, and leading to nutritional deficiencies. It has been postulated (Hill & Matrone, 1970) that elements with similar properties will act antagonistically to each other biologically, as a result of their competition for binding sites on proteins that require metals as cofactors. The interaction of zinc with other metals, such as copper, iron and calcium, has been reviewed in some detail elsewhere (Walsh et al., 1994; Bremner & Beattie, 1995).

7.8.1 Zinc and copper

Copper deficiency induced by excess zinc intake in experimental animals is manifested by reduced copper concentrations in liver, serum and heart, and decreased activities of copper metalloenzymes (Duncan et al., 1953; Van Reen, 1953; Cox & Harris, 1960; L'Abbe & Fischer, 1984).

Excessive zinc intake has been shown to inhibit intestinal absorption, hepatic accumulation and placental transfer of copper, as well as to induce clinical and biochemical signs of copper deficiency (Campbell & Mills, 1974; Bremner et al., 1976; Hall et al., 1979; L'Abbe & Fischer, 1984). Results of an isotope experiment suggest that zinc interferes with copper metabolism by decreasing utilization and increasing excretion of copper in the rat, but has little effect on copper absorption (Magee & Matrone, 1960). High levels of zinc in the diet have been shown to induce *de novo* synthesis of metallothionein in a dose-related fashion. It has been suggested that the induced metallothionein sequesters copper, reducing its bioavailability (Hall et al., 1979). Animals deficient in copper are infertile (Mertz, 1987). Richmond (1992) decreased the mortality of pups delivered of copper-deficient dams by injection of oxytocin at term. Atrophy of the exocrine pancreas in copper deficiency (Alvarez et al., 1989) may be secondary to vascular changes (Weaver, 1989). Allen et al. (1982) found that copper-deficient rats responded poorly to injection of thyrotropin-releasing hormone. Deficient, non-anaemic rats at 24 °C became hypothermic with, *inter alia*, decreased concentration of triiodothyronine in plasma. Mice deficient in copper (Lynch & Klevay, 1992) have a bleeding tendency characterized, *inter alia*, by increased activated partial thromboplastin time, prothrombin time.

Findings of infertility, thyroid abnormalities, pancreatic changes, coagulation defects and bone pathology in experiments using increasingly high doses of zinc may impair copper utilization. Characterization of the copper contents of diets and the copper levels in organs is important in understanding the relevance of these effects. If the effects of high doses of zinc are not accompanied by decreased copper in target organs, it seems likely that the they are related to zinc intoxication.

In a study designed to measure the level of zinc at which copper metabolism begins to be affected, Wistar rats (10 per group) were fed diets containing zinc as zinc sulfate at 0, 15, 30, 60, 120 or 240 mg/kg. Ceruloplasmin activity is significantly reduced at doses of 30 mg/kg and greater, and the number of rats with extremely low ceruloplasmin activity increases with increased zinc levels in the diet. The level of zinc at which 50% of animals would have abnormally low ceruloplasmin activity was calculated to be 125–129 mg/kg. Liver superoxide dismutase and heart cytochrome *c* oxidase activities were significantly reduced at 120 and 240 mg/kg respectively, as compared to controls (L'Abbe & Fischer, 1984).

Mink (11 females and 3 males per group) were exposed to a basal diet or to a diet supplemented with zinc as zinc oxide at 1000 µg/g throughout the mating, gestation and lactation periods. The basal diet contained zinc at 20.2 µg/g and copper at 3.1 µg/g. Supplementation of the basal diet with zinc had no significant effect on the body, liver, spleen or kidney weights of the adult female mink. No significant differences from control females were seen in the haematological parameters measured. Clinical signs consistent with copper deficiency (alopecia, anaemia or achromotrichia) were not observed in the adult mink. All females on the basal diet whelped, but only 8 females on the zinc-supplemented diets produced offspring. The body weights of male kits born to dams consuming the zinc supplemented diet were significantly lower than those of controls at 12 weeks of age. No significant differences were noted in erythrocyte or leukocyte count, haemoglobin concentration, mean corpuscular haemoglobin concentration, mean corpuscular volume or the leukocyte differential count between the zinc-treated and control kits bled at 8 weeks of age. There was a significant decrease in haematocrit value in the zinc-exposed kits. The T-cell mitogenic response was significantly reduced in the zinc-treated mink kits; however, the immunosuppression was reversible, as a normal response was seen approximately 14 weeks after the kits were weaned and placed on basal diets. In 3- to 4-week-old kits, whelped and nursed by females, that were fed a zinc-supplemented diet, greying of the fur developed around the eyes, ears, jaws and genitals, with a concomitant hair loss and dermatosis in these areas. The condition was stated to be consistent with copper deficiency; it spread over much of the body within a few weeks and persisted for several weeks after the kits were removed from the supplemented diets (Bleavins et al., 1983).

Pregnant sheep (5–12 per group) were fed diets containing zinc as zinc sulfate at 0, 30, 150 and 750 mg/kg for approximately 110 days. The diet contained copper at 2.5 mg/kg. Food consumption, weight gain and efficiency of food utilization were reduced in ewes in the 750-mg dose group. The reduction in feed intake began within 10 days of the beginning of the treatment. Some 20 days prior to parturition, copper status in the highest dose group was severely depressed, with reductions in plasma copper, ceruloplasmin and amine oxidase activity when compared to the group on the basal diet. The concentration of zinc in plasma was greatly increased in the 750 mg dose group only. Reproductive performance was severely impaired in the highest dose group, with increased incidence of non-viable lambs, defined as lambs which were aborted, stillborn or died within 7 days of birth. The cause of death in these lambs was not determined. Lambs born alive in the high dose group were weak, did not suckle, displayed ataxia and died following convulsions within 48 h of birth. Of 20 lambs conceived in the high-dose group, only one survived longer than 5 weeks. Two findings common to all non-viable lambs from the high-dose group were high tissue zinc concentrations and low tissue copper concentrations; radiographs of these lambs revealed arrested growth in the long bones. Addition of copper (10 mg/kg of diet) to another group of pregnant sheep fed diets containing zinc at 750 mg/kg, prevented the development of copper deficiency, but failed to prevent the adverse effect of high zinc on weight gain, feed consumption, efficiency of feed utilization and lamb viability. The doses of zinc in pregnant ewes were calculated to be 20 mg/kg body weight per day at the start of the study, declining to 10 mg/kg per day with reduced food intake. It was postulated that the reduced viability of lambs may have been due to fetal malnutrition caused by the reduced maternal food intake and food utilization, or alternatively to direct toxicity of zinc to the fetus (Campbell & Mills, 1979).

The reverse interaction, namely the effect of copper on zinc status, is less clear. Excessive copper can affect zinc metabolism in some species, but zinc absorption does not appear to be seriously affected. The intestinal absorption of zinc in the rat was decreased by 20% when the dietary copper level was increased from 3 to 24 mg/kg, with no further decreases seen at copper levels of 300 mg/kg (Hall et al., 1979). However, there is some evidence for competition and/or inhibition of copper or zinc uptake into intestinal cells when the luminal concentration of the respective metal is very high (Oestreicher & Cousins, 1985).

7.8.2 Zinc and other metals

High levels of zinc (0.5–1%) fed to rats have been shown to interfere directly with iron metabolism (Magee & Matrone, 1960). The occurrence of hypochromic, macrocytic anaemia in rats following the ingestion of excessive zinc and the reversal of this anaemia by iron supplementation demonstrate the interaction between iron and zinc (Cox & Harris, 1960; Magee & Matrone, 1960). Zinc intoxication affects iron metabolism by increasing the iron turnover, decreasing the life span of erythrocytes and decreasing the hepatic accumulation of iron as ferritin (Settlemire & Matrone, 1967a,b).

Zinc appears to be a less effective inhibitor of iron absorption than iron is of zinc absorption. In iron-deficient mice and rats, the oral absorption of zinc is greatly increased, which has been interpreted to indicate a shared transport pathway (Pollack et al., 1965; Forth & Rummel, 1973; Hamilton et al., 1978). Iron absorption and distribution is altered by zinc deficiency. A marked increase in iron and a decrease in zinc concentration in various organs, such as the liver, bone, pancreas and testes, have been observed in zinc-deficient animals in comparison to pair-fed controls. These changes are reversed following zinc supplementation (Prasad et al., 1967; Prasad et al., 1969; Petering et al., 1971).

Excess dietary zinc administered to pregnant rats and also to weanling and adult rats lowers the tissue iron content of the treated animals (Duncan et al., 1953; Cox & Harris, 1960; Magee & Matrone, 1960; Cox et al., 1969). In another study, reduced tissue iron and copper levels in weanling rats and reductions in calcium and phosphorus deposition in bones of young rats were observed following feeding with excess zinc (Sadasivan, 1951). High levels of dietary zinc have been also been shown to interfere with the metabolism of calcium and to increase total calcium and concentrations of calcium in the liver, but to decrease these levels in the body of exposed fetuses (Cox et al., 1969). Elevated magnesium concentrations (mg/kg) but not total magnesium content were detected in the liver and body of fetuses from mothers fed excess zinc (Cox et al., 1969).

Interaction between zinc and cadmium in animals has been reviewed elsewhere (IPCS, 1993). Supplementation with zinc has been shown to prevent the teratogenic and carcinogenic effects of cadmium: the induction of severe facial abnormalities in hamster embryos induced by cadmium (2–4 mg/kg administered intravenously) was prevented by the simultaneous administration of zinc (as zinc sulfate at 992 mg/kg) (Ferm & Carpenter, 1968); and the induction of interstitial cell tumours in rats and mice was prevented by concurrent zinc supplementation (Gunn et al., 1963).

7.9 Zinc deficiency in animals

Zinc is essential for DNA replication, RNA polymerases, protein synthesis and many metabolic processes. All cell replication, protein synthesis and growth processes are partially dependent upon zinc. Systemic depletion of this element therefore inevitably leads to deleterious effects.

The essentiality of zinc for growth has been described elsewhere (Todd et al., 1934; Hove et al., 1937; Hove et al., 1938). In experimental animals, restriction of zinc in the diet leads to an immediate decline in plasma zinc levels, followed by a loss of appetite and poor growth, which are evident within a few days of zinc depletion. Further symptoms can include dermatitis, alopecia and testicular atrophy (Macapinlac et al., 1966, 1968; Chesters & Quarterman, 1970; Wallwork et al., 1981). Zinc deficiency in experimental animals is characterized by rash, alopecia, hyperkeratosis, parakeratosis and hypopigmentation (O'Dell et al., 1959; Oberleas et al., 1962). In monkeys, as the deficiency progresses, animals stand in a hunched position, have an unsteady gait and unkempt fur, become emaciated and eventually die (Macapinlac et al., 1967; Sandstead et al., 1978; Swenerton & Hurley, 1980). It has been observed that the healing of wounds is retarded in zinc-deficient rats and that healing can be accelerated with zinc supplementation (Sandstead et al., 1970).

Zinc deficiency has an adverse effect on the pancreas of experimental animals. *In vitro* assays in pancreatic preparations from rats fed a zinc-deficient diet showed a marked impairment of the insulin response to glucose , which was directly proportional to the degree of zinc deficiency (Huber & Gershoff, 1973; Jhala & Baly, 1991). Plasma insulin levels in response to glucose injection were decreased in obese but not in lean rats fed a zinc-deficient diet over 8 weeks (Zwick et al., 1991). Additionally, a markedly zinc-deficient diet in rats produced a significant reduction in the total pancreatic content of zinc within 2 days and was associated with a more than 50% reduction in the activity of γ-glutamyl hydrolase (an indicator of pancreatic activity) in pancreatic tissue (Canton & Cremin, 1990). Rapid loss of pancreatic carboxypeptidase activity has been demonstrated under similar conditions (Mills et al., 1967).

Zinc deficiency has been shown to be correlated with a diminished activity of some enzymes. The level of a serum enzyme alkaline phosphatase decreased in zinc-deficient animals and increased with zinc replenishment (Sadasivan, 1952; Van Reen, 1953). It has been postulated that the promoter region of the gene for intestinal alkaline phosphatases contains a metal-responsive element, and that zinc deficiency leads to suboptimal transcription of this type of enzyme (Stuart et al., 1985; Millan, 1987). Zinc deficiency has also been reported to impair the activity of intracellular hepatic enzymes. The biotransformation of some pharmacological agents was reduced in zinc-deficient rats and was also associated with a decrease in the cytochrome P450 content of microsomes (Becking & Morrison, 1970).

Serum lipid concentrations were shown to be lower than normal in zinc-deficient rats, and this was postulated to be caused by the impairment of intestinal absorption of lipids by zinc deficiency (Koo et al., 1987).

Adverse reproductive effects were seen in rats when their diets were low in zinc (Hurley & Swenerton, 1966; Hurley & Shrader, 1974). Spermatogenesis was shown to be arrested in weanling rats and the germinal epithelium of the testes was atrophic (Barney et al., 1968). The menstrual cycle of rats (Apgar, 1970) and monkeys (Swenerton & Hurley, 1980) was also reported to be impaired, and ovarian follicular development appeared to be retarded (see also Evenson et al., 1993 for another study in rats).

Zinc deficiency is lethal or injurious to the embryos and fetuses of experimental animals. Evidence indicates that adequate levels of zinc are essential for conception (Swenerton & Hurley, 1980), blastula development and implantation (Hurley & Shrader, 1974), organogenesis (Blamberg et al., 1960; Kienholz et al., 1961; Hurley & Swenerton, 1966), fetal growth (McKenzie et al., 1975; Fosmire et al., 1977), prenatal survival (Hurley & Swenerton, 1966) and parturition (Apgar, 1973). Severe zinc deficiency results in high fetal resorption, with malformation of the skeleton, nervous system and viscera found in most of the surviving fetuses (Hurley & Swenerton, 1966; Hurley et al., 1971; Hurley & Shrader, 1972). Impaired synthesis and/or metabolism of DNA is postulated to cause these abnormalities (Swenerton et al., 1969; Dreosti et al., 1972; Dreosti & Hurley, 1975).

Zinc deficiency impairs development of the brain and has been shown to cause long-term behavioural consequences in rats. Evidence for the essentiality of zinc for the maturation of brain was provided by studies in rats (Hurley & Swenerton, 1966; Warkany & Petering, 1972), which demonstrated a variety of malformations in the brains of offspring that had been deprived of zinc early in gestation. Inhibition of DNA synthesis in neural crest cells is postulated to be one of the causes of such malformations (Swenerton et al., 1969). Zinc-deprived 10-day-old suckling rats showed suppression of incorporation of thymidine into their DNA (Sandstead et al., 1972). The cerebellum of a 21-day-old rat showed histological evidence of retarded maturation (Buell et al., 1977), and impaired division and migration of external granular cell neurons (Dvergsten et al., 1983; Dvergsten et al., 1984). The long-term functional significance of zinc deficiency in the fetus and neonate was studied in rats deprived of zinc during late gestation and/or suckling. Severe maternal deprivation (zinc at < 1 mg/kg in the diet) on days 14–20 of gestation caused stunting and a decrease in brain cell number in fetuses (McKenzie et al., 1975). Nutritionally rehabilitated offspring subsequently showed active avoidance of shock and an increased aggressive response to shock (Halas et al., 1975, 1976, 1977). Severe maternal zinc deprivation throughout nursing impaired growth of suckling pups and subsequently increased errors by nutritionally rehabilitated offspring in maze tests (Lokken et al., 1973; Halas et al., 1983). Reference to or long-term memory of shock on days 18–21 of nursing was also impaired (Halas et al., 1979). In rats that were mildly zinc-deprived during gestation and lactation (zinc at 10 mg/kg) where there was only a minimum effect on the growth of pups, it was subsequently revealed that the zinc-rehabilitated offspring had deficits in working memory (Halas et al., 1986). Maternal zinc deprivation of rhesus monkeys throughout most of the third trimester (Sandstead et al., 1978) and throughout gestation and lactation (Golub et al., 1985) caused acrodermatitis in the dam and subsequent reduction of exploration and play in infants during weaning. Later study of these animals found impaired ability to solve complex learning sets at 300 and 700, but not at 1000 days (Strobel & Sandstead, 1984).

Immune function was shown to be adversely affected by zinc deficiency. Calves with an inborn error in zinc absorption display thymic hypoplasia, an increased susceptibility to infection, growth failure, diarrhoea, dermatitis and death. Treatment with zinc can prevent and cure the illness (Brummerstedt et al., 1977). However, it may be difficult to separate immune deficiency from malnutrition in this case. In rats and mice, zinc deficiency was reported to impair the growth of the thymus and to retard both cellular and humoral immunity (Fraker et al., 1978; Luecke et al., 1978; Fernandes et al., 1979; Pekarek et al., 1979; Lennard, 1980). Mice fed diets deficient in zinc for 30 days developed thymic atrophy, had markedly depleted numbers of lymphocytes and macrophages in the spleen, and showed a markedly reduced ability to produce antibody-mediated responses to T-cell dependent and T-cell independent antigens. Delayed-type hypersensitivity responses, cell-mediated responses to tumour antigens and the function of natural killer cells were also significantly reduced (Fraker et al., 1978, 1986; Fernandes et al., 1979). However, in another study, it was shown that the ability of lymphocytes to proliferate and to produce interleukins and mitogenic-stimulated antibody responses was normal in zinc-deficient mice (Cook-Mills & Fraker, 1993). The reasons for this discrepancy are unclear, but mitogenic responses are a less reliable indicator of immune reactivity than antigen-specific responses. It has been established that, although the T-cell:B-cell ratio is unaffected, the total number of lymphocytes is significantly reduced in zinc-deficient mice (King & Fraker, 1991). The incidence of oesophageal tumours induced by methylbenzylnitrosamine (MBN) was higher in rats fed diets low in zinc, at 3 mg/kg compared to rats fed diets containing 60 mg/kg (Fong et al., 1978). This effect may arise through the oesophageal epithelium being damaged by zinc deficiency, making it sensitive to MBN and/or its activated metabolite (Fong et al., 1984). The mechanism appears to be via the activation of a specific cytochrome P450 by zinc deficiency, with a resultant increase in MBN-induced formation of O^6-methylguanine in oesophageal DNA (Barch & Fox, 1987). Studies investigating the effect of dietary zinc deficiency in oesophageal carcinogenesis are reported in section 8.3.7.

8. EFFECTS ON HUMANS

In the general population, essential elements have a range of acceptable exposures at which there are no untoward effects. Below this range there is the potential for effects associated with deficiency, and above it, effects associated with toxicity. The curve describing this concept of acceptable intake is shown in Chapter 10 (Fig. 1). As zinc is an essential component in a multiplicity of enzymatic reactions (see section 6.5.2), there is a need to define the range of acceptable intake to provide for biological requirements that balance the consequences of deficit and excess. In the position of balance, there is homeostasis, with optimum health. An additional factor is the consequence of interactions of zinc with other elements, which can introduce a toxicity mediated by zinc excess (Hill & Matrone, 1970). In this Chapter, the effects associated with zinc deficiency are described, along with the adverse effects associated with zinc excess, including those mediated by interaction with other elements.

8.1 Human dietary zinc requirements

8.1.1 Estimation of zinc requirements

There are inherent difficulties in estimating zinc requirements for humans, with a number of physiological, dietary and environmental factors affecting various populations. Estimates have been made using metabolic balance studies, in which zinc intake was compared with zinc excretion in the urine and faeces (Sandstead, 1984, 1985; Sandstead et al., 1990), and using additional factorial calculations that account for the zinc required for growth, losses (including zinc lost in sweat, shed hair and skin, semen and milk) and bioavailability (Sandstead, 1973; King & Turnlund, 1989). Growing infants, children, growing adolescents, and pregnant and lactating mothers require more zinc per kilogram of body weight than do mature adults (WHO, 1996b). The factorial estimates for zinc requirements are outlined in Tables 27 and 28.

A major factor affecting zinc requirements is the variation in the percentage absorption of zinc from differing dietary sources; this is discussed in section 6.1.2.

Table 27. Provisional dietary requirements for zinc in relation to estimates of retention, losses and availability[a]

Age	Peak daily retention (mg)	Urinary excretion (mg)	Sweat excretion (mg)	Total required (mg)	Intake necessary (mg) in daily diet for available zinc content of		
					10%	20%	40%
Infants							
0–4 months	0.35	0.4	0.5	1.25	12.5	6.3	3.1
5–12 months	0.2	0.4	0.5	1.1	11.0	5.5	2.8
Males							
1–10 years	0.2	0.4	1.0	1.6	16.0	8.0	4.0
11–17 years	0.8	0.5	1.5	2.8	28.0	14.0	7.0
18+ years	0.2	0.5	1.5	2.2	22.0	11.0	5.5
Females							
1–9 years	0.15	0.4	1.0	1.55	15.5	7.8	3.9
10–13 years	0.65	0.5	1.5	2.65	26.5	13.3	6.6
14–16 years	0.2	0.5	1.5	2.2	22.0	11.0	5.5
17+ years	0.2	0.5	1.5	2.2	22.0	11.0	5.5

Table 27 (contd.)

Age	Peak daily retention (mg)	Urinary excretion (mg)	Sweat excretion (mg)	Total required (mg)	Intake necessary (mg) in daily diet for available zinc content of		
					10%	20%	40%
Pregnant women							
0–20 weeks	0.55	0.5	1.5	2.55	25.5	12.8	6.4
20–30 weeks	0.9	0.5	1.5	2.9	29.0	14.5	7.3
30–40 weeks	1.0	0.5	1.5	3.0	30.0	15.0	7.5
Lactating women	3.45	0.5	1.5	5.45	54.5	27.3	13.7

[a] WHO, 1973. The above estimates were based on the assumption that the fat-free tissue concentration of zinc in humans is approximately 30 µg/g. This figure is equivalent to 2.0 g of zinc in the soft tissues of an adult male and 1.2 g in the soft tissues of an adult female, as determined from lean body mass. The zinc requirement at various ages was determined from the change in lean body mass with age. Bone zinc was not included in these calculations, because zinc in bone is relatively sequestered from the metabolically active pool of body zinc. The zinc content of sweat is based on an assumed zinc surface loss of 1mg/litre. The estimated requirement for lactation is based on a zinc content in milk of 5 mg/litre and a daily milk secretion of 650 ml. The urinary excretion of zinc is based on reported levels.

159

Table 28. Dietary reference values for zinc (mg/day)

Age	United Kingdom[a]			USA RDA[b]	WHO[c]	European DRI[d]
	LNRI	EAR	RNI			
Infants						
0–3 months	2.6	3.3	4.0	5.0		
4–6 months	2.6	3.3	4.0	5.0		
7–12 months	3.0	3.8	5.0	5.0	5.6	4.0
1–3 years	3.0	3.8	5.0	10.0	5.5	4.0
4–6 years	4.0	5.0	6.5	10.0	6.5	6.0
7–10 years	4.0	5.4	7.0	10.0	7.5	7.0
Males						
11–14 years	5.3	7.0	9.0	15.0	12.1	9.0
15–18 years	5.5	7.3	9.5	15.0	13.1	9.5
19–50+ years	5.5	7.3	9.5	15.0	9.4	9.5
Females						
11–14 years	5.3	7.0	9.0	12.0	10.3	9.0
15–18 years	4.0	5.5	7.0	12.0	10.2	7.0
19–50+ years	4.0	5.5	7.0	12.0	6.5	7.1
Pregnancy	c	c	c	15.0	7.3–13.3	b
Lactation						
0–4 months				19.0	12.7	+5.0
4+ months				16	11.7	+5.0

DRI = dietary reference intake; EAR = estimated average requirement;
LNRI = lower reference nutrient intake; RDA = recommended daily allowance;
RNI = recommended nutrient intake
[a] UK (1991).
[b] US National Academy of Sciences (1989).
[c] WHO (1996b); normative requirement for diet of moderate zinc availability
[d] EU (1993); no increment.

The effects of dietary supplementation on humans have recently been reviewed (Gibson, 1994). Tables 29, 30 and 31 provide a summary, taken from this review, of the effects of supplementation in infants, children and lactating women.

Table 29. Double-blind zinc supplementation studies in infants

| Country | Subjects and treatment | Mean plasma zinc levels (µmol/litre) | | | | Growth effects and other responses | Reference |
| | | Zinc supple- mentation | | Control | | | |
		Start	End	Start	End		
USA	68 normal healthy full-term male infants at birth; studied for 6 months double-blind study; formula with zinc at 1.8 mg/litre or 5.8 mg/litre		119		110	improved weight and length in males only	Walravens & Hambidge (1976)
France	57 normal healthy infants at 5.4 months old, studied for 3 months double-blind study; zinc at 5 mg/day (25) or placebo (32)					improved weight gain; improved length in males only	Walravens et al. (1992)
USA	50 failure-to-thrive infants, 8–27 months old, studied for 6 months randomized double-blind trial, pair matched; zinc at 5.7 mg/day as syrup (25) or placebo (25)	10.7	9.8	10.7	10.4	improved weight especially in males:. tendency to increased activity of serum alkaline phosphate in zinc group	Walravens et al. (1989)

Table 29 (contd.)

Country	Subjects and treatment	Mean plasma zinc levels (µmol/litre)				Growth effects and other responses	Reference
		Zinc supplementation		Control			
		Start	End	Start	End		
Chile	32 marasmic infants, 7–8 months old, studied for 90 days randomized double-blind trial; zinc at 2 mg/kg daily in solution (16) or placebo (16)	14.7	15.6	16.1	15.6	weight-for-length effect; decrease in percentage of anergic infants; increase in serum IgA in zinc group	Castillo-Duran et al. (1987)
Chile	39 severely malnourished infants studied for 104 days double blind trial; zinc at 1.9 mg/kg (19) or 0.35 mg/kg in daily formula (20)	19.4	18.6	23.4	18.0	linear length effect; improved immune function	Schlesinger et al. (1993)

Table 29 (contd.)

Bangladesh	60 severely malnourished infants 5–60 months old studied for 3 weeks rice-based diet *ad lib* with vitamins and minerals; zinc at 10 mg/kg daily if < 6 kg or 50 mg/day if > 6 kg as zinc sulfate; non-supplemented group (30)	8.2	18.5	7.9	10.6	improved weight gain and weight for length	Khanum et al. (1988)
Bangladesh	65 children with AD and 152 with PD 3–24 months old supplemented for 2 weeks followed for 2 or 3 months in a double-blind randomized study after supplementation for 2 weeks with zinc at 15 mg /kg daily or placebo					improved length gain in AD group, and in PD with < 90% weight/age and 90% height/age; reduced no. of episodes of diarrhoea in AD and PD groups and attack rate of respiratory tract infections in AD group only	Roy et al. (1992)

AD = acute diarrhoea; PD = persistent diarrhoea; SGA = small for gestational age

163

Table 30. Double-blind zinc supplementation studies in children

Country, date	Subjects and treatment	Dietary zinc intake (mg)	Mean plasma zinc levels (μmol/litre)				Growth effects and other responses	Reference
			Zinc supplementation		Control			
			Start	End	Start	End		
Egypt 1965–1966	90 growth-retarded schoolboys, 11–18 years old studied for 5.5 months randomized trial; zinc at 14 mg (30) or placebo (30); capsules given at school	14	1–7	19–2	11–7	13–3	no weight or height effects; no difference in sexual maturation; no effect on serum alkaline phosphatase	Carter et al. (1969)
Iran 1967–1968	60 growth-retarded schoolboys, 12–18 years old, studied for 17 months (5 months trial, 7 months rest, 5 months trial); first 5 months, 28 mg of zinc (20), 67 mg of iron (20) or placebo (20); second 5 months, micro-nutrients (20), micronutrients + 40 mg of zinc (20) or placebo (20); capsules given at school	12	17–2	14–7	11–6	14–1	no weight or height effects; difference in sexual maturation	Ronaghy et al. (1968)

Table 30 (contd.)

Iran 1969–1971	50 growth-retarded schoolboys, 13 years old, studied for 17 months (5 months trial, 7 months rest, 5 months trial) non-randomized trial; micronutrients (20), micronutrients + 40 mg of zinc (20) or placebo (10); capsules given at school	12	8-2	10-2	10-5	10-7	weight and height effects; difference in bone age; tendency for faster sexual development; no effect on serum alkaline phosphatase	Ronaghy et al. (1974)
USA, Colorado	40 growth-retarded, low-zinc-status children, 2–6 years old, studied for 1 year randomized pair-matched trial; zinc at 10 mg/day (20) or placebo (20); syrup given by parents at home	4-6	10-7	10-8	11-3	11-3	height effect (especially in boys); increase in appetite	Walravens et al. (1983)
Canada 1986	60 growth-retarded boys, 5–7 years old, studied for 12 months randomized pair-matched trial; zinc at 10 mg/day (30) or with placebo (30); fruit juice drink given by parents at home	6-4	15-6	16-2	16-5	16-4	height effect only in subjects with low hair zinc (<1.68 µmol/g); increase in appetite perceived by parents	Gibson et al. (1989b)

Table 30 (contd.)

Country, date	Subjects and treatment	Dietary zinc intake (mg)	Mean plasma zinc levels (µmol/litre)				Growth effects and other responses	Reference
			Zinc supplementation		Control			
			Start	End	Start	End		
Thailand 1989–1990	133 children, 6–13 years old, with suboptimal zinc and vitamin A nutriture studied for 6 months randomized pair-matched trial; zinc at 25 mg/day (33), vitamin A (33), vitamin A + zinc (32) or with placebo (35); capsules taken on school days	4.3	13.2	19.0	13.2	14.3	no weight or height effects; increase in serum alkaline phosphatase activity; improved dark adaptation; improved conjunctival integrity	Udomkesmalee et al. (1992)
Gambia 1989–1990	109 apparently healthy children, 0.5–3 years old, studied for 15 months randomized group-matched trial; 70 mg of zinc (55) or placebo (54); drink given twice a week at clinic						no weight or height effects; increase in arm circumference; less malaria; improved intestinal permeability	Bates et al. (1993)

Table 30 (contd.)

Guatemala 1989	162 schoolchildren, 6–8 years old, studied for 25 weeks randomized pair-matched trial; micronutrients (82), micronutrients + zinc at 10 mg/day (80); chewable tablet given at school on weekdays	10	14.2	16.2	14.4	14.9	no weight or height effects; increase in triceps skinfold; smaller decrease in mid-arm circumference; no increase in serum alkaline phosphatase	Cavan et al. (1993)
Chile 1991	46 short-stature schoolchildren, 6–12 years old, consuming diets providing 50–60% of normal daily zinc intake, studied for 12 months randomized study; zinc at 10 mg/day or placebo						no weight effect; height effect in males only; no difference in plasma zinc	Castillo-Duran et al. (1995)
Chile 1993	98 healthy pre-school children studied for 14 months zinc at 10 mg/day or placebo						height effect in males; trend towards improved immune function and reduced giardiasis	Ruz et al. (1997)

Table 31. Double-blind zinc supplementation studies in lactating women

Country	Subjects and treatment	Dietary zinc intake (mg)	Response	Reference
USA, Colorado	53 middle-income lactating women, studied for varying durations up to 9 months controlled trial; zinc at 15 mg/day (14), placebo (39) or control (8); tablets taken at home	12.2	decreased fall in milk zinc levels	Krebs et al. (1985)
USA, Indiana	49 middle-income mothers studied for first 6 months of lactation controlled trial; micronutrients (25) or micronutrients + zinc at 25 mg/day (24); different commercial supplements taken at home	11.2	higher milk zinc levels	Karra et al. (1986)
USA, Maryland	40 middle-income women studied for first 6 months of lactation randomized double-blind trial; micronutrients (20) or micronutrients + zinc at 25 mg/day (20); tablets taken at home	12	no effect on milk zinc levels	Moser-Veillon & Reynolds (1990)

Methods for the assessment of zinc status in humans are discussed in section 6.5.1.

8.2 Zinc deficiency

8.2.1 Clinical manifestations

Cases of severe zinc deficiency are now rare, but mild deficiency during periods of rapid growth, pregnancy, synthesis of new tissue, and in persons consuming plant-based diets, is not uncommon. Zinc deficiency also occurs in the presence of certain disease states such as malabsorption syndromes, renal and hepatic diseases, and in association with burns and alcoholism. Two genetic disorders, acrodermatitis enteropathica and sickle-cell disease, are associated with suboptimal zinc status.

The first cases of human zinc deficiency were reported in the Middle East among adolescent dwarves in the 1960s (section 8.2.4). Since those first reports, mild zinc deficiency has been reported in infants and younger children living both in developing and in industrialized countries.

The health effects associated with zinc deficiency in humans have been extensively reviewed (Prasad, 1988; Aggett, 1989; Clegg et al., 1989; Hambidge, 1989; Keen & Hurley, 1989; Walsh et al., 1994). Zinc deficiency has been classified into three syndromes (Henkin & Aamodt, 1983): acute, chronic and subacute zinc deficiency. The clinical symptoms range from neurosensory changes, oligospermia in males, decreased thymulin activity, decreased interleukin-2 production, hypogeusia and impaired neuropsychological functions (Prasad, 1988; Penland, 1991) in mild or marginal deficiency, through to growth retardation, male hypogonadism, and delayed wound healing with moderate deficiency, and alopecia, mental disturbances, cell-mediated immune disorders and pustular dermatitis in patients with severe zinc deficiency (Prasad, 1988). These conditions are generally reversible when the deficiency is corrected by zinc supplementation.

8.2.2 Brain function

In an experimental study (Henkin et al., 1975b) in which severe, acute zinc deficiency was induced in eight patients with scleroderma

169

by treatment with large doses of oral histidine, severe zincuria was produced, and plasma zinc levels decreased from 60–105 µg/dl to 40–60 µg/dl. Signs of zinc deficiency included anorexia, dysosmia, ataxia, tremor, loss of memory, impaired higher intellectual processes, paranoid ideation and receptive aphasia. Treatment with zinc by mouth improved signs within 24 h.

The effects of less severe zinc deficiency are less easily characterized and include reduced growth and impaired immune function (WHO, 1996b). In a study in which 14 men were fed diets providing zinc at a rate of 1, 2, 3, 4 or 10 mg/day for periods of 35 days in a 7-month study (Johnson et al., 1993), impaired neuromotor and cognitive function was observed (Penland, 1991), with significant decreases in sensory motor, attention, visual memory and spatial and perceptual tasks.

8.2.3 Immune function

In patients suffering from acrodermatitis enteropathica — a rare genetic defect affecting the assimilation of zinc — an increased incidence of secondary infections is seen, and T-cell numbers, thymic hormone levels and T-cell mediated cellular and humoral immunities are deficient (Aggett, 1989). Similar changes have been noted in other patients with zinc deficiency and with sickle-cell anaemia (Fraker et al., 1986; Endre et al., 1990), and patients with suboptimal zinc intakes have been reported to be at greater risk of infection and disease (Bogden et al., 1987). In an experimental study in which male volunteers with experimentally induced mild zinc deficiency had decreased interleukin-2 activity, a decreased T4+:T8+ ratio and increased T101 cells and serum immunoglobulin (Ig), these changes were corrected upon zinc repletion (Prasad et al., 1988). Immune function related to zinc deficiency has been reviewed by Keen & Gershwin (1990). It has been suggested that zinc may act as an antiviral agent. Possible mechanisms by which this could be achieved are inhibition of virus protein coat synthesis and prevention of virus entry into the cell (Korant & Butterworth, 1976; Prasad, 1996).

8.2.4 Growth

Growth retardation and hypogonadism were reported in adolescents in the Middle East, and these effects were believed to be related to inadequate dietary zinc intake (Prasad et al., 1961, 1963b). The principal features of this syndrome were growth failure and delayed sexual maturation, giving 16- to 18-year-olds a physical appearance resembling that of prepubertal 9-year-olds, commonly associated with hepatosplenomegaly and iron deficiency. Zinc deficiency appears to be a major contributing factor in this syndrome. Administration of zinc, along with a balanced diet, produced accelerated growth, and enlargement of the penis and testes in males, and of breasts in females; a well-balanced diet alone was not followed by rapid improvement (Prasad et al., 1963a; Sandstead et al., 1967; Halsted et al., 1972). A subsequent series of zinc supplementation studies in Iran gave mixed results (Ronaghy et al., 1974; Mahloudji et al., 1975): there was a clear stimulation of growth after supplementation, but no significant stimulation of gonadal development (Ronaghy et al., 1974). Supplementation with zinc plus iron did not stimulate growth (Mahloudji et al., 1975).

Details of more recent double-blind zinc supplementation studies conducted on infants and children are reviewed in Gibson (1994).

8.2.5   Dermal effects

Severe zinc deficiency resulting from total parenteral nutrition without zinc (Arakawa et al., 1976; Kay et al., 1976), and in patients suffering from acrodermatitis enteropathica (Aggett, 1989) leads to dermatological effects, including erythematous scaling eruptions in the naso-labial and retro-auricular folds, with the dermatitis extending to the trunk and becoming exudative upon continued zinc deficiency (total parenteral nutrition), and bullous pustular dermatitis of the extremities and the oral, anal and genital areas, combined with paronychia and generalized alopecia (acrodermatitis enteropathica).

8.2.6   Reproduction

An association between low serum zinc levels and reproduction was made when one of 83 infants in a series of studies (Jameson, 1976) showed a congenital cardiac malformation, with a ventricular septum defect and coarctation of the aorta. The infant's mother had shown the lowest serum zinc level (12.2 µmol/litre) in the 13th week of gestation, but all other laboratory findings were normal. In women with complications such as abnormal labour or atonic bleeding, serum zinc concentrations had been significantly reduced during early pregnancy. Additionally, of 316 pregnancies, a high proportion (60%) of the women who gave birth to infants with congenital defects had shown low serum zinc concentrations in the first trimester.

In a study in which 450 women were followed during and after pregnancy (Mukherjee et al., 1984), plasma zinc was reported to be an indicator of feto-maternal complications, including fetal distress and maternal infections, for those women in the lowest quartile for plasma zinc. In a study in low-income women, there was a significantly higher prevalence of low birth weight in the infants of mothers in the lowest quartile for plasma zinc at 16 weeks gestation than in those born to the other mothers (Neggers et al., 1990).

Studies to examine whether maternal zinc status is a useful predictor of pregnancy outcome have produced mixed results. Scholl et al. (1993), in a cohort study of pregnant girls and women of low socioeconomic status, reported that low dietary intakes of zinc (< 6.0 mg/day) were associated with increased risk of low-birth-weight infants, after controlling for energy intake and other variables known to influence outcome. Some studies (including Hunt et al., 1984; Cherry et al., 1989; Goldenberg et al., 1995), but not all double-blind supplementation trials have provided further support for this suggestion. In a study by Tamura & Goldenberg (1996) of 580 indigent African-American pregnant women, those randomly assigned to a zinc-supplemented group (25 mg of zinc daily as zinc sulfate) at 19.2 weeks of gestation had infants with a significantly higher birth weight (126 g; $P = 0.03$) and head circumference (0.4 cm; $P = 0.04$) than infants born to mothers in the placebo group. The results suggested that, by increasing the zinc intakes of pregnant women with suboptimal zinc nutriture, pregnancy outcomes could be improved. Recent reviews of this subject appear in Gibson (1994) and Tamura & Goldenberg (1996), and a summary of some of these findings is provided in Table 32.

8.2.7 Carcinogenicity

In a study in Chinese men aged between 45 and 75 years, the zinc levels in serum and hair were lower in those patients with oesophageal cancer (Lin et al., 1977). The results of these studies do

Table 32. Zinc supplementation studies in pregnant women

Country, date	Subjects	Treatment	Dietary zinc intake (mg)	Responses	References
United Kingdom 1985–1986	494 middle-class women studied for last 4 months of pregnancy	randomized double-blind trial; zinc at 20 mg/day (246) or placebo (248); capsules taken at home	9	no effect on birth weight; no differences in leukocyte zinc	Mahomed et al. (1989)
USA, New Orleans	556 low-income adolescent women studied for last 3 months of pregnancy	randomized double-blind trial; zinc at 30 mg/day (268) or placebo (288); tablets taken at home	30	no effect on birth weight; reduced rates of pre-maturity and neonatal morbidity	Cherry et al. (1989)
USA, Los Angeles 1981–1982	138 Hispanic teenagers studied for last 4 months of pregnancy	randomized double-blind trial; micronutrients (68) or micronutrients + zinc at 20 mg/day (70); capsules taken at home	9–8	no effect on birth weight	Hunt et al. (1995)

173

Table 32 (contd.)

Country, date	Subjects	Treatment	Dietary zinc intake (mg)	Responses	References
USA, Los Angeles	213 Hispanic low-income women enrolled at gestation age < 27 weeks	randomized double-blind trial; micronutrients (106) or micronutrients + zinc at 20 mg/day (107)	9–3	no effect on birth weight; reduced incidence of pregnancy-induced hypertension	Hunt et al. (1995)
United Kingdom	56 pregnant women at risk of small-for-gestational age infants studied for last 15–25 weeks of pregnancy	randomized double-blind trial; zinc at 22.5 mg/day (30) or placebo (26)	22.5	lower incidence of Intra-uterine growth retardation; reduction in induced labours and Caesarean sections	Simmer et al. (1991)
USA	46 pregnant middle-income women studied for 7–9 months	double-blind study; zinc at 15 mg/day (10) or placebo (36); tablet taken 2 h after dinner	11	no effect on birth weight; no other effects observed	Hambidge et al. (1983)

not provide evidence for any causal relationship between low plasma/serum zinc levels and an increased incidence of cancer in humans. Similarly, in another study by Lipman et al. (1987), mean plasma zinc and mean plasma vitamin A in the 21 oesophageal cancer patients were significantly lower than in the 17 patients with oesophagitis, or the 12 normal controls. However, there were no differences in oesophageal zinc content between the cancerous tissue and adjacent normal tissue, the oesophagitis tissue and adjacent normal tissue, and normal oesophageal tissue.

8.3 Zinc toxicity: general population

8.3.1 Poisoning incidents

A number of reports outline the effect of acute exposure to zinc in humans. However, these reports are generally old and poorly documented, with inadequate characterization of the actual exposure levels, although some estimates of exposure have been made. For example, high concentrations of zinc in drinks (up to 2500 mg/litre) have been linked with effects such as severe abdominal cramping, diarrhoea, tenesmus, bloody stools, nausea, and vomiting in 300 people, and symptoms of dryness of the mouth, nausea, vomiting and diarrhoea in more than 40 people (Brown et al., 1964). The amount of zinc ingested was estimated to be approximately 325–650 mg. Lethargy, along with drowsiness, unsteady gait, and increased serum lipase and amylase levels, was seen in an individual who had ingested 12 g of elemental zinc, equivalent to 150 mg/kg body weight, resulting in increases in blood zinc concentrations (Murphy, 1970). No gastrointestinal distress was reported and chelation therapy was effective in achieving clinical improvement and reducing blood zinc levels. Severe local burns, metabolic acidosis, hepatic damage, hyperamylasaemia, lethargy and hypertension resulting from the ingestion of zinc chloride/ammonium chloride soldering flux were reported in a 16-month-old boy who developed pancreatic exocrine insufficiency 5 months later (Knapp et al., 1994).

Excess hepatic copper and zinc levels in a small number of Cree and Ojibwa-Cree children were associated with severe chronic cholestatic liver disease progressing to end-stage biliary cirrhosis in these children (Phillips et al., 1996). It was postulated that the effects might have been due to an inborn error of metal metabolism, secondary dietary or environmental factors, or genetic factors. Zinc and copper also appeared to be accumulated in transplanted livers, but these findings were not quantitative and there were no detectable histological effects following transplantation. There were no data to indicate that any exposure to excess zinc had occurred in these children.

8.3.2 Dermal effects

Contact dermatitis has been reported following use of shampoos containing zinc pyrithione (Nigam et al., 1988). The specific etiological role for zinc was not clear, and the dermal application of zinc as zinc oxide has not been associated with any adverse dermal effects in humans.

8.3.3 Immune function

An adverse lymphocytic response was reported in 11 healthy adult men who ingested 150 mg of elemental zinc twice a day for 6 weeks; the subjects also showed a reduction in the lymphocytic stimulation response to phytohaemagglutinin (up to 70% reduction at 6 weeks), chemotaxis (50% reduction) and phagocytosis of bacteria by polymorphonuclear leukocytes (50% reduction). There were no control groups in this study and the copper status of the subjects was not measured. The absolute number of lymphocytes and the proportions of T- and B-cells were not altered. However, the measurement of immune status conducted *in vitro* may not be a true reflection of the immune responses in the subjects, in whom a two-fold elevation in serum zinc was measured (Chandra, 1984).

Conversely, when 103 apparently healthy elderly subjects were randomly assigned to one of three treatments and given supplementary daily doses of placebo, or 15 or 100 mg of zinc for 3 months, none of the treatments significantly altered delayed dermal hypersensitivity to a panel of seven recall antigens or *in vitro* lymphocyte proliferative responses to mitogens and antigens. A modest increase in plasma zinc was not accompanied by a decrease in plasma copper levels (Bogden et al., 1988). Subjects also received a daily supplement of 2 mg of copper above dietary intake.

Phagocytic fungicidal capacity was evaluated in a double-blind study in which marasmic infants received formula fortified with zinc and iron at 15 mg/litre for up to 105 days, with a mean daily zinc intake of 1.9 mg/kg (Schlesinger et al., 1993). A decrease in the number of infants whose monocyte phagocytic activity was above the median was observed after 60 days of zinc supplementation (63% upon admission compared to 32% after 60 days; $P < 0.05$). There was also a decrease in the number of infants whose monocyte fungicidal activity was above the median after 105 days of zinc supplementation (61% upon admission compared to 39% after 105 days; $P < 0.04$). The number and duration of impetigo episodes were greater in the group of infants fed the zinc-supplemented formula (1.31 ± 1.1 infectious episodes/infant compared to 0.55 ± 0.8 in controls).

However, in another study in marasmic infants (Castillo-Duran et al., 1987) in which 16 subjects received a daily elemental zinc supplement as zinc acetate of 2 mg/kg and 16 subjects received a placebo for 90 days, the incidence of infection, especially pyoderma, was significantly decreased in the zinc-supplemented group (3/16 in the supplemented group compared to 10/16 in controls; $P < 0.025$). The proportion of anergic infants decreased (from 50% to 25% between days 0 and 90) and serum IgA increased significantly (from 81 ± 32 to 111 ± 26 mg/100 ml) only in the zinc-supplemented group.

8.3.4 Reproduction

Dietary supplementation with zinc at a rate of 20 mg/day did not result in adverse effects on pregnancy progress or outcomes in healthy pregnant women in a number of large, controlled trials (Hunt et al., 1984; Kynast & Saling, 1986; Mahomed et al., 1989). In a double-blind trial in low-income pregnant adolescents thought to be at risk for poor zinc nutriture, supplementation with zinc at 30 mg/day did not result in adverse pregnancy outcomes (Cherry et al., 1989). Of the women, one-third received the zinc for the first trimester and the remainder from the second trimester. Similarly, dietary supplementation for the last 15–25 weeks of pregnancy with 22.5 mg/day to women at risk of delivering a small-for-gestational-age baby did not result in adverse reproductive effects (Simmer et al., 1991).

When seven pregnant women with low serum zinc levels (< 11.5 µmol/litre) were given a zinc supplement of 90 mg/day for the last 13–25 weeks of pregnancy, no adverse effects were associated with the supplementation (Jameson, 1976). In a follow-up study (Jameson, 1982) in which 133 women with low serum zinc levels (< 10 µmol/litre) were randomly assigned zinc supplementation at 45 mg/day or no supplementation, no adverse effects were associated with zinc supplementation, and serum copper levels were unaffected.

8.3.5 Zinc-induced copper deficiency

Elevated intakes of zinc have been shown to induce copper deficiency in humans (Prasad et al., 1978a; Fischer et al., 1984; Hoffman et al., 1988). The level of intestinal metallothionein may be important in the development of this zinc-induced hypocupraemia. As metallothionein has a greater affinity for copper than zinc, and zinc induces high levels of metallothionein in the intestinal mucosa (Richards & Cousins, 1975), the proposed mechanism for this copper deficiency is a reduction in copper absorption followed by sequestration of copper in a stable copper-metallothionein complex, which is returned to the intestinal lumen by the desquamation of the intestinal mucosal cells (Richards & Cousins, 1976a; Fischer et al., 1983). Balance studies indicate that, as the amount of zinc in the diet increases, so does the amount of dietary copper required, so that persons on a diet high in zinc may have an increased risk of copper deficiency (Sandstead, 1982b). The ingestion of zinc at levels near the recommended daily allowance of 15 mg (see section 5.2.2) may result in increased copper requirements, increased copper excretion, impaired copper status, or reduced copper retention (Greger et al., 1978a,b,c; Burke et al., 1981; Festa et al., 1985). The effect of dietary zinc on copper utilization depends markedly on the amount of dietary copper and the copper status of the individual (see section 8.3.5.1).

8.3.5.1 Controlled human studies

In a study in which adolescent females received dietary copper at a rate of 1.2 mg/day, faecal copper excretion was increased by approximately 14% (0.9 compared to 0.79 mg/day) in subjects receiving zinc at 14.7 mg/day compared with those fed 11.5 mg/day during 10-day periods (Greger et al., 1978c), with all subjects in positive copper balance. The standard error in the estimate for zinc was 2.18 mg/day. In another study in adolescent females, no increased copper excretion was reported in subjects given dietary zinc at 7.4 or 13.4 mg/day and copper at 2.8 mg/day for 18 days, again with all subjects in positive copper balance (Greger et al., 1978b). Other studies are cited in Table 33.

A group of 18 female volunteers participated in a 10-week, single-blind dietary supplementation study designed to investigate the effect of zinc supplementation on iron, copper and zinc status. When subjects were given zinc at 50 mg/day (administered as two gelatin tablets daily, each containing 25 mg of zinc as zinc gluconate), there was a significant reduction ($P < 0.05$) compared with pretreatment levels in serum ferritin (23%), haematocrit (4%) and erythrocyte copper,zinc-superoxide dismutase (ESOD; 47%). Serum zinc was increased by approximately 25% ($P < 0.01$), but there were no changes in serum ceruloplasmin or haemoglobin. When subjects received iron at 50 mg/day in addition to the zinc, similar reductions in ESOD were observed (47%), while there were increases in serum ferritin (25%) and serum zinc (21%); there were no changes in haemoglobin, haematocrit or ceruloplasmin (Yadrick et al., 1989). No indication of dietary intake of zinc or copper was noted in this study.

The effects of zinc supplementation on the copper status of two groups of healthy adult men were investigated for 6 weeks. A test group containing half of the subjects received 25 mg of elemental zinc as zinc gluconate twice daily in gelatin capsules (50 mg/day), while the control group received placebo capsules (Fischer et al., 1984). No significant differences in plasma copper levels or ferroxidase activities were observed between the groups. Increases in plasma zinc (approximately 20%) and decreases in ESOD (approximately 20%) were observed in the zinc-supplemented group, the differences becoming statistically significantly ($P < 0.05$) in week 6 of the study. No indication of dietary intake of zinc or copper was noted in this study.

In a 12-week, double-blind cross-over study (Samman & Roberts, 1988), 47 healthy adult volunteers received 50 mg of elemental zinc (as 220 mg of zinc sulfate) or placebo, in a capsule,

Table 33. Summary of effects of zinc on copper homeostasis in humans[a]

Copper intake (mg/day)	Subjects, duration	Zinc dose (mg/day)	Effects	Described health effects	Reference
1.2	adolescents 10 days	14.7	increase in faecal copper excretion; positive copper balance		Greger et al. (1987c)
2.8	adolescents (females) 18 days	7.4 or 13.4	no effect on copper excretion; positive copper balance; copper intake 2.8 mg/day		Greger et al. (1978b)
2.0	healthy adult females 12 days	9.5 or 19.9	no effect on faecal copper excretion; positive copper balance; copper intake 2 mg/day		Colin et al. (1983)
2.0	adult females 12 days	8 or 24	no effect on copper excretion; negative copper balance; copper intake 2 mg/day		Taper et al. (1980)
2.6	adult males 0.5 weeks	1.8–18.5	no effect on serum copper concentration; sudden increase in zinc intake from 4–8 mg/day to 18.5 mg/day resulted in a temporary increase in faecal copper excretion	none	Festa et al. (1985)
2.33	elderly adults (5 male, 6 female) 30 days	7.8 or 23.26	reduced copper retention and increased faecal copper excretion at higher zinc dose compared with lower dose; most subjects in positive copper balance	none	Burke et al. (1981)

[a] The interpretation of studies is difficult because in many supplemental studies the total intake of diet and supplement of zinc is not given.

three times a day for 6 weeks. The zinc supplementation resulted in reductions in the ferroxidase activity of serum ceruloplasmin (from 13.0 to 11.3 U/ml), and ESOD activity (from 2184 to 1672 U/g of haemoglobin), but only in females. The change in plasma zinc levels was greater in females (8.4 ± 1.5 μmol/litre) than males (5.5 ± 1.1 μmol/litre), and no changes were reported in plasma copper or haematocrit. No indication of dietary intake of zinc or copper was noted in this study.

A study of five men and six women aged 56–83 years showed that zinc intakes of about 23 mg daily (about 6 mg from the diet and 17 mg from zinc sulfate given in a beverage consumed at breakfast) significantly lowered retention of copper (about 1 mg from the diet and 1.3 mg from copper sulfate given in the breakfast beverage) compared to copper retention when the diet plus zinc sulfate in the beverage provided 7.8 mg of zinc daily (Burke et al., 1981). When the amount of copper lost in sweat (≥ 0.3 mg daily) is considered in the interpretation of the data (Jacob et al., 1979; Milne et al., 1983), it is evident that intakes of 23 mg of zinc daily placed the subjects at risk of negative copper retention. A zinc:copper ratio of about 10 had an adverse effect on copper retention that was not evident with a ratio of about 3.5. Of note in this study is that the increased intake of zinc was from a zinc sulfate supplement.

8.3.5.2 Case reports

In case studies, effects associated with long-term, excessive zinc intakes (ranging from 150 mg/day to 1–2 g/day) have included sideroblastic anaemia, hypochromic microcytic anaemia, leukopenia, lymphadenopathy, neutropenia, hypocupraemia and hypoferraemia. Patients recovered to normal blood patterns after cessation of zinc intake with or without copper supplementation (Porter et al., 1977; Prasad et al., 1978b; Hoffman et al., 1988; Simon et al., 1988; Broun et al., 1990; Forman et al., 1990; Gyorffy & Chan, 1992; Ramadurai et al., 1993).

8.3.6 Serum lipids and cardiovascular disorders

Following the induction of hypercholesterolaemia in rats by administration of a high ratio of ingested zinc to copper (Klevay, 1973) and the identification of an association between the mortality rate for coronary heart disease and the zinc:copper ratio in cows' milk in 47 cities in the USA (Klevay, 1975), it was hypothesized that the zinc:copper ratio has important influences on processes related to coronary heart disease (Klevay, 1975, 1980, 1983).

As a partial investigation of these concepts, a number of studies have been conducted to examine the effects of zinc intake on blood lipid levels. The lowest dose of zinc that affects lipid metabolism is ill-defined, but it was approximately twice the US recommended daily allowance. Doses of zinc of 50–300 mg in excess of dietary amounts generally have potentially harmful effects on lipid metabolism.

Effects resulting from zinc-induced copper deficiency are discussed in section 8.3.5.

In a 12-week, double-blind study in adult males, subjects received daily a placebo tablet ($n = 9$), or tablets containing 50 ($n = 13$) or 75 ($n = 9$) mg of elemental zinc, as zinc gluconate. Dietary analysis revealed that subjects in the 75 mg group consumed significantly less total fat, saturated fatty acids and protein than those in the other groups. Serum total cholesterol, low-density lipoprotein (LDL) cholesterol, very-low-density lipoprotein (VLDL) cholesterol and triglycerides were not affected by zinc supplementation. Serum high-density lipoprotein (HDL) cholesterol was significantly decreased ($P \leq 0.05$) at zinc doses of 75 mg/day (reductions of 11% and 15% at weeks 6 and 12, respectively) and 50 mg/day (15% at week 12) compared with placebo, and was also lower than baseline values ($P \leq 0.05$) at weeks 6, 8 and 12 at 75 mg/day (reductions of 13%, 15% and 13%, respectively) and week 12 at 50 mg/day (11%). Serum copper levels did not change with zinc supplementation (Black et al., 1988). The dietary intake of nutrients including copper and zinc were monitored throughout the study.

A study was conducted to investigate the relationship between level of exercise, zinc supplementation, and serum HDL cholesterol in men and women over the age of 60 years (Goodwin et al., 1985). There was a significant positive correlation between levels of exercise and serum HDL cholesterol in the 180 subjects not supplemented with zinc ($r = 0.26$; $P = 0.005$), but not for those subjects taking supplemental zinc. Following discontinuation of zinc supplementation (24 mg/day; median 17–52 mg/day), there was a significant increase in HDL cholesterol levels (2.0 mg/100 ml; approximately 4%; $P = 0.04$) after 8 weeks in 22 subjects. This change was positively correlated with the level of exercise of the subjects. The authors noted that in young runners, HDL is unchanged by zinc administered at 50 mg/day for 8 weeks. They suggested that age and sex differences may be important in the relationship between zinc and lipid metabolism in humans, but no data were provided to support this hypothesis.

Reiser et al. (1985) described a diet mainly of conventional food but low in copper, and containing an amount of fructose similar to that consumed by many Americans. The effects of this diet on more than 20 male subjects have been described in a number of papers (Reiser et al., 1985, 1987; Bhathena et al., 1986; Holbrook et al., 1989). Prominent among these effects were decreased plasma encephalins and dyslipidaemia characterized by decreased HDL cholesterol and increased LDL cholesterol. The experiment was interrupted because of fear of adverse cardiac effects. Evidence of copper deficiency assessed by traditional means was minimal, but included decreased activity of ESOD.

In a study in which 12 healthy male subjects received 440 mg of zinc sulfate (160 mg of elemental zinc) daily for 5 weeks, HDL cholesterol levels decreased to 25% below baseline values (30.1 compared with 40.5 mg/100 ml; $P = 0.0001$), while total cholesterol, triglyceride and LDL cholesterol levels remained unchanged (Hooper et al., 1980). No indication of dietary intake of zinc or copper was noted in this study.

When 11 healthy male subjects ingested 150 mg of elemental zinc twice a day, serum HDL concentrations decreased significantly compared with baseline values after 4 weeks (20% reduction; $P < 0.01$) and 6 weeks (30% reduction; $P < 0.001$), while LDL levels increased slightly (by 10–15% at 4–6 weeks; $P < 0.05$); however, this study lacked a placebo control (Chandra, 1984). Dietary zinc estimates were made using 24-h dietary recall interviews.

Not all studies show that zinc supplementation affects serum HDL levels. In a double-blind, cross-over trial involving 26 women and 21 men, the diets of healthy volunteers were supplemented with zinc at a rate of 150 mg/day for 6 weeks. Plasma total cholesterol and HDL levels remained unchanged in both sexes, while in women only, the LDL level decreased by 9%. There was also a trend for HDL to be redistributed in women, with slight increases in HDL_2 and slight decreases in HDL_3 (Samman & Roberts, 1988). When groups of eight women were given dietary supplementation of zinc at doses of 0, 15, 50 or 100 mg/day for 2 months, a transient 8% decrease in HDL cholesterol was seen at 4 weeks at the highest zinc level, but no uniform or sustained response of plasma cholesterol or HDL cholesterol was observed (Freeland-Graves et al., 1980). Records of the dietary nutrients, including zinc and copper, were obtained from the 3-day dietary records kept throughout this study.

8.4 Occupational exposure

8.4.1 Acute toxicity

Inhalation exposure to zinc chloride following the military use of "smoke bombs" has been reported to result in various effects, including interstitial oedema, interstitial fibrosis, pneumonitis, bronchial mucosal oedema and ulceration, and changes in the mucous membrane of the larynx and trachea (Pare & Sandler, 1954; Johnson & Stonehill, 1961; Milliken et al., 1963; Schenker et al., 1981; Matarese & Matthews, 1986). Acute injury has been associated with mortality under extreme exposure conditions, sometimes attributed to the effects upon the respiratory tract mucosa due to the hygroscopic and astringent nature of the zinc chloride particles released by such devices (Evans, 1945; Milliken et al., 1963; Hjortso et al., 1988; Homma et al., 1992).

8.4.2 Short-term exposure

The term "metal-fume fever" describes an acute industrial illness characterized by a variety of symptoms, including fever, chills, dyspnoea, muscle soreness, nausea and fatigue, which occur in workers following the inhalation of finely dispersed particulate matter formed when certain metals are volatilized. The oxides of a number of metals, including zinc, can cause this acute, reversible syndrome (Drinker et al., 1927a–d; Rohrs, 1957; Doig & Challen, 1964; Gordon et al., 1992). The description of the effects has been cited extensively, and the condition has variously been called brassfounder's ague, zinc chills, zinc fever, Spelter's shakes and metal shakes (Batchelor et al., 1926; McCord & Friedlander, 1926; Mueller & Seger, 1985; Blanc et al., 1991).

Metal-fume fever is common in welders who work on various types of non-ferrous metals or ferrous metals alloyed with or coated with other metals. Zinc fume from galvanized coatings is a common cause. While the disease is generally short, transient and severe, serious complications are not common and individuals tend to develop a tolerance (Drinker et al., 1927a–d). Symptoms might occasionally be followed by pulmonary oedema or pneumonia (Doig & Challen, 1964). The size of the ultrafine zinc oxide particles appears to be critical in the development of the syndrome, with the particles needing to be small enough to reach the alveoli when inhaled (Brown, 1988). Recent studies in humans following occupational exposure to zinc oxide fumes have demonstrated some changes in pulmonary function and/or radiological abnormalities, which are reversible following cessation of exposure.

A cross-sectional analysis, conducted on spirometric lung-function parameters in zinc welders, non-welders with exposure to welding fumes and control subjects (Marquart et al., 1989), revealed no differences in lung function between groups, and changes in lung function over five consecutive work shifts were not related to the exposure level. The highest measured concentrations of welding fumes were 5.1 and 8.0 mg/m^3 for an 8-h time-weighted average.

In a study designed to examine the pathogenesis of metal-fume fever in humans (Blanc et al., 1991), 14 subjects welded galvanized mild steel over a period of 15–30 min in special environmental exposure chambers, with controlled ventilation, humidity and temperature, designed to produce an exposure level in excess of 10 mg/m^3 over 15 min. The mean cumulative exposure to zinc oxide for the 14 participants was reported to be 2.3 ± 1.7 g/min per m^3 (range 0.6–5.1 g/min per m^3), resulting in a range of mean exposure levels of 77–153 mg/m^3, and a minimum exposure of 20–40 mg/m^3, depending upon whether duration was 15 or 30 min. Pulmonary function and airway responsiveness were measured after 1 h ($n = 14$), 6 h ($n = 5$) and 20 h ($n = 9$), while bronchoalveolar lavage was conducted 8 h or 22 h after welding. A marked, dose-dependent inflammatory response was observed in the lungs, with a positive correlation between cumulative zinc exposure and polymorpho-nuclear leukocyte count in bronchoalveolar lavage fluid at early ($r = 0.93$; $P < 0.05$) and late ($r = 0.87$; $P < 0.01$) follow-up. The proportion of polymorphonuclear leukocytes in the late follow-up sample, 37% (range 19–63%), was increased compared with the early follow-up figure of 9% (2–21%). There was only a minimal effect on pulmonary function, and no statistically significant correlation was observed between cumulative zinc exposure and pulmonary function. In the late follow-up group, the four participants with the highest cumulative exposures (> 3.5 g/min per m^3) all had myalgia. Two of the participants (with exposure of > 5 g/min per m^3, i.e., approximately 150 mg/m^3) also had fever (38 °C).

In a subsequent paper (Blanc et al., 1993), further information from the same subjects was reported together with additional data from a total of 23 volunteers adding a 3-h post-exposure time-point for bronchoalveolar lavage fluid (zinc exposure 1.8 ± 0.2 mg/m^3). Increased concentration of tumour necrosis factor (TNF) in bronchoalveolar lavage fluid was prominent at 3 h, and less marked at 8 h or 22 h after exposure, exhibiting a statistically significant exposure–response relationship to airborne zinc at each time-point ($P < 0.05$). There were also significant changes in the concentrations of interleukin-6 and interleukin-8, but not of interleukin-1. The findings are consistent with a role of these cytokines in the pathogenesis of the inflammatory changes in metal-fume fever. Although these short-term exposures (15–30 min) were to zinc concentrations well above 10 mg/m^3, it should be noted that they do not exceed an 8-h time-weighted average of 5 mg/m^3 if recalculated to an 8-h time interval; however, it is unlikely that an acute reaction of the type observed would occur if the same cumulative exposures were given over 8 h.

A number of case reports have demonstrated the acute effects of zinc fume inhalation in occupational settings. Reversible clinical signs and radiological effects, including aches and pains, dyspnoea, dry cough, lethargy, neutrophil leukocytosis, pyrexia, and widespread abnormality of both lung fields, with multiple nodules measuring 3–4 mm and becoming confluent and ill-defined in some areas, were seen when an individual was exposed to zinc fumes in a shipyard over a 3-week period (Brown, 1988). A systemic reaction and a self-limiting response in the periphery of the lung were reported when a patient with a clinical history of recurring zinc fume fever underwent experimental welding exposures of 1 h using zinc-coated tubing (Vogelmeier et al., 1987). An acute lung reaction was also seen in an individual working with heated zinc who experienced chills, muscle ache and dyspnoea; radiographic examination revealed diffuse nodular infiltrates, which cleared after 10 days away from the job (Malo et al., 1990).

8.4.3 Long-term exposure

The complex environment encountered by workers in galvanizing and metal plating plants results in exposure to a variety of compounds, including zinc and zinc compounds.

A causal association between the exposure to zinc and any occupational asthma is difficult to establish. Occasional cases of occupational asthma have been reported among workers using soft solder fluxes containing ammonium chloride and zinc chloride. A causative relationship with zinc could not be concluded. The most suggestive case was a subject who developed asthma symptoms 2.5 years after being employed at a plant where metals were galvanized in heated zinc (Malo et al., 1993). Positive immediate skin tests to zinc sulfate at concentrations of 1 and 10 mg/ml were obtained, although no specific IgE antibodies to zinc were observed. An immediate asthmatic reaction was elicited after the subject inhaled nebulized zinc sulfate at a concentration of 10 mg/ml for 6 min.

The exposure of groups of volunteers to a polydisperse aerosol of zinc ammonium sulfate in an environmental control chamber at a nominal concentration of 20 µg/m^3 produced minimal or no short-term respiratory effects, even in subjects diagnosed as asthmatics prior to the study (Linn et al., 1981).

8.4.4 Epidemiological studies

In general, well-conducted epidemiological studies in the workplace with adequate characterization of zinc exposure values are lacking, and there are inadequate data available to make an association between occupational exposure to zinc and disease states.

8.5 Subpopulations at special risk

8.5.1 Dialysis patients

Acute zinc toxicity has been reported in patients following kidney dialysis (Gallery et al., 1972; Petrie & Row, 1977). A patient who, for home dialysis, used rainwater draining from a painted galvanized iron roof, which had been stored in a galvanized iron tank, developed severe nausea and vomiting within 2 h of starting the procedure, with similar symptoms at subsequent dialyses. The tank water contained zinc at a concentration of 625 µg/100 ml. The patient's plasma and red cell zinc concentrations were 700 and 3500 µg/100 ml, respectively, haemoglobin was 3.5 g/100 ml, and a blood film showed moderate polychromatophilia; 6 weeks after rehospitalization, plasma zinc was still moderately raised. Intercurrent hospital dialyses were uneventful, and subsequent deionization of the patient's home water supply resulted in asymptomatic dialyses (Gallery et al., 1972). The use of water drawn through galvanized iron piping resulted in a fall in the haemoglobin levels of two home dialysis patients; these effects were eliminated after the installation of carbon filtration of the dialysis water (Petrie & Row, 1977). Severe anaemia was also seen in 9/10 patients dialysed in a hospital dialysis unit, following the installation of a new galvanized iron water softener in the dialysate water supply system. The dialysate contained zinc at a concentration of 4.89 µmol/litre (32 µg/100 ml). The installation of an activated carbon filter in the system reduced the zinc concentration to < 0.15 µmol/litre (< 1 µg/100 ml), resulting in a rise in haemoglobin levels in the patients towards previous values (Petrie & Row, 1977).

8.5.2 People with diabetes

Non-infective furunculoid skin lesions were reported in an insulin-dependent diabetic subject, apparently induced by the zinc acetate component of an intermediate-acting insulin preparation. This rare complication of insulin therapy was attributed to a reparative granulomatous phase arising from tissue damage caused by the zinc in the preparation (Jordaan & Sandler, 1989; Sandler & Jordaan, 1989). In two patients using insulin preparations containing zinc, pruritic, erythematous, papular lesions were observed at the injection site. Intradermal skin tests for zinc were positive in both patients. Zinc-free insulin did not produce any allergic reactions in the patients (Feinglos & Jegasothy, 1979).

8.5.3 Hospital patients

An elderly woman died after she received 46 mmol (7.4 g) of zinc sulfate intravenously over 60 h, owing to a prescribing error; her serum zinc concentration was 640 µmol/litre (4184 µg/100 ml). Zinc intoxication was characterized by hypotension, pulmonary oedema, diarrhoea, vomiting, jaundice and oliguria (Brocks et al., 1977). In another incident, seven hospital patients undergoing intravenous feeding with fluid containing elemental zinc at a concentration of 227 µg/100 ml were inadvertently given fluid containing 10 times that amount (2270 µg/100 ml) for 26–60 days. Mortality was high (5/7). While the clinical manifestation of the zinc overdose was hyperamylasaemia (unaccompanied by clinical signs of pancreatitis), the authors concluded that all deaths had resulted from septic complications already present before the appearance of this symptom (Faintuch et al., 1978).

To investigate the effects of zinc administration on the healing of chronic leg ulcers, a double-blind trial was conducted in 27 subjects; 13 patients received 200 mg of zinc sulfate three times a day (approximately 135 mg of elemental zinc daily) for 18 weeks, while 14 patients received placebo. No signs of toxicity associated with the zinc treatment were reported in the study (Hallbook & Lanner, 1972). Similarly, in another study investigating the effect of oral zinc treatment on leg ulcers, no clinical signs of toxicity were reported in 18 patients administered 220 mg of zinc sulfate three times daily (approximately 150 mg of elemental zinc daily) for 16–26 weeks (Greaves & Skillen, 1970). Mild diarrhoea was reported in 3/52 patients receiving three daily doses of 220 mg of zinc sulfate for up to 71 days (Husain, 1969), while diarrhoea was reported in 6/16 geriatric patients receiving a similar zinc dose for 24 weeks (Czerwinski et al., 1974).

8.5.4 Other populations

No adverse effects were observed as a result of ingestion of 300–1200 mg of zinc sulfate heptahydrate daily for 3 years or 150 mg of zinc as zinc acetate daily for several weeks to 2 years by

189

Wilson disease patients (Hoogenraad et al., 1979, 1983, 1984; Brewer et al., 1983, Hill et al., 1987), doses of zinc of 1–2 mg/kg daily by infants and children with acrodermatitis enteropathica (Hambidge & Walravens, 1982), and 68–102 mg of zinc daily during pregnancy by a woman with acrodermatitis enteropathica (Jones & Peters, 1981). Few long-term studies of the effects of high oral zinc in healthy adults have been reported. In 11 female and 13 male patients with Wilson disease, the administration of 50 mg of elemental zinc as zinc acetate three times a day for about 2 years resulted in a decrease in total cholesterol of about 10% in both sexes and a reduction of HDL cholesterol of about 20% in male patients. The authors concluded that the coronary heart disease risk factor was not changed significantly in either sex (Brewer et al., 1991).

Recently, a controlled, randomized double-blind study showed that oral zinc therapy, 100 mg of zinc sulfate twice daily taken with food, significantly reduced visual loss in individuals with macular degeneration (Newsome et al., 1988).

8.6 Interactions

8.6.1 Copper

Impaired copper nutriture in humans has been noted following chronic elevated intake of zinc; these effects are reported in section 8.3.5.

8.6.2 Iron

The effect of inorganic zinc on the absorption of inorganic iron from a solution was investigated in two studies in healthy male volunteers (Crofton et al., 1989). Simultaneous administration of 344 µmol of zinc had no effect on the absorption of 842 µmol of radiolabelled iron (^{59}Fe) in the first study, based upon the area under the plasma iron concentration–time curve at 3 h and 6 h, the whole body retention of ^{59}Fe, and plasma content of ^{59}Fe. However, the authors noted a reduction in 4/9 subjects of the areas under the curve at 3 h and 6 h for iron, and suggested that there was a trend (not statistically significant) for zinc to inhibit the intestinal absorption of iron. The second study was conducted without a radiolabel, and the results indicated that the simultaneous administration of iron with

190

zinc at molar ratios of 1:1 (421 µmol) and 2.5:1 (1048 µmol) significantly reduced increments in the concentrations of iron in the plasma.

In a study described in section 8.3.5.1, in which women were supplemented with zinc at 50 mg/day for 10 weeks, competitive interactions between iron and zinc were suggested by the authors (Yadrick et al., 1989). Serum ferritin, the level of which is proportional to tissue iron stores, was reduced following zinc supplementation alone, but when iron at 50 mg/day was administered together with the zinc, serum zinc and serum ferritin increased.

In human subjects, the presence of inorganic iron in solution with ionic zinc at molar ratios of between 2:1 and 3:1 resulted in significant inhibition of zinc absorption (Solomons & Jacob, 1981; Solomons et al., 1983; Valberg et al., 1984; Sandstroem et al., 1985), while the presence of haem iron in the same molar excess did not inhibit the absorption of zinc (Solomons & Jacob, 1981). In healthy, non-pregnant woman, a progressive decrease in plasma zinc was seen as the ratio of iron to zinc was increased from 0.1 to 3.1, while the intake of zinc remained constant at 25 mg (Solomons & Jacob, 1981).

In studies in which iron was given with food, no inhibitory effect on zinc uptake was observed when the iron intake was not unusually high. The consumption of 54 mg of "organic" zinc in oysters with 100 mg of ferrous iron did not alter plasma uptake of zinc (Solomons & Jacob, 1981). Neither the addition of ferrous iron (at an iron:zinc ratio of 25) to a composite meal containing 2.6 mg of zinc (Sandstroem et al., 1985), nor the consumption of turkey meat containing 4 mg of zinc with ferric iron (17 or 34 mg) (Valberg et al., 1984) significantly changed the absorption of zinc. No effects of iron-fortified infant foods on zinc absorption of zinc from natural sources were demonstrated in adults or children (Fairweather-Tait, 1995) or in healthy infants given an iron supplement (30 mg of iron as ferrous sulfate) before a meal. However, dietary supplementation with large amounts of iron may impair zinc absorption, and this was observed in four human volunteers fed a zinc-deficient (zinc at 3.5 mg/day), protein-based, semisynthetic soy diet for 4 months (Prasad et al., 1978a); the two subjects receiving 130 mg of iron daily displayed a more rapid reduction in plasma zinc than did the two volunteers fed 20.3 mg of iron daily.

The effect of iron on zinc absorption may depend upon zinc status. For example, serum copper or zinc levels were not affected in healthy infants who were fed a zinc- sufficient diet supplemented with 30 mg of iron as ferrous fumarate daily given 30 min before a meal (Yip et al., 1985).

Three pregnant women, whose daily diets were supplemented with iron at rates of 100 mg/day or more, had lower plasma zinc than other pregnant women whose iron supplementation was less than 100 mg/day (Campbell-Brown et al., 1985), and daily multivitamin supplements containing 60–65 mg of iron inhibited zinc absorption in first-trimester pregnant women, compared with pregnant women receiving no iron supplementation, or with iron supplementation of less than 30 mg/day (Breskin et al., 1983). It is not known whether the iron supplements in these studies were taken in the presence or absence of food.

8.6.3 Calcium

Human subjects with a constant zinc intake of 14.5 mg/day and calcium intakes of 200–2000 mg/day showed no changes in zinc absorption (Spencer et al., 1983). Conversely, the intake of high zinc levels (140 mg/day) reduced calcium absorption in men with low calcium intakes (200 mg/day) but calcium absorption was not affected when calcium intake was 800 mg/day (Spencer et al., 1987).

9. EFFECTS ON OTHER ORGANISMS
IN THE LABORATORY AND FIELD

Zinc is an essential micronutrient in all biota owing to its involvement in many physiological processes. It is essential in the maintenance of plasma membrane stability (Bettger & O'Dell, 1981; Cakmak & Marschner, 1988), in the activation of more than 300 enzymes, in transcription factors and in hormone receptors (see section 6.5.2).

Generally, organisms growing in natural terrestrial environments do not show symptoms of zinc deficiency. However, species introduced by humans into the environment may show these deficiencies. Zinc toxicity is observed in organisms exposed to anthropogenic zinc enrichment (Ernst, 1972) and in crops grown in naturally enriched environments (Chaney, 1993). More often than toxicity, zinc deficiency is reported from environments where humans have grown plants that are not adapted and/or have not been properly selected, ranging from crops and pastures in Australia (Donald & Prescott, 1975), Africa (Cottenie et al., 1981), Asia (Katyal & Ponnamperuma, 1974) and North America (Lingle & Holmberg, 1957), to fruit trees (SSSA, 1990) and forest trees. Application of various types of zinc fertilizers to soil or onto leaves can help to overcome these problems (Takkar & Walker, 1993). Another approach is to increase the zinc efficiency of cultivated plant species (El Bassam et al., 1990). Animals fed or feeding on zinc-deficient plants will also show symptoms of zinc deficiency (Blamberg et al., 1960, Elinder & Piscator, 1979).

Nutritional zinc deficiency is relatively rare for aquatic organisms. A possible exception may be the low zinc environments that characterize open oceans. Extremely low concentrations of zinc, iron and copper have been observed in open oceans and it has been suggested that these are rate limiting for phytoplankton growth (Anderson et al., 1978; Reuter & Morel, 1981; Bruland, 1993). In most other circumstances, organisms appear to have developed appropriate physiological mechanisms to ensure adequate uptake of zinc from the concentrations present in their native environment. Organisms not capable of doing this would of course have disappeared from a particular ecosystem. Information concerning zinc deficiency in aquatic organisms must thus be obtained primarily from laboratory experiments. There are several reports of zinc deficiency under experimental conditions in protozoa (Falchuk, 1988), algae (Vymazal, 1986), daphnids (Keating & Caffrey, 1989), fish (Spry et al., 1988) and amphibians (Herkovits et al., 1989). White & Rainbow (1985) calculated theoretical estimates for the minimum metabolic requirements of zinc in molluscs and crustaceans. Enzymatic requirements for zinc in both groups were estimated to be 34.5 mg/kg dw. The possession of haemocyanin as a respiratory pigment adds a further non-enzymatic metabolic requirement of 58.3 mg/kg for certain gastropod molluscs and 36.3 mg/kg for some crustaceans such as decapods. However, Depledge (1989) recalculated the amount of zinc required by decapod crustaceans to be 67.9 mg/kg dw.

9.1 Laboratory experiments

Many experiments performed in laboratories give insufficient information on the speciation of zinc, especially when zinc is added to a medium rich in complexing agents such as sewage sludge and agar (for example, Codina et al., 1993). In the case of soils, there is a lack of information on the time period between the zinc application and the start of the experiments, i.e., the time necessary for an equilibrium to be reached between the metal application and the soil solution (Spurgeon & Hopkin, 1996). The lack of this information adversely affects the reliability and utility of toxicity determinations. A similarly inadequate procedure is followed in many experiments with animals in which zinc added to the feed is only adsorbed, whereas in the natural situation it is processed by the organism and incorporated into organic compounds. The difference between adsorbed and metabolically processed zinc has clearly been shown in experiments with Japanese quail fed spinach and lettuce (McKenna et al., 1992).

To be useful, toxicity testing requires, at a minimum, the following information: actual exposure concentrations (nominal concentrations are unacceptable); acceptable control results (i.e., an acceptably low level of mortalities and/or effects); physicochemical conditions (at a minimum, temperature, pH, dissolved oxygen and hardness); and a concentration–response relationship. Studies that met these criteria are so indicated where appropriate in the text and tables that follow.

9.1.1 Microorganisms

9.1.1.1 Water

Studies on the effect of zinc on microorganisms in the aquatic environment generally measure either growth or survival. However, the zinc concentrations added in these tests are often too high to be of environmental relevance (Codina et al., 1993, 50–432 mg/litre; Tijero et al., 1991, 200–600 mg/litre). Values for the EC_{50} (the concentration producing effects in 50% of the tested organisms) and LC_{50} (the concentration killing 50% of the tested organisms) in other experiments varied in a species-specific manner (Table 34).

Table 34. Zinc toxicity (LC_{50} or EC_{50} values in mg/litre) for microorganisms in the aquatic environment

Species	Duration of exposure (h)	LC_{50}	Reference
Drepanomonas revoluta	24	0.25	Madoni et al. (1994)
Spirostomum teres	24	0.67	Madoni et al. (1994)
Blepharisma americanum	24	1.05	Madoni et al. (1994)
Tetrahymena pyriformis	56	5.77	Carter & Cameron (1973)
Tetrahymena pyriformis	8	< 1.00	Chapman & Dunlop (1981)
Zoogloea ramigera	24	approximately 3.0[a]	Norberg & Molin (1983)
Euplotes patella	24	50.0	Madoni et al. (1992)

[a] EC_{50}

9.1.1.2 Soil

Laboratory experiments are often carried out without equilibrium between the added zinc and the soil, which is a critical drawback in short-term experiments (< 3 weeks). Three parameters of microbial activities in soil have been studied: mineralization of macronutrients (N, S); soil respiration as a parameter of the mineralization of organic compounds; and general soil activity (dehydrogenase). Microbial activity is less affected by zinc in soils rich in organic materials than in sandy and loamy soils (this situation was found for N-mineralization (Doelman & Haanstra, 1984), soil respiration (Frostegård et al., 1993) and dehydrogenase activity (Rogers & Li, 1985). These results can be explained by differences in zinc speciation.

More recent literature confirms the importance of organic matter in reducing the effects of zinc in microbial processes, such as the breakdown of glutamic acid, and phosphatase activity. Increasing exposure time lowers the EC_{50} value (Table 35).

9.1.2 Aquatic organisms

9.1.2.1 Plants

Acute toxicity of zinc is often determined in short-term experiments of 24–96 h (Table 36). In the case of unicellular algae, these experiments cover 1–4 cell-division cycles. EC_{50} values range from 0.058 to 10 mg/litre (nominal concentration) in a species-specific manner. The toxicity of zinc depends on the external concentration, the zinc speciation, and the pH and hardness of the water (Starodub et al., 1987, Stauber & Florence, 1989). Aquatic macrophytes are generally insensitive to zinc.

Most of the data for unicellular algae were obtained using culture media as the assay solutions. These results should be used with caution, since complexing agents, e.g., EDTA in the culture media, may reduce zinc bioavailability and lower its toxicity (Stauber, 1995). Crucial information with respect to physicochemical parameters (e.g., water hardness, dissolved organic carbon, dissolved oxygen) is not generally provided in most of the cited references.

Acute toxicity values tend to be lower for marine unicellular algae than for freshwater species. Only one set of experiments satisfies the ideal criteria as previously specified. In these tests, no-observed-effect concentrations, obtained under standardized test conditions (OECD 201 algae growth-inhibition test) for *Selenastrum capricornutum*, ranged between 30 µg/litre and 50 µg/litre (measured as dissolved concentration; hardness 16 mg/litre, $CaCO_3$).

Table 35. Impact of zinc (mg/kg) on nitrogen mineralization in relation to soil types

Process	Duration	EC$_{50}$				Reference
		Sand	Sandy loam	Silty loam	Clay	
Urease	6 weeks	420	480	1030	1780	Doelman & Haanstra (1986)
	18 months	230	110	–	90	
Nitrification	–	–	100[a]	–	ca. 80	Wilson (1977)
			1000			

[a] No-observed-effect concentration.

197

Table 36. Toxicity of zinc to algae and aquatic plants in static conditions[a]

Organism	Temp (°C)	Zinc compound tested	Hardness (CaCO$_3$ mg/litre)	Parameter	End-point	EC$_{50}$ (mg/litre)	Reference
Unicellular algae: freshwater							
Green algae							
Chlorella vulgaris	15.5	sulfate	n.g.	96-h EC$_{50}$	culture growth	2.4 (n)	Rachlin & Farran (1974)
Scenedesmus quadricauda	20	sulfate	n.g.	24-h EC$_{50}$	photosynthesis	> 0.225 (n)	Starodub et al. (1987)
Selenastrum capricornutum	25	zinc powder	16	72-h EC$_{50}$	culture growth	0.15 (m,d)	van Woensel (1994)
				NOEC	culture growth	0.05 (m,d)	van Woensel (1994)
Selenastrum capricornutum	25	oxide	16	72-h EC$_{50}$	culture growth	0.17 (m,d)	van Ginneken (1994)
				NOEC	culture growth	0.03 (m,d)	van Ginneken (1994)
Diatoms							
Navicula incerta	19	chloride	n.g.	9-h EC$_{50}$	culture growth	10.0 (n)	Rachlin et al. (1983)
Unicellular algae: marine							
Marine diatoms							
Asterionella japonica	23	sulfate	n.g.	72-h EC$_{50}$	culture growth	0.058 (n)	Fisher & Jones (1981)
Nitzschia closterium	15.5	sulfate	n.g.	96-h EC$_{50}$	culture growth	0.271 (n)	Rosko & Rachlin (1975)
Nitzschia closterium	21	chloride	n.g.	96-h EC$_{50}$	culture growth	0.065 (n)	Stauber & Florence (1990)

Table 36 (contd.)

Macrophytes: freshwater

Acute toxicity						
Elodea canadensis (segments)	24	sulfate	10	photosynthetic O_2	8.1 (n)	Brown & Rattigan (1979)
Prolonged tests						
Elodea canadensis	n.g.	sulfate	n.g.	28-d EC_{50}	22.5 (n)	Brown & Rattigan (1979)
Lemna minor	n.g.	sulfate	n.g.	28-d EC_{50}	67.7 (n)	Brown & Rattigan (1979)
Lemna minor	25–28	chloride	n.g.	7-d EC_{50}	10 (n)	Dirilgen & Inel (1994)
				frond growth inhibition		
Elodea nuttallii	21	sulfate	n.g.	14 d	32.7 (n)	Van der Werff & Pruyt (1982)
				no toxic symptoms		
Callitriche platycarpa	21	sulfate	n.g.	28 d	32.7 (n)	Van der Werff & Pruyt (1982)
				no toxic symptoms		
Callitriche platycarpa	21	sulfate	n.g.	73 d	0.654 (n)	Van der Werff & Pruyt (1982)
				no toxic symptoms		
Spirodela polyrhiza	21	sulfate	n.g.	73 d	0.654 (n)	Van der Werff & Pruyt (1982)
				no effects observed		
Lemna gibba	21	sulfate	n.g.	73 d	0.654 (n)	Van der Werff & Pruyt (1982)
				no effects observed		

d = measurements expressed as dissolved zinc; m = measured concentrations; n = nominal concentrations; n.g. = not given

[a] Many of the older test results should be regarded with caution because the assays were carried out in culture media containing complexing agents like EDTA, which could affect the bioavailability of zinc. Crucial information concerning physicochemical factors such as hardness, DOC and DO is lacking in most of the papers.

199

Floating aquatic plants can take up zinc by the roots and shoots (the lower surface with water contact). Zinc uptake is governed not only by the zinc concentration in the water but also by evapotranspiration, which is not taken into account in most experiments with duckweed (*Lemna minor*) (Hutchinson & Czyrska, 1975; Brown & Rattigan, 1979; Dirilgen & Inel, 1994). EC_{50} values vary from 10 to 67.7 mg/litre depending on the test period and conditions. Submerged aquatic plants, e.g., pondweeds (*Elodea* sp.), are more sensitive than floating aquatic plants (Brown & Rattigan, 1979).

Permanent high exposure to zinc gives rise to the selection of zinc-tolerant genotypes, e.g., *Lemna minor* (Van Steveninck et al., 1990) which detoxifies zinc as zinc phytate in vacuoles, and in several algal species (Say et al., 1977; Harding & Whitton, 1981). Zinc tolerance in plants and other organisms is discussed further in section 9.2.

9.1.2.2 Invertebrates and vertebrates

Information on the acute toxicity of zinc to freshwater and marine invertebrates is summarized in Tables 37 and 38, respectively, and to freshwater and marine fish is summarized in Tables 39 and 40, respectively. Studies that meet the criteria specified above so indicated in these tables.

The toxicity of zinc can be influenced both by intrinsic and by extrinsic factors. Numerous studies with aquatic animals have demonstrated that zinc toxicity decreases with increasing water hardness (Sinley et al., 1974; Bradley & Sprague, 1985; Winner & Gauss, 1986; Paulauskis & Winner, 1988; Everall et al., 1989) and decreasing temperature (McLusky & Hagerman, 1987; Hilmy et al., 1987; Zou & Bu, 1994). However, Berglind & Dave (1984) reported that, hardness over the range 50–300 mg/litre, $CaCO_3$, had no significant effect on the toxicity of zinc to daphnids. Similarly, Rehwoldt et al. (1972) found no effect of temperature (15–28 °C) on the toxicity of zinc to freshwater fish. Smith & Heath (1979) reported that the effect of temperature on zinc toxicity was species specific. While increased temperature resulted in an increase in toxicity of zinc to goldfish (*Carassius auratus*) and bluegill (*Lepomis macrochinus*), it had no effect on the toxicity of zinc to golden shiners (*Notemigonus crysoleucas*) or rainbow trout (*Oncorhynchus mykiss*).

Table 37. Toxicity of zinc to freshwater invertebrates[a]

Organism	Size/age	Stat/flow	Temp (°C)	Hardness (mg/litre)	pH	Zinc compound tested	Parameter	Concentration (mg/litre)	Reference
Snail	eggs	stat	17	50	7.6		24-h LC$_{50}$	28.1 (m)	Rehwoldt et al. (1973)
Amnicola sp.	eggs	stat	17	50	7.6		96-h LC$_{50}$	20.2 (m)	(1973)
	adult	stat	17	50	7.6		24-h LC$_{50}$	16.8 (m)	Rehwoldt et al. (1973)
	adult	stat	17	50	7.6		96-h LC$_{50}$	14 (m)	(1973)
Mollusc	<2 mm	stat	10			sulfate	96-h LC$_{50}$	3.2 (n)	Willis (1988)
Ancylus fluviatilis	>3 mm	stat	10			sulfate	96-h LC$_{50}$	4.5 (n)	Willis (1988)
Annelid	<4 mg	stat	10			sulfate	96-h LC$_{50}$	2.05 (n)	Willis (1989)
Erpobdella oculata	>15 mg	stat	10			sulfate	96-h LC$_{50}$	8.8 (n)	Willis (1989)
Bristle worm		stat	17	50	7.6		24-h LC$_{50}$	21.2 (m)	Rehwoldt et al. (1973)
Nais sp.		stat	17	50	7.6		96-h LC$_{50}$	18.4 (m)	(1973)
Water flea		stat	17–19	44–53	7.4–8.2	chloride	48-h EC$_{50}$	0.1 (n)	Biesinger & Christensen (1972)
Daphnia magna		stat	17–19	44–53	7.4–8.2	chloride	48-h EC$_{50}$	0.28 (n)	Christensen (1972)
	<48 h	stat	20	175	6.0	sulfate	48-h LC$_{50}$	0.24 (n)	LeBlanc (1982)
	<24 h	stat	20		6.5	sulfate	48-h LC$_{50}$	0.151 (n)	Oikari et al. (1992)
	<24 h	stat	20		6.5	sulfate	48-h LC$_{50}$	0.244 (n, hw)	Oikari et al. (1992)
	<24 h	stat		45	7.2–7.4	sulfate	48-h LC$_{50}$	0.068 (n)	Mount & Norberg (1984)

Table 37 (contd.)

Organism	Size/age	Stat/flow	Temp (°C)	Hardness (mg/litre)	pH	Zinc compound tested	Parameter	Concentration (mg/litre)	Reference
D. magna (contd.)	<24 h	stat*	20			sulfate	48-h LC$_{50}$	0.75 (n)	Arambasic et al. (1995)
		stat				bromide	48-h LC$_{50}$	1.22 (m)	Magliette et al. (1995)
D. pulex	<24 h	stat		45	7.2–7.4		48-h LC$_{50}$	0.107 (n)	Mount & Norberg (1984)
Ceriodaphnia dubia		stat	25			bromide	48-h LC$_{50}$	0.50 (m)	Magliette et al. (1995)
C. reticulata	<24 h	stat		45	7.2–7.4		48-h LC$_{50}$	0.076 (n)	Mount & Norberg (1984)
D. hyalina	1.27 mm	stat	10		7.2	sulfate	48-h LC$_{50}$	0.04 (n)	Baudouin & Scoppa (1974)
D. lumholtzi		stat*	28.5	200	7.9		48-h LC$_{50}$	2.29 (n)	Vardia et al. (1988)
		stat*	28.5	200	7.9		96-h LC$_{50}$	0.44 (n)	

Table 37 (contd.)

	<24 h			<5					
Moina irrasa		stat	20		8.0	chloride	48-h LC$_{50}$	0.059 (n)	Zou & Bu (1994)
M. macrocopa		stat*	24–27		6.5	sulfate	48-h LC$_{50}$	1.17 (n)	Wong (1992)
Copepod Cyclops abyssorum	1.27 mm	stat	10		7.2	sulfate	48-h LC$_{50}$	5.5 (n)	Baudouin & Scoppa (1974)
Eudiaptomus padanus	1.27 mm	stat	10		7.2	sulfate	48-h-LC$_{50}$	0.50 (n)	Baudouin & Scoppa (1974)
Parastenocaris germanica	adult	stat	10.5	10*	6.8	sulfate	48-h LC$_{50}$	4.5 (m)	Notenboom et al. (1992)
	adult	stat	10.5		6.8	sulfate	96-h LC$_{50}$	1.7 (m)	
Amphipod Gammarus sp.		stat	17	50	7.6		24-h LC$_{50}$	10.2 (m)	Rehwoldt et al. (1973)
		stat	17	50	7.6		96-h LC$_{50}$	8.1 (m)	
Crangonyx pseudogracilis	4 mm	stat	13	50	6.75	sulfate	48-h LC$_{50}$	121 (n)	Martin & Holdich (1986)
Isopod	4 mm	stat	13	50	6.75	sulfate	96-h LC$_{50}$	19.8 (n)	
Asellus aquaticus	7 mm	stat	13	50	6.75	sulfate	96-h LC$_{50}$	18.2 (n)	
Ostracod Cypris subglobosa		stat*	28.5	200	7.9		48-h LC$_{50}$	34.99 (n)	Vardia et al. (1988)
		stat*	28.5	200	7.9		96-h LC$_{50}$	8.35 (n)	Vardia et al. (1988)

203

Table 37 (contd.)

Organism	Size/age	Stat/flow	Temp (°C)	Hardness (mg/litre)	pH	Zinc compound tested	Parameter	Concentration (mg/litre)	Reference
Harpacticoid Nitocra spinipes	adult	stat	21	7	7.8	sulfate	96-h LC$_{50}$	4.3 (n)	Lindén et al. (1979)
Rotifer Brachionus calyciflorus	juvenile	stat	20	36.2	7.3	chloride	24-h LC$_{50}$	1.32 (n)	Couillard et al. (1989)
		stat	25				24-h LC$_{50}$	1.3 (n)	Snell et al. (1991)
Midge Chironomus sp.		stat	17	50	7.6		24-h LC$_{50}$	21.5 (m)	Rehwoldt et al. (1973)
		stat	17	50	7.6		96-h LC$_{50}$	18.2 (m)	
C. tentans	3rd instar	stat	13	25	6.3	sulfate	48-h EC$_{50}$	8.2 (n)	Khangarot & Ray (1989)
Caddis fly Unidentified		stat	17	50	7.6		24-h LC$_{50}$	62.6 (m)	Rehwoldt et al. (1973)
		stat	17	50	7.6		96-h LC$_{50}$	58.1 (m)	
Damsel fly		stat	17	50	7.6		24-h LC$_{50}$	32 (m)	Rehwoldt et al.
Unidentified		stat	17	50	7.6		96-h LC$_{50}$	26.2 (m)	(1973)

hw = humic water; m = measured concentrations; n = nominal concentrations; stat = static conditions (water unchanged for duration of test); stat* = static renewal conditions (water changed at regular intervals)

[a] EC$_{50}$ values based on immobilization; hardness expressed as mg/litre O$_3$.

204

Table 38. Toxicity of zinc to marine invertebrates[a]

Organism	Size/age	Stat/flow	Temp (°C)	Salinity (‰)	pH	Zinc salt	Parameter	Concentration (mg/litre)	Reference
Starfish Asterias forbesi	11.2 g	stat	20	20	7.8	chloride	96-h LC_{50}	39 (n)	Eisler & Hennekey (1977)
American oyster Crassostrea virginica	embryo	stat	26	25		chloride	48-h LC_{50}	0.31 (n)	Calabrese et al. (1973)
Mussel Mytilus edulis		stat	12	7		chloride	24-h LC_{50}	20.8 (n)	Hietanen et al. (1988)
M. edulis planulatus		stat flow	20.6 17.6	34	8.0 7.8	chloride chloride	96-h LC_{50} 96-h LC_{50}	2.5 (m) 3.6 (m)	Ahsanullah (1976) Ahsanullah (1976)
Bay scallop Argopecten irradians	juvenile	stat*	20	25		chloride	96-h LC_{50}	2.25 (n)	Nelson et al. (1988)
Surf clam Spisula solidissima	juvenile	stat*	20	25		chloride	96-h LC_{50}	2.95 (n)	Nelson et al (1988)
Soft-shell clam Mya arenaria	4.6 g	stat	20	20	7.8	chloride	96-h LC_{50}	7.7 (n)	Eisler & Hennekey (1977)

Table 38 (contd.)

Organism	Size/age	Stat/flow	Temp (°C)	Salinity (‰)	pH	Zinc salt	Parameter	Concentration (mg/litre)	Reference
Squid *Loligo opalescens*	larvae	stat	8.6	30	8.1	chloride	96-h LC_{50}	>1.92 (m)	Dinnel et al. (1989)
Cabezon *Scorpaenichthys marmoratus*	larvae	stat	8.3	27	7.9	chloride	96-h LC_{50}	0.191 (m)	Dinnel et al. (1989)
Eastern mud snail *Nassarius obsoletus*	0.4 g	stat	20	20	7.8	chloride	96-h LC_{50}	50 (n)	Eisler & Hennekey (1977)
Amphipod *Allorchestes compressa*	0.06 g	stat	20.5	34.5	7.9	chloride	96-h LC_{50}	0.58 (m)	Ahsanullah (1976)
Harpacticoid copepod *Nitocra spinipes*						chloride chloride sulfate sulfate	96-h LC_{50} 96-h LC_{50} 96-h LC_{50} 96-h LC_{50}	0.85 (n) 1.3 (n) 2.4 (n) 2.8 (n)	Bengtsson & Bergström (1987) Bengtsson & Bergström (1987)
Ragworm *Nereis virens*	7.6 g	stat	20	20	7.8	chloride	96-h LC_{50}	8.1 (n)	Eisler & Hennekey (1977)

206

Table 38 (contd.)

Species	Size/stage	Method				Compound	Endpoint	Value	Reference
Sandworm *Neanthes vaalii*	0.33 g	stat	18.7	34.2	7.9	chloride	96-h LC$_{50}$	5.5 (m)	Ahsanullah (1976)
Dungeness crab *Cancer magister*	larvae	stat	8.5	30	8.1	chloride	96-h LC$_{50}$	0.586 (m)	Dinnel et al. (1989)
Fiddler crab *Uca annulipes*	24–29 mm	stat	29	25		sulfate	96-h LC$_{50}$	31.9 (n)	Devi (1987)
	24–29 mm	stat	29	25		sulfate	96-h LC$_{50}$	77 (n) (polluted)	Devi (1987)
U. triangularis	24–29 mm	stat	29	25		sulfate	96-h LC$_{50}$	39.1 (n)	Devi (1987)
	24–29 mm	stat	29	25		sulfate	96-h-LC$_{50}$	66.4 (n) (polluted)	Devi (1987)
Hermit crab *Pagurus longicarpus*	0.5 g	stat	20	20	7.8	chloride	96h LC$_{50}$	0.4 (n)	Eisler & Hennekey (1977)
Grapsid crab *Paragrapsus quadridentatus*	1.44 g	stat	19.6	34.2	8.1	chloride	96-h LC$_{50}$	11 (m)	Ahsanullah (1976)
Crab *Portunus pelagicus*	zoeae	stat*	25–27	35		chloride	48-h LC$_{50}$	0.56–0.77 (n)	Greenwood & Fielder (1983)
P. sanguinolentus	zoeae	stat*	25–27	35		chloride	48-h LC$_{50}$	0.62 (n)	Greenwood & Fielder (1983)
Charybdis feriatus	zoeae	stat*	25–27	35		chloride	48-h LC$_{50}$	0.96 (n)	Greenwood & Fielder (1983)

207

Table 38 (contd.)

Organism	Size/age	Stat/flow	Temp (°C)	Salinity (‰)	pH	Zinc salt	Parameter	Concentration (mg/litre)	Reference
Copepod *Tisbe holothuriae*		stat	22	38		sulfate	48-h LC_{50}	0.62 (n)	Verriopoulos & Dimas (1988)
Grass shrimp *Palaemonetes pugio*	juvenile	stat*	20	10		chloride	48-h LC_{50}	11.3 (m)	Burton & Fisher (1990)
Shrimp *Palaemon* sp.	0.28 g	stat	19.5	35.5	7.8	chloride	96-h LC_{50}	9.5 (m)	Ahsanullah (1976)
Mysid *Holmesimysis costata*	juvenile juvenile	stat stat*	13–15.5 13–16	34–36 34–40		sulfate sulfate	48-h LC_{50} 96-h LC_{50}	0.458 (m) 0.097 (m)	Martin et al. (1989) Martin et al. (1989)
Prawn *Metapenaeus dobsoni*	30–50 mm 30–50 mm	stat* stat*	27.5 27.5		7.5 7.5	sulfate sulfate	48-h LC_{50} 96-h LC_{50}	3 (n) 0.84 (n)	Sivadasan et al. (1986)

flow = flow-through conditions (zinc concentration in water continuously maintained); m = measured concentrations; n = nominal concentrations; stat = static conditions (water unchanged for duration of test); stat* = static renewal conditions (water changed at regular intervals)

[a] EC_{50} values based on immobilization; hardness expressed as mg/litre O_3.

208

Table 39. Toxicity (96-h LC$_{50}$) of zinc to freshwater fish[a]

Organism	Size/age	Stat/flow	Temp (°C)	Hardness (mg/litre)	pH	Zinc salt	Concentration (mg/litre)	Reference
Chinook salmon *Oncorhynchus tshawytscha*	1.03 g	stat	12	211	7.4–8.3	chloride (47.3%)	1.27 (n)	Hamilton & Buhl (1990)
	juvenile	flow	11–13	20–21	7.1–7.2	sulfate	0.084 (m)	Finlayson & Verrue (1982)
	alevin	flow	12	23	7.1		>0.66 (n)	Chapman (1978b)
	swim-up	flow	12	23	7.1		0.097 (n)	Chapman (1978b)
	parr	flow	12	23	7.1		0.46 (n)	Chapman (1978b)
	smolt	flow	12	23	7.1		0.7 (n)	Chapman (1978b)
Coho salmon *O. kisutch*	alevin	stat	12	41	7.1–8.0	chloride	0.73 (n)	Buhl & Hamilton (1990)
	0.47 g	stat	12	41	7.1–8.0	chloride	0.82 (n)	Buhl & Hamilton (1990)
	0.63 g	stat	12	41	7.1–8.0	chloride	1.81 (n)	Buhl & Hamilton (1990)
	2.7 kg	flow	14	25	7.4	chloride	0.91 (n)	Chapman & Stevens (1978)
Rainbow trout *O. mykiss*	alevin	stat	12	41	7.1–8.0	chloride	2.17 (n)	Buhl & Hamilton (1990)
	0.60 g	stat	12	41	7.1–8.0	chloride	0.17 (n)	Buhl & Hamilton (1990)
	juvenile	flow			6.4–8.3	acetate	0.550 (m)	Hale (1977)
	alevin	flow	12	23	7.1		0.815 (n)	Chapman (1978b)
	swim-up	flow	12	23	7.1		0.093 (n)	Chapman (1978b)

Table 39 (contd.)

Organism	Size/age	Stat/flow	Temp (°C)	Hardness (mg/litre)	pH	Zinc salt	Concentration (mg/litre)	Reference
Rainbow trout (contd.)	parr	flow	12	23	7.1		0.136 (n)	Chapman (1978b)
	smolt	flow	12	23	7.1		>0.651 (n)	Chapman (1978b)
	2.7 kg	flow	10	83	7.45	chloride	1.76 (n)	Chapman & Stevens (1978)
	juvenile	flow	15	26	6.8	sulfate	0.43 (n)	Sinley et al. (1974)
	juvenile	flow	15	333	7.8	sulfate	7.21 (n)	Sinley et al. (1974)
	25–70 g	flow	12.7	137	7.3	sulfate	2.6 (m)	Meisner & Quan Hum
	160–290 g	flow	12.9	143	7.1	sulfate	2.4 (m)	(1987)
Cutthroat trout	0.6 g	stat	10	38	7.5	sulfate	0.152	Mayer & Ellersieck (1986)
Salmo clarki	0.9 g	stat	15	43	7.5	sulfate	0.600	Mayer & Ellersieck (1986)
	0.9 g	stat	10	40	7.8	sulfate	0.130	Mayer & Ellersieck (1986)
	1.0 g	stat	10	40	8.5	sulfate	0.061	Mayer & Ellersieck (1986)
	1.0 g	stat	10	38	6.5	sulfate	0.100	Mayer & Ellersieck (1986)
	1.0 g	stat	5	38	7.5	sulfate	0.074	Mayer & Ellersieck (1986)
Fathead minnow	79 mg	flow	25	220	7.8	sulfate	2.61 (n)	Broderius & Smith (1979)
Pimephales promelas	1–2 g	stat	25	20	7.5	sulfate	0.77–0.96 (n)	Pickering & Henderson
	1–2 g	stat	25	360	8.2	sulfate	33.4 (n)	(1966)

Table 39 (contd.)

	1–2 g	stat	25	20	7.5	acetate	0.88 (n)	Pickering & Henderson (1966)
	1–2 g	stat	15	20	7.5		2.33 and 2.55 (n)	
Arctic grayling Thymallus arcticus	fry	stat	12	41	7.1–8.0	chloride	0.32 (n)	Buhl & Hamilton (1990)
	alevin	stat	12	41	7.1–8.0	chloride	2.92 (n)	Buhl & Hamilton (1990)
	0.20 g	stat	12	41	7.1–8.0	chloride	0.14 (n)	Buhl & Hamilton (1990)
	0.85 g	stat	12	41	7.1–8.0	chloride	0.17 (n)	Buhl & Hamilton (1990)
Bluegill Lepomis macrochirus	1–2 g	stat	25	20	7.5	sulfate	4.85–5.82 (n)	Pickering & Henderson (1966)
	1–2 g	stat	25	360	8.2	sulfate	40.9 (n)	
	1–2 g	stat	15	20	7.5		6.44 (n)	
Pumpkinseed Lepomis gibbosus		stat	28	55	8.0		20.1 (m)	Rehwoldt et al. (1972)
Banded killifish Fundulus diaphanus		stat	28	55	8.0		19.2 (m)	Rehwoldt et al. (1972)
Striped bass Roccus saxatilis		stat	28	55	8.0		6.8 (m)	Rehwoldt et al. (1972)

Table 39 (contd.)

Organism	Size/age	Stat/flow	Temp (°C)	Hardness (mg/litre)	pH	Zinc salt	Concentration (mg/litre)	Reference
White perch _Roccus americanus_		stat	28	55	8.0		14.4 (m)	Rehwoldt et al. (1972)
American eel _Anguilla rostrata_		stat	28	55	8.0		14.5 (m)	Rehwoldt et al. (1972)
Carp _Cyprinus carpio_		stat	28	55	8.0		7.8 (m)	Rehwoldt et al. (1972)
	3.2 cm	stat*	15		7.1	sulfate	0.45–1.34 (n)	Alam & Maughan (1992)
	6.0 cm	stat*	15		7.1	sulfate	1.64–2.25 (n)	Alam & Maughan (1992)
	47–62 mm	stat*	15	19	6.3	sulfate	3.12 (n)	Khangarot et al. (1983)
Goldfish _Carassius auratus_	1–2 g	stat	25	20	7.5	sulfate	6.44 (n)	Pickering & Henderson (1966)
Guppy _Poecilia reticulata_	0.1–0.2 g	stat	25	20	7.5	sulfate	1.27 (n)	Pickering & Henderson (1966)

Table 39 (contd.)

Species		Conditions			pH	Salt	Concentration	Reference
Flagfish *Jordanella floridae*	juvenile	flow	25	44	7.1–7.8	sulfate	1.5 (n)	Spehar (1976)
Channelfish *Nuria denricus*	500 mg	stat	4		6.1		6.06 (n)	Abbasi & Soni (1986)
Tilapia *Tilapia zilli*	subadult	stat	9.3	20–22	6.7	sulfate	33 (n)	Hilmy et al. (1987)
	subadult	stat	25	20–22	6.7	sulfate	13 (n)	Hilmy et al. (1987)
Catfish *Clarius lazera*	subadult	stat	9.3	20–22	6.7	sulfate	52 (n)	Hilmy et al. (1987)
	subadult	stat	25	20–22	6.7	sulfate	26 (n)	Hilmy et al. (1987)

flow = flow-through conditions (zinc concentration in water continuously maintained); m = measured concentrations; n = nominal concentrations; stat = static conditions (water unchanged for duration of test); stat* = static renewal conditions (water changed at regular intervals)

[a] Hardness expressed as $CaCO_3$ in mg/litre.

Table 40. Toxicity of zinc to marine fish

Organism	Size/age	Stat/flow	Temp (°C)	Salinity (‰)	pH	Zinc salt	Parameter	Concentration (mg/litre)	Reference
Chinook salmon Oncorhynchus tshawytscha	2.6 g	stat	11–13	brackish	7.6–8.1	chloride (47.3%)	96-h LC$_{50}$	2.88 (n)	Hamilton & Buhl (1990)
Atheriniform fish Rivulus marmoratus	0.03–0.1 g	flow	26–27	14			96-h LC$_{50}$	119.3–176.6	Lin & Dunson (1993)
Mummichog Fundulus heteroclitus	0.02–0.1 g juvenile	flow stat*	26–27 20	14 10		chloride	96-h LC$_{50}$ 48-h LC$_{50}$	129.5 (n) 96.5 (m)	Lin & Dunson (1993) Burton & Fisher (1990)
	1.3 g	stat	20	20	7.8	chloride	96-h LC$_{50}$	60 (n)	Eisler & Hennekey (1977)
Grey mullet Chelon labrosus	0.87 g	flow	12	34.6	7.7	nitrate	96-h LC$_{50}$	21.5 (m)	Taylor et al. (1985)

Table 40 (contd.)

English sole *Parophrys vetulus*	larvae	stat	12			sulfate	96-h LC_{50}	14.5 (n)	Shenker & Cherr (1990)
Bleak *Alburnus alburnus*	8 cm	stat	10	7	7.8	chloride	96-h LC_{50}	32 (n)	Lindén et al. (1979)
	8 cm	stat	10	7	7.8	sulfate	96-h LC_{50}	41.9 (n)	Lindén et al. (1979)
Tidewater silverside *Menidia peninsulae*	larvae	stat	25	20		sulfate	9-6h LC_{50}	5.6 (n)	Mayer (1987)
Spot *Leiostomus xanthurus*	adult	stat	26	25		sulfate	96-h LC_{50}	38 (n)	Mayer (1987)

flow = flow-through conditions (zinc concentration in water continuously maintained); m = measured concentrations; n = nominal concentrations; stat = static conditions (water unchanged for duration of test); stat* = static renewal conditions (water changed at regular intervals)

Zinc toxicity is also influenced by water pH and salinity, although the dose–response relationship is not necessarily monotonic (McLusky & Hagerman, 1987; Meinel & Krause, 1988; Reader et al., 1989). Notenboom et al. (1992) found no effect of reducing dissolved oxygen concentration (5.4 mg/litre to 0.1 mg/litre) on the toxicity of zinc to the copepod, *Parastenocanis germanica.* Paulauskis & Winner (1988) reported that the toxicity of zinc to *Daphnia magna* decreased with increasing concentrations of humic acids.

Bengsston (1974a) reported that yearling minnow (*Phoxinus phoxinus*) were more sensitive to zinc than adults, and Naylor et al. (1990) reported that juvenile *Gammarus pulex* and *Asellus aquaticus* were more sensitive than large adults. However, other studies have found little effect of organism age on zinc toxicity (Martin et al., 1989; Collyard et al., 1994).

Acute and short-term toxicity

Of the studies reported in Table 37, the results from five freshwater crustaceans meet the minimal data requirements. For four species the LC_{50} values for zinc at 48–96 h range from 0.5 to 10 mg/litre; *Asellus aquaticus* was less sensitive to zinc (194–575 mg/litre). Other acute toxicity test results reported range from 0.04 to 2.29 mg/litre zinc for daphnids (*Daphnia, Ceriodaphnia* and *Moina*) to 28.1 and 62.6 mg/litre for a snail species and a caddisfly, respectively.

Acute toxicity results for eight marine invertebrate species were acceptable, in accordance with the minimal data requirements (Table 38). The 96-h LC_{50} values for four species (including cabezon, amphipod, crab and mysid species) ranged from 0.191 to 0.586 mg/litre; those for the remaining species ranged from 2.5 to 11.3 mg/litre. Other results ranged from 0.31 (American oyster) to 77 mg/litre (fiddler crab).

Dinnel et al. (1989) reported on short-term zinc toxicity tests with the early life stages of echinoderms. Threshold values (EC_{50}) for the purple sea urchin (*Strongylocentrotus purpuratus*) were 23 and 262 µg/litre for embryo development (120 h) and gamete fertilization (80 min), respectively. Using the latter end-point, these authors also report EC_{50} values of 383 and 28 µg/litre for the green sea urchin (*S. droebachiensis*) and the sand dollar (*Dendraster excentricus*).

Baird et al. (1991) found that the 48-h EC_{50} for zinc for different clones of *Daphnia magna* ranged from 0.76 to 1.83 mg/litre. Hietanen et al. (1988) exposed the common mussel *Mytilus edulis* to increased zinc concentrations in brackish water (salinity 7‰) at a temperature of 12 °C. The 24-h EC_{50} values, based on an increased opening response and on byssal attachment, were found to be 1.35 and 0.64 mg/litre respectively. Kraak et al. (1994a) calculated the 48-h EC_{50}, based on filtration rate, to be 1.35 mg/litre for the zebra mussel (*Dreissena polymorpha*). The no-observed-effect concentration (NOEC) for the same parameter was 0.19 mg/litre.

Acute zinc toxicity data for two species of freshwater fish met the minimal requirements (Table 39). The 96-h LC_{50} values for *Oncorhynchus tshawytscha* and *O. mykiss* were 1.27 and 2.6 mg/litre, respectively. Other results for freshwater fish ranged from 0.061 to 52 mg/litre.

Data on three marine fish species were acceptable (Table 40). The 96-h LC_{50} values ranged from 21.5 mg/litre for grey mullet (*Chelon labrosus*) to 176.6 mg/litre for *Rivulus marmoratus*. A 96-h LC_{50} range of 2.88 to 129.5 mg/litre was found in the other reported data.

Norberg & Mount (1985) calculated the 7-day LC_{50} for the fathead minnow (*Pimephales promelas*) to be 0.238 mg/litre in Lake Superior water (hardness 48 mg/litre, $CaCO_3$). No zinc-induced growth inhibition was observed at 0.18 mg/litre but survival was significantly lower at that concentration. The maximum acceptable toxicant concentration was estimated to be 0.125 mg/litre. Magliette et al. (1995) exposed larval fathead minnow (*Pimephales promelas*) to zinc bromide in 7-day static renewal tests. The 7-day LC_{50} and EC_{50} (growth), based on measured concentrations, were 0.78 and 0.76 mg/litre respectively. The lowest-observed-effect concentration (LOEC) for growth was 0.63 mg/litre.

Reader et al. (1989) found that mortality in brown trout (*Salmo trutta*) exposed to zinc at a concentration of 281 µg/litre at pH 6.5 in soft water (calcium 22 µmol/litre) remained low during 30-day exposures, while in fish exposed to 0.316 mg/litre at pH 4.5, mortality was greater than 80%.

Mount et al. (1994) fed rainbow trout (*Oncorhynchus mykiss*) on a brine shrimp (*Artemia* sp.) diet containing zinc at concentrations of 920, 930 or 1900 mg/kg dw for up to 60 days. No significant mortality or effect of zinc on growth was observed during the experiment. Spry et al. (1988) fed rainbow trout (*O. mykiss*) on a purified diet containing zinc concentrations of 1, 90 and 590 mg/kg, which ranged from deficient to excessive. Fish were simultaneously exposed to zinc concentrations in water of up to 0.5 mg/litre for 16 weeks. There was no significant difference in the physical condition of fish in any treatment compared with controls.

Chronic and long-term toxicity

Data that meet the selection criteria are presented in Table 41 for freshwater invertebrates and in Table 42 for freshwater fish. No data are presented for marine and estuarine species. Table 41 contains data for four invertebrate species, two crustaceans, an insect and a snail, tested under a variety of experimental conditions and in waters of different pH (6.9–8.39), hardness (15–197 mg/litre, $CaCO_3$) and humic acid concentration. The threshold zinc concentrations range from 25 to 225 µg/litre (both values for *D. magna*) and clearly illustrate the influence of water hardness and humic acid concentration on zinc toxicity.

With respect to freshwater fish (Table 42), there are primary chronic toxicity data for six species covering water hardness ranging from 35 to 374 mg/litre, $CaCO_3$. For water hardness of ≥ 100 mg/litre, $CaCO_3$, all NOECs are ≥ 500 mg/litre, except in one behavioural study (Korver & Sprague, 1989), which reported a NOEC of 60 mg/litre. For studies in which water hardness was ≤ 100 mg/litre, $CaCO_3$, all NOECs were ≤ 50 mg/litre.

Freshwater studies

Farris et al. (1989) studied growth and cellulase activity in the Asiatic clam (*Corbicula* sp.) during a 30-day exposure to zinc sulfate concentrations ranging from 0.034 to 1.1 mg/litre. The cellulase

Table 41. Long-term and chronic toxicity to freshwater invertebrates[a]

Species	Life stage/age	End-point	pH	Hardness (mg/litre)	Humic acid (mg/litre)	Temp (°C)	Duration (days)	Threshold (zinc in µg/litre)	Reference
Water flea *Daphnia magna*	< 24 h	survival (LC$_{50}$)	7.74	45.3	–	18	21	158	Biesinger & Christensen (1972)
		production of young (EC$_{50}$)	7.74	45.3	–	18	21	102	(1972)
		production of young (MATC)	8.39	51.9	–	20	50	25	Paulauskis & Winner (1988)
			8.32	101.8	–	20	50	87.5	(1988)
			8.29	197	1.5	20	50	175	Paulauskis & Winner (1988)
			8.29	197	1.5	20	50	225	
			8.39	51.9		20	50	100	
Ceriodaphnia dubia	< 24 h	production of young (MATC)	8	97.6	–	25	7	22	Belanger & Cherry (1990a)
				113.6				71	
				182				71	
Midge *Tanytarsus dissimilis*	eggs	survival (LC$_{50}$)	7.5	46.8	–	22	10	36.8	Anderson et al. (1980)
Snail *Ancylus fluviatilis*	adults	eggs per capsule (NOEC–LOEC)	6.9	15–15.3	–	–	31	105–187	Willis (1988)

LOEC = lowest-observed-effect concentration; MATC = maximum acceptable toxicant contamination; NOEC = no-observed-effect concentration

[a] Measured zinc concentrations were ± 15% at nominal concentrations.

Table 42. Long-term and chronic toxicity to freshwater fish

Species	Life stage/ age	End-point	pH	Hardness (mg/litre)	Humic acid (mg/litre)	Temp (°C)	Duration (days)	Threshold (µg zinc/litre)	Reference
Brachydanio rerio	embryo-larval	hatchability (NOEC)	7.5	100	–	25	16	500	Dave et al. (1987)
Phoxinus phoxinus	yearling	growth (MATC)	7.5	(3.9 dH; alkalinity; 64 mg/litre)	–	12	150	80.6	Bengtsson (1974a)
Oncorhyncus mykiss	yearling (45 g)	growth and hypo-glycaemia (MATC)	7.3	374	–	10	100	763	Watson & McKeown (1976)
Pimephales promelas	males (4.6 g)	avoidance (MATC)	8.1	318	–	20	7.5	130.5	Korver & Sprague (1989)

Table 42 (contd.)

P. promelas	full life cycle	critical end-point (MATC)	7–8	46	–	25	154	106	Benoit & Holcombe (1978)
O. nerka	adult embryo–juvenile	survival, fertility, fecundity, growth, osmoregulation, acclimation (MATC)	7.2	35	–	9–14	21 months	164.6	Chapman (1978a)
Salvelinus fontinalis	3-generation life-cycle	all life-cycle parameters (egg fragility was critical end-point) (MATC)	7.0–7.7	45	–	9	3 generations	852	Holcombe et al. (1979)

MATC = maximum acceptable toxicant contamination ; NOEC = no-observed-effect concentration index declined following weight and shell loss between days 20 and 30 at the lower dose and by day 30 the growth rate was only 50% of controls. At the higher dose, animals did not grow after 5 days and had a rapidly declining cellulase index; 50% of the clams at this exposure concentration died.

Münzinger & Guarducci (1988) exposed the freshwater snail *Biomphalaria glabrata* to increased zinc concentrations (0.5 to > 5.0 mg/litre) for 33 days. At a zinc concentration of 1.5 mg/litre, 60% of young snails and 20% of adults died; at concentrations of ≥ 3.0 mg/litre no snails survived. Egg capsules were produced at zinc concentrations of up to 1.5 mg/litre. The number of eggs per capsule and the fecundity of the molluscs were significantly reduced by zinc exposure.

Mirenda (1986) calculated the 2-week LC_{50} for zinc for the crayfish (*Orconectes virilis*) to be 84 mg/litre in soft water (26 mg/litre, $CaCO_3$). Bodar et al. (1989) exposed parthenogenetic eggs of *Daphnia magna* to zinc concentrations of 10, 50 and 100 mg/litre. Exposure at 10 and 50 mg/litre had no significant effect on death rates in the six early life stages. There was no effect of zinc on survival, even at the highest exposure concentration, during developmental stages 1 and 2 (these stages take about half of the development time from egg to juvenile). The toxicity of zinc at 100 mg/litre was exerted during stages 3–6. Winner (1981) studied the toxicity of zinc to *Daphnia magna* in lifetime exposure tests. Zinc caused a significant reduction in body length of primiparous animals and longevity at concentrations of ≥ 0.1 mg/litre; however, mean brood sizes of animals reared at 0.2 mg/litre were not significantly different from those of control animals. Winner & Gauss (1986) found that an increase in water hardness from 52 to 102 mg/litre ($CaCO_3$) resulted in a significant reduction in zinc toxicity as estimated from survival curves over a 50-day exposure to zinc at 0.125 mg/litre. The addition of humic acid (1.5 mg/litre) to soft water (52 mg/litre, $CaCO_3$) significantly increased survival.

Paulauskis & Winner (1988) studied the effect of zinc on the brood size of *Daphnia magna* in chronic (50-day) toxicity tests. An increase in water hardness from 50 to 200 mg/litre ($CaCO_3$), and the addition of humic acid (1.5 mg/litre) significantly reduced the toxic effect of zinc on brood size. NOEC values were 0.1 and

0.025 mg/litre in soft water with and without humic acid and 0.225 and 0.175 mg/litre in hard water with and without humic acid, respectively.

Belanger & Cherry (1990) exposed *Ceriodaphnia dubia* to zinc in reproductive toxicity tests at three pH levels (6, 8 and 9) in three different surface waters from Virginia and Louisiana, USA. In New River water (hardness, 97.6 mg/litre, $CaCO_3$) significant reproductive impairment, as measured by the number of young per female, was found at a zinc concentration of 0.025 mg/litre at pH 6 and 8, while in Amy Bayou water (hardness, 113.6 mg/litre, $CaCO_3$) significant reproductive impairment was noted at 0.1 mg/litre. Reproductive impairment was found at 0.05 mg/litre in Clinch River water at pH levels of 6 and 9 but not 8.

Biesinger & Christensen (1972) exposed *Daphnia magna* to zinc chloride for 3 weeks. The 3-week LC_{50} was 0.16 mg/litre; the 3-week EC_{50}, based on reproductive impairment, was 0.10 mg/litre. Enserink et al. (1991) calculated the 21-day LC_{50} for *Daphnia magna* to be 0.84 mg/litre in Lake Ijssel water (background zinc concentration < 0.01 mg/litre; hardness, 225 mg/litre, $CaCO_3$). An EC_{50} based on population growth was 0.57 mg/litre. Münzinger & Monicelli (1991) carried out 21-day tests on *Daphnia magna* at added zinc concentrations of 0.05, 0.10 or 0.15 mg/litre in lake water (total zinc < 6 µg/litre). No significant effects on survival or reproduction were reported at the two lower concentrations. At 0.15 mg/litre, mortality was 80%, the number of progeny was reduced by more than 50%, and primiparous individuals were significantly smaller and produced significantly fewer eggs.

Wong (1993) studied the effect of zinc on the longevity and reproduction of the cladoceran *Moina macrocopa* reared in aquarium water with a zinc content of less than 1 µg/litre. A significant reduction in survival was observed at < 0.5 mg/litre within 1 day. The LT_{50} (time taken for 50% of animals to die) was reduced by more than 2 days at a zinc concentration of > 0.45 mg/litre and the average life span was reduced by more than 50% at > 0.70 mg/litre compared to controls. The net reproductive rate decreased abruptly at 0.7 mg/litre.

Maltby & Naylor (1990) exposed *Gammarus pulex* to zinc concentrations of 0.1, 0.3 or 0.5 mg/litre in 7-day tests. Zinc had no significant effect on either the number or size of offspring released from the current brood or on the number of offspring released from the subsequent brood, incubated under non-stressed conditions. The metal did cause a significant reduction in the size of offspring released from the subsequent brood and a positive correlation between zinc concentration and the number of broods aborted.

Anderson et al. (1980) calculated a 10-day LC_{50} for zinc for the midge *Tanytarsus dissimilis* reared in unfiltered Lake Superior water (background zinc concentration 5.1 µg/litre) to be 0.037 mg/litre. The midges were exposed to zinc during embryogenesis, hatching and larval development to the 2nd or 3rd instar. In flow-through life-cycle tests with caddisfly (*Clistoronia magnifica*), the highest zinc concentration tested, 5.2 mg/litre, had no significant effect on any life stage (Nebeker et al., 1984).

Dave et al. (1987) reported the results of a ring test of the 16-day embryo-larval toxicity test on zebrafish (*Brachydanio rerio*) using zinc sulfate as the toxicant. Hatching time delay was found to be the most sensitive parameter, with an NOEC of 0.5 mg/litre. Dawson et al. (1988) studied the effect of zinc on fathead minnow (*Pimephales promelas*) and South African clawed toad (*Xenopus laevis*) in embryo-larval assays. Static renewal tests were conducted for 6 days with minnow embryos to allow for hatching to take place and for 4 days with toad embryos. LC_{50} values were found to be 3.6 mg/litre for fathead minnows and 34.5 mg/litre for toad embryos. EC_{50} values, based on malformation, were 0.8 and 3.6 mg/litre for the two species, respectively; the minimum concentrations that significantly inhibited growth were 0.6 and 4.2 mg/litre, respectively.

Sayer et al. (1989) exposed yolk-sac fry of brown trout (*Salmo trutta*) to zinc concentrations of 4.9, 9.8 and 19.5 µg/litre (75, 150 and 300 nmol/litre) at pH 4.5 and calcium concentrations of 20 or 200 µmol for 30 days. Mortalities were high (70–100%) at the lower calcium concentration for all three zinc concentrations. No deaths or significant effects on mineral uptake were observed for zinc at the higher calcium exposure.

Bengtsson (1974a) exposed both yearling and adult minnows (*Phoxinus phoxinus*) to zinc as zinc nitrate in freshwater (< 0.02 mg/litre) over a 150-day period. Yearlings were the most sensitive, with growth significantly reduced at 0.13 mg/litre. Suppressed growth was associated with reduced feeding activity.

Chapman (1978a) studied the chronic toxicity of zinc to sockeye salmon (*Oncorhynchus nerka*) in a 22-month adult-to-smolt toxicity test. Fish were exposed to zinc concentrations ranging from 30 to 242 µg/litre in well-water (background zinc concentration 2 µg/litre; hardness, 35 mg/litre, $CaCO_3$). No adverse effects on survival, fertility, fecundity, growth or the subsequent survival of smolts transferred to seawater were observed.

Spehar (1976) exposed flagfish (*Jordanella floridae*) to zinc concentrations ranging from 28 to 267 µg/litre during a complete life-cycle test in untreated Lake Superior water (background zinc concentration 10 µg/litre; hardness, 44 mg/litre, $CaCO_3$). The 30-day survival of larvae previously exposed to zinc as embryos was significantly reduced at 267 µg/litre, while growth (100 days) was significantly reduced at 139 µg/litre. Reproduction was unaffected at zinc concentrations of up to 139 µg/litre. In a second experiment, fish were not exposed as embryos and significant reductions were observed in survival at 85 µg/litre after 30 days. The growth of female fish (100 days) was significantly reduced at 51 µg/litre. It should be noted that background zinc concentrations were less than 1 µg/litre in the second experiment. Spehar et al. (1978) found that cadmium (4.3–8.5 µg/litre) did not influence the mode of action of zinc under the same experimental conditions. The joint action of the toxicants on survival was little different from the toxicity of zinc alone.

Benoit & Holcombe (1978) carried out fathead minnow (*Pimephales promelas*) life-cycle tests in Lake Superior water (mean total zinc 2 µg/litre) at total zinc concentrations ranging from 44 to 577 µg/litre. The most sensitive parameters were egg adhesiveness and fragility, which were significantly affected at 145 µg/litre but not at 78 µg/litre. Hatchability and survival of larvae were significantly reduced, and deformities at hatching significantly increased at ≥ 295 µg/litre.

Holcombe et al. (1979) found no significant harmful effects on brook trout (*Salvelinus fontinalis*) exposed to zinc concentrations ranging from 2.6 (control) to 534 µg/litre for three generations in Lake Superior water. In a second experiment, a zinc concentration of 1368 µg/litre significantly reduced both the survival of embryos and 12-week-old larvae.

Kumar & Pant (1984) studied the toxic effects of zinc on the gonads of the fish *Puntius conchonius* exposed to one-third of the 96-h LC_{50} for zinc (which is 33.26 mg/litre) for up to 4 months. Male fish showed dilation in the testicular blood capillaries with necrosis and disintegration of the seminiferous tubules. Significant atresia in the ovary and damage to younger oocytes was found in female fish.

Watson & McKeown (1976) exposed yearling rainbow trout (*Oncorhynchus mykiss*) to zinc concentrations ranging from < 0.1 (control) to 1.12 mg/litre for up to 63 days. Growth was significantly inhibited at 1.12 mg/litre. Significant hyperglycaemia was found at all zinc exposure concentrations after 7 days but the condition remained significant by the end of the experiment only at 1.12 mg/litre.

Nemcsók et al. (1984) found that zinc chloride (1, 10 or 50 mg/litre) did not decrease acetylcholinesterase activity in serum, brain, heart or muscle of common carp (*Cyprinus carpio*) exposed for 2 h.

Korver & Sprague (1989) analysed the ability of male fathead minnows (*Pimephales promelas*) to avoid zinc concentrations ranging from 0.02 (control) to 13.5 mg/litre for up to 180 min. The LOEC was found to be 0.284 mg/litre; however, when fish were exposed in the presence of a shelter, the LOEC was 1.83 mg/litre.

Bengtsson (1974b) studied the effect of zinc on the ability of minnows (*Phoxinus phoxinus*) to compensate for a rotating water mass. Fish were exposed for approximately 100 days and significant adverse effects were found at 0.06 mg/litre for under yearlings, 0.16 mg/litre for yearlings and 0.2 mg/litre for adults.

Seawater studies

Calabrese et al. (1973) exposed eggs of American oyster (*Crassostrea virginica*) from within one hour of fertilization for 42–48 h to zinc chloride under static conditions (26 °C; salinity 25‰). An EC_{50} based on embryonic development was calculated to be 0.31 mg/litre. In similar tests, Calabrese & Nelson (1974) found the EC_{50} for the hard clam (*Mercenaria mercenaria*) to be 0.166 mg/litre. Calabrese et al. (1977) found the 8–10 day LC_{50} and an EC_{50}, based on growth, for hard clam larvae (*Mercenaria mercenaria*) to be 195.4 and 61.6 µg/litre, respectively, when tested in natural seawater. The values do not include the background zinc concentration of 17.7 µg/litre. Strömgren (1982) studied the effects of zinc on growth of the common mussel *Mytilus edulis* in tests of 10–22 days. Zinc concentrations ranging from 0.01 to 0.20 mg/litre were added to local seawater (background zinc concentration 5 µg/litre). Significant reductions in growth were observed at 0.01 mg/litre; an EC_{50} of 0.06 mg/litre was calculated for days 2–6.

Hunt & Anderson (1989) exposed the red abalone *Haliotis rufescens* to increased zinc concentrations in natural seawater. A 48-h EC_{50}, based on larval development, and a 9-day EC_{50}, based on metamorphosis, were found to be 0.068 and 0.050 mg/litre, respectively; NOEC values for the two parameters were 0.037 and 0.019 mg/litre, respectively.

Dinnel et al. (1989) exposed the purple sea urchin (*Strongylocentrotus purpuratus*), green sea urchin (*Strongylocentrotus droebachiensis*) and red sea urchin (*Strongylocentrotus franciscanus*) to zinc in 120-h sperm/fertilization tests. The sand dollar (*Dendraster excentricus*) was exposed to zinc in a 72-h test. EC_{50} values were found to be 0.26, 0.38, 0.31 and 0.028 mg/litre in the sperm test for the four species, respectively.

Reish & Carr (1978) found that the reproduction of the polychaetous annelids *Ctenodrillus serratus* and *Ophryotrocha diadema* was significantly inhibited at zinc concentrations of ≥ 0.5 mg/litre in 21-day tests.

Bengtsson & Bergström (1987) found that the 13-day EC_{50}, based on fecundity, for the harpacticoid copepod *Nitocra spinipes* ranged from 0.17 to 0.43 mg/litre. There was no significant effect of salinity, which ranged from 7‰ to 25‰.

Price & Uglow (1979) found the LT_{50} for the marine shrimp *Crangon crangon* to be 130 h at a zinc concentration of 14.4 mg/litre. When the test was carried out on various moult stages, LT_{50} values were 64, 140 and 152 h for the post-moult, inter-moult and pre-moult stages respectively.

Macdonald et al. (1988) exposed embryos of yellow crab (*Cancer anthonyi*) to zinc in 7-day tests. Zinc concentrations of ≥ 0.1 mg/litre significantly reduced survival. Hatching of embryos and larval survival were significantly reduced at 0.01 mg/litre; no embryos hatched at zinc concentrations of ≥ 1.0 mg/litre.

Redpath & Davenport (1988) reported that the pumping rate in the common mussel (*Mytilus edulis*) decreased with increasing zinc concentration and stopped completely at zinc concentrations in the range 0.47–0.86 mg/litre.

Weeks (1993) found a significant reduction in the feeding rate of the talitrid amphipod *Orchestia gammarellus* at dietary zinc concentrations ranging from 63 to 458 mg/kg during 48-h tests. However, no significant effect was found in 20-day exposures.

Somasundaram et al. (1984a) exposed the Atlantic herring (*Clupea harengus*) to zinc concentrations ranging from 0.1 to 6 mg/litre for up to 408 h. Zinc concentrations of ≥ 2 mg/litre significantly decreased total egg and yolk volumes throughout the study. At zinc concentrations of 0.1, 0.5 and 2.0 mg/litre, the development rate of eggs was faster than controls but at 6 mg/litre the rate was slower.

Somasundaram et al. (1984b) incubated eggs of Atlantic herring (*Clupea harengus*) at four zinc concentrations (0.5, 2.0, 6.0 and 12.0 mg/litre). The ultrastructural changes in the trunk muscle tissue of larvae hatched from the eggs were examined by morphometric analysis. The mean relative volumes of mitochondria, sarcoplasmic reticulum and muscle fibre were significantly increased and the surface:volume ratio of the mitochondrial cristae was significantly reduced. The ultrastructural changes in brain cells of larvae were also examined (Somasundaram et al., 1984c). All zinc exposures caused significant swelling of the nuclear membranes and rough endoplasmic reticulum, an increase in intracellular spaces and a decrease in the relative volumes of mitochondria. Somasundaram et al. (1985) studied the ultrastructural changes in the posterior gut and pronephric ducts of the herring larvae. Significant changes were observed only at zinc concentrations of 6.0 and 12.0 mg/litre; the endoplasmic reticulum, perinuclear space and mitochondria were swollen and there was a reduction in the surface:volume ratio of the mitochondrial cristae. At the highest zinc concentration, the posterior gut cells showed signs of necrosis. Examination of the epidermal structure revealed more vesicles and intracellular spaces in the epidermal cells, swollen mitochondria and signs of necrosis at zinc concentrations of 6.0 or 12.0 mg/litre (Somasundaram, 1985).

9.1.2.3 Effects on communities

Mesocosms

Belanger et al. (1986) exposed the Asiatic clam (*Corbicula* sp.) to zinc concentrations ranging from 0.025 to 1 mg/litre for 30 days in outdoor artificial stream systems. Background total zinc concentrations ranged from 0.02 to 0.094 mg/litre. Zinc concentrations of ≥ 0.05 mg/litre significantly reduced weight gain between days 20 and 30. Exposures to zinc at 1 mg/litre resulted in mortality of 10–50% by day 30.

Genter et al. (1988) added zinc at a concentration of 0.5 mg/litre to a flow-through stream mesocosm and studied the effects on an established periphyton community for 30 days. Seven diatoms and a coccoid green alga were significantly inhibited by zinc exposure. The algal total biovolume-density was reduced to < 5% of control levels by zinc from days 5–30. Zinc addition reduced protozoan numbers by more than 50%.

Marshall et al. (1983) conducted *in situ* experiments in Lake Michigan (background zinc concentration ~ 1 µg/litre) to determine the responses of the plankton community to added zinc for 2 weeks. Total zinc concentrations of 17.1 µg/litre significantly reduced chlorophyll *a*, primary productivity, dissolved oxygen, specific zooplankton populations and zooplankton species diversity.

Niederlehner & Cairns (1993) studied the effect of zinc on a naturally-derived periphyton community collected from a 195-ha lake (pH 7.1; hardness, 12.6 mg/litre; background zinc concentration 13.3 µg/litre). Toxicity tests with added zinc at concentrations of 73 and 172 µg/litre were carried out in dechlorinated tap water (pH 7.78; hardness, 73.8 mg/litre; zinc 1.3 µg/litre) for 21 days. Species richness was significantly impaired at the higher zinc exposure; primary production and community respiration were impaired at both zinc concentrations. The community was then exposed to pH levels ranging from 3 to 4.5 for 48 h. The pH stress significantly reduced species richness from the initial levels in controls and at both zinc concentrations. No significant differences between zinc treatments were observed at pH < 4.0.

Colwell et al. (1989) studied the effect of zinc on epilithic communities in artificial streams. Zinc concentrations of 0.05 and 1 mg/litre were added to the streams (background zinc concentration 0.02 mg/litre). After 30 days, greater biomass and lower protein:carbohydrate ratios were evident in epilithon exposed to the highest zinc concentration compared to the controls. Metal-tolerant populations had replaced metal-sensitive organisms by the end of the experiment at the higher zinc exposure.

Kiffney & Clements (1994) exposed benthic macroinvertebrate communities from two different sites to zinc (130 µg/litre) for 7 days. The background zinc concentrations at the two collection sites and in the artificial stream were below detection limits. Significant effects were observed at the community and population level following the addition of zinc. Specifically, mayflies from both sites were sensitive to zinc, but the magnitude of the response varied between sites. The results indicated that benthic macroinvertebrate communities from different stream orders may vary in sensitivity to zinc.

Field observations

Etxeberria et al. (1994) reported that increasing environmental levels of bioavailable zinc are associated with enlarged digestive lysosomes in mussels.

Solbé (1977) found that macroinvertebrates and fish were adversely affected by effluent from a steel works entering a hard water stream that had its source in the neighbouring limestone hills. The observed concentration of dissolved zinc in the river was 25 mg/litre; ammonia was the only other contaminant found at concentrations toxic to aquatic life, and it was quickly oxidized.

Graham et al. (1986) found increased mortality of rainbow trout (*Oncorhynchus mykiss*) at zinc-contaminated sites on the Molonglo River, New South Wales, Australia when compared with non-contaminated sites on the same river. Zinc concentrations of up to 2.32 mg/litre in the water and 1016 mg/kg dw in gill tissue were reported. The authors concluded that the concentrations of copper measured in the water were not sufficiently high to be lethal to the fish, although copper could have acted synergistically with zinc.

Hogstrand et al. (1989) reported that hepatic levels and metallothionein in perch (*Perca fluviatilis*) caught downstream from a brass works in Sweden reflected the water concentration of zinc (0.56–59 μg/litre). A significant correlation was found between hepatic zinc and metallothionein levels.

Clements & Kiffney (1995) examined benthic macroinvertebrate community responses to heavy metals at 33 sites in six Colorado streams (USA) in which zinc concentrations ranged from 2 to 691 μg/litre. The number of taxa and species richness of mayflies (Ephemeroptera), and the abundance of most mayfly and stonefly taxa were significantly reduced at sites where the zinc concentration exceeded the hardness-based criterion .

Van Tilborg (1996) reported on a freshwater stream in Belgium (the Kleine Nete) which contained total zinc at an average of 60 μg/litre (range <20–140 μg/litre). According to the Belgian Biotic Index, this stream has a high quality ecosystem.

9.1.3   Terrestrial organisms

9.1.3.1   Plants

Geochemical differences in zinc concentrations in soils and autonomic selection processes during the evolution of plants result in a great variation in zinc demand and zinc content between plant species and between plant genotypes of the same species. As a general rule, plants from environments poor in zinc are characterized by low zinc concentrations, those from zinc-enriched environments by high concentrations (Ernst, 1996). Within each ecosystem, biodiversity can only be maintained if species differ in their various ecological niches; zinc-demand is one variable. There is no convincing explanation as to why certain plant species have a higher uptake rate and accumulation pattern of zinc than others, although one possible reason may be to develop a defence against herbivores by accumulating high metal levels (Ernst et al., 1990). A great variation of zinc content is well known in forest ecosystems growing on soils with a normal zinc content. This variation is due to a number of factors including changes in the degree of infestation with endomycorrhizal fungi during a growing season and changes in ectomycorrhizal partners during the life history of the plant. When comparing zinc-sensitive and zinc-tolerant genotypes, it was found that, in zinc-tolerant genotypes only, the rapid compartmentation of zinc in the vacuole is one reason for an increased demand for zinc (Mathys, 1977). Zinc-activated enzymes, such as carbonic anhydrase, therefore reach the same activity in tolerant plants at higher external zinc concentrations than in zinc-sensitive plants. This response pattern has to be interpreted as a decrease in zinc efficiency in zinc-tolerant plants given the same level of zinc uptake by both genotypes (Harmens et al., 1993a). Zinc tolerance in plants is coded by only two major genes (Schat et al., 1996). Whether these genes are related to the zinc-efficiency genes reported from soybean varieties (Hartwig et al., 1991) remains to be investigated.

Therefore critical zinc levels cannot be established by analysing only the zinc content of leaves or other plant tissues; it is necessary to test the zinc demand of the plant genotype, its potential for physiological flexibility (allocation and retranslocation) (Ernst, 1995), siderophore exudation (Von Wirén et al., 1996), and cellular zinc compartmentation. The aim should be to establish the range of effects on the physiological processes under consideration from no effect up to 100% effect.

Toxicity to plants grown hydroponically and in soil

All such experiments involve acute toxicity, defined as toxicity over less than one life cycle in duration (seed to seed: cereals, rape, trees) or a harvest cycle (spinach, lettuce, cabbage). Zinc toxicity is first expressed in reduced root growth, a parameter that is used routinely in testing zinc-resistance in plants (Antonovics et al., 1971; Wainwright & Woolhouse, 1977; Schat et al., 1996). In higher plants the toxicity of zinc increases with exposure time, and therefore increasing zinc concentration in the plant and translocation from root to shoot (Mitchell & Fretz, 1977; Rauser, 1978; Dijkshoorn et al., 1979; Davies, 1993; Sheppard et al., 1993).

Zinc toxicity affects general physiological processes, e.g., transpiration, respiration and photosynthesis, and plant development in general can be visibly inhibited. Stunted growth, leaf epinasty and chlorosis of the younger leaves are striking symptoms of strong zinc toxicity. However, at lower degrees of zinc toxicity, these visible symptoms are less pronounced or can even be absent, whereas at the cellular level several processes are affected, owing to increases in local metal concentrations. Several mechanisms of metal action at the physiological and biochemical level have been described (for a review see Chaney, 1993; Vangronsveld & Clijsters, 1994), ranging from disturbance of cell division (Powell et al., 1986a,b; Davies et al., 1991) and ion balance (Ernst, 1996) to inhibition of photosynthesis (Van Assche & Clijsters, 1986). In *Phaseolus vulgaris*, growth inhibition and stress enzyme induction were both observed to occur when exactly the same internal zinc concentration was exceeded (Van Assche et al., 1988).

The critical leaf tissue concentration of zinc at which growth is affected was found for many plant species to be between 200 and 300 mg/kg dry matter (Davis & Beckett, 1978; Van Assche et al., 1988; Balsberg Påhlsson, 1989; Vangronsveld & Clijsters, 1992; Mench et al., 1994; Marschner, 1995). However, zinc phytotoxicity in leaves can depend to a large extent on the plant species, the age of the leaf and other factors, such as exposure period and exposure concentration.

Evaluations of phytotoxicity of zinc-polluted substrata are generally made by chemical analysis of the substratum itself. These results give rise to misinterpretations since availability of zinc to plants in and consequently metal uptake from the substratum are functions of the chemical form of the element in the soil, several soil parameters (e.g., pH, organic matter content, soil type) and plant species. Moreover, soils are frequently contaminated by a mixture of metals. Each of these materials separately can be phytotoxic, or they can interact in a synergistic, antagonistic or cumulative way (Beckett & Davis, 1978) (see section 9.3). The physiological and metabolic responses of test plants can be considered as a biological criterion for the total phytotoxic effect, since they are the result of the interactions of the metals present in the soil with other soil factors (biotic or abiotic) and with the plant. Phytotoxicity responses of test plants grown under controlled environmental conditions only reflect further the interference with metabolic processes of metals assimilated through the roots. Morphological responses (e.g., root growth, stem elongation, leaf expansion, biomass) and physiological and biochemical parameters (respiration, photosynthesis, capacities of enzymes and isozyme patterns, but not phytochelatin levels) can be used for the evaluation of phytotoxicity (Van Assche & Clijsters, 1990; Vangronsveld & Clijsters, 1992; Harmens et al., 1993b) (Tables 43–44).

Table 43. Impact of zinc-enriched sewage sludge added to non-dried sassafras sandy loam on crop plants (after Chaney, 1993)

Crop species	Yield as % of control	Geometric mean of zinc shoot (mg/kg dry weight)	Chlorosis[a]
Red fescue	17.2	965	1.8
Tall fescue	69.8	1060	1.7
Canadian blue grass	16.4	898	2.3
Cyperus	33.6	580	3.1
Barley	57.6	1060	1.4
Soybean	11.4	1140	4.2
Lettuce	1.2	3620	4.6

[a] Range of 1–5 with 5 being severe chlorosis.

Concentrations of zinc that are subtoxic or non-toxic to plants may have metabolic effects higher up the food chain. The disappearance of herbivorous insects on zinc-tolerant plants is one example of differences in species-specific tolerances (Ernst et al., 1990). Similarly, the zinc-content of zinc-efficient plants may be

Table 44. Acute toxicity of zinc to plants grown in hydroponic culture or soil

Plant species	Exposure		End-point	Toxicity data	Parameter	Reference
	Concentration (mg/litre)	Time (days)				
Hydroponic culture Hordeum vulgare	0–450	16	biomass, concentration	NOEC LD$_{100}$	8.3–27.2 mg/litre 168–460 mg/kg dw 150–800 mg/litre 1000–8000 mg/kg dw	Davis & Beckett (1978)
Phaseolus vulgaris	0.975–26	4	biomass, starch content	NOEC EC$_{50}$	0.975 mg/litre 26 mg/litre	Rauser (1978)
Phaseolus vulgaris	0–1990	16	primary leaf physiology and zinc concentration	NOEC EC$_{50}$	189–266 mg/kg dw 500 mg/kg dw (for shoot length)	Van Assche et al. (1988)
Allium cepa	6.5–65	2	root length	NOEC EC$_{50}$	6.5 mg/litre 25.9 mg/litre	Arambasic et al. (1995)
Lepidium sativum	65–164	2	root length	NOEC EC$_{50}$	65 mg/litre 547 mg/litre	Arambasic et al. (1995)

Table 44 (contd.)

Plant species	Exposure		End-point	Toxicity data	Parameter	Reference
	Concentration (mg/litre)	Time (days)				
Lolium perenne	0–30	14	root length	NOEC EC₅₀	< 0.1 mg/litre 1.6 mg/litre	Wong & Bradshaw (1982)
Festuca rubra zinc-sensitive zinc-tolerant	0–0.2	4	mitosis	EC₁₇ +NOEC	0.2 mg/litre 0.2mg/litre	Powell et al. (1986a)
Acer rubrum	0–0.4	78	root growth, zinc content	EC₅₀ zinc root zinc leaf	0.05 mg/litre 2190 mg/kg dw 381 mg/kg dw	Mitchell & Fretz (1977)
Picea abies	0–0.4	66	root growth, zinc content	EC₅₀ zinc root zinc needle	0.2 mg/litre 4125 mg/kg dw 1440 mg/kg dw	Mitchell & Fretz (1977)
Pinus strobus	0–0.4	78	root growth, zinc content	EC₅₀ zinc root zinc needle	0.2 mg/litre 9375 mg/kg 1005 mg/kg	Mitchell & Fretz (1977)

Table 44 (contd.)

Soil					
Acer rubrum	0–165 mg/kg pH 6.7	root growth, zinc content	NOEC zinc root zinc leaf	165 mg/kg in soil 618 mg/kg dw 209 mg/kg dw	Mitchell & Fretz (1977)
Picea abies	0–165 mg/kg pH 6.7	root growth, zinc content	NOEC zinc root zinc needle	165 mg/kg in soil 615 mg/kg dw 127 mg/kg dw	Mitchell & Fretz (1977)
Pinus strobus	0–165 mg/kg pH 6.7	root growth, zinc content	+NOEC zinc root zinc needle	165 mg/kg in soil 1430 mg/kg 314 mg/kg	Mitchell & Fretz (1977)
Plantago lanceolata	9.5–614 mg/kg pH 4.4, 4% organic matter	6 weeks	zinc content	EC_{50}	>1010 mg/kg in leaf	Dijkshoorn et al. (1979)
Trifolium repens	9.5–614 mg/kg pH 4.4, 4% organic matter	6 weeks	zinc content	EC_{50}	800 mg/kg leaf	Dijkshoorn et al. (1979)
Lolium perenne	9.5–614 mg/kg pH 4.4, 4% organic matter	6 weeks	zinc content	EC_{50}	500–600 mg/kg leaf	Dijkshoorn et al. (1979)

NOEC = no-observed-effect concentration; dw = dry weight

insufficient for optimum performance of herbivorous animals and humana, especially if all the cellular zinc is present in a form which is not readily bioavailable.

9.1.3.2 Invertebrates

Haight et al. (1982) calculated 24-h, 48-h and 72-h LC_{50} values for zinc of 82, 39.4 and 20 mg/litre (added as zinc sulfate to the growth medium), respectively, for juvenile free-living nematodes (*Panagrellus silusiae*) and 255, 95.1 and 47.5 mg/litre for adults.

Neuhauser et al. (1985) exposed earthworms (*Eisenia fetida*) to zinc in contact and artificial soil toxicity tests. In 48-h contact tests LC_{50} values were 13 µg/cm^2 for zinc acetate, 12 µg/cm^2 for zinc chloride, 10 µg/cm^2 for zinc nitrate and 13 µg/cm^2 for zinc sulfate. There were no significant differences between the toxicities of the different zinc salts. In an artificial soil test, the 2-week LC_{50} was found to be 662 mg/kg. Spurgeon et al. (1994) reported the 14-day LC_{50} for *E. fetida* to be 1010 mg/kg. The 56-day LC_{50} and NOEC were 745 and 289 mg/kg respectively; the EC_{50} and NOEC based on cocoon production were 276 and 199 mg/kg respectively.

Neuhauser et al. (1984) exposed earthworms (*E. fetida*) to zinc concentrations of 1000, 2500, 5000 and 10 000 mg/kg of manure (dry weight) for 6 weeks. Zinc at ≥ 5000 mg/kg significantly reduced growth and cocoon production. Similar results were obtained with four different zinc salts (acetate, chloride, nitrate and sulfate). The growth rate and reproduction had returned to normal after a subsequent 6-week period without zinc. Malecki et al. (1982) exposed earthworms (*E. fetida*) to six different zinc salts for 8 weeks. Significant reductions in cocoon production were observed at zinc carbonate and sulfate concentrations of 500 mg/kg dw. A zinc concentration of 2000 mg/kg adversely affected reproduction (acetate, chloride and nitrate) and growth (chloride, nitrate and sulfate). Zinc oxide significantly affected both growth and reproduction at 4000 mg/kg. Zinc carbonate did not adversely affect growth at the highest exposure (40 000 mg/kg). Long-term studies (20 weeks) with zinc acetate revealed significant reductions in cocoon production at 5000 mg/kg. Van Gestel et al. (1993) exposed earthworms (*E. andrei*) to zinc as zinc chloride at concentrations in dry artificial soil of 100–1000 mg/kg. Zinc significantly reduced reproduction at soil concentrations of 560 and 1000 mg/kg and induced the production of malformed cocoons. EC_{50} values for the effect of zinc on cocoon production and the number of juveniles per worm per week were 659 and 512 mg/kg dry soil, respectively. At the end of a 3-week recovery period, reproduction had returned to normal.

Marigomez et al. (1986) fed terrestrial slugs (*Arion ater*) for 27 days on a diet containing zinc concentrations ranging from 10 to 1000 mg/kg. No treatment-related effect on mortality was observed. Zinc concentrations of 1000 mg/kg significantly reduced feeding activity and growth.

Beyer & Anderson (1985) fed woodlice (*Porcellio scaber*) for 64 weeks on soil litter containing zinc at concentrations of up to 12 800 mg/kg. Soil litter containing ≥ 1600 mg/kg had adverse effects on reproduction; adult survival was reduced at ≥ 6400 mg/kg. Woodlice fed diets containing up to 20 000 mg/kg for 8 weeks showed decreased survival at concentrations of ≥ 5000 mg/kg (Beyer et al., 1984).

9.1.3.3   *Vertebrates*

High dietary levels of zinc are frequently fed to poultry to force moulting and reduce egg deposition (Hussein et al., 1988). Stahl et al. (1990) fed hens on a diet containing zinc at concentrations of 48, 228 or 2028 mg/kg for 12 or 44 weeks. Zinc treatments had no effect on overall egg production, feed conversion, feed consumption, hatchability, or progeny growth to the age of 3 weeks. Zinc levels were elevated in eggs from hens fed the diet containing 2028 mg/kg, but chick performance and tissue zinc content were unaffected by maternal zinc nutritional status. Stahl et al. (1989) fed chicks on a diet containing zinc at 37 (control), 100 or 2000 mg/kg for 21 days. There were no zinc-related deaths; at the highest exposure growth rate was decreased, anaemia was evident, tissue copper and iron decreased and tissue zinc increased.

Japanese quail (*Coturnix coturnix japonica*) fed a diet containing zinc (as zinc oxide) at 15 000 mg/kg for 7 days showed significant reductions in body weight. Egg production approached zero on day 3, eggshell breaking strength was reduced and moulting was induced (Hussein et al., 1988).

Dewar et al. (1983) fed 2-week-old chicks on a diet containing zinc at 74 (controls), 2000, 4000 or 6000 mg/kg for 4 weeks. High mortality was noted at the highest dose; all groups receiving zinc-supplemented food showed an increased incidence of gizzard and pancreatic lesions. Similar results were found when 1-day-old chicks were fed a diet containing zinc at 1000, 2000 or 4000 mg/kg for 4 weeks and when hens were fed diets containing 10 000 or 20 000 mg/kg for only 4 days. No lesions were found in hens exposed for 4 days to 10 000 mg/kg followed by 28 days on a control diet. Dean et al. (1991) fed day-old male chicks a diet containing zinc at 73 (controls) or 5280 mg/kg for 4 weeks. The zinc-supplemented feed significantly decreased body weight but did not affect food consumption compared with controls. Serum cholesterol, thyroxine and triiodothyronine levels were significantly reduced; serum growth hormone was significantly reduced but had recovered by the end of the experiment.

Gasaway & Buss (1972) maintained young mallard duck (*Anas platyrhynchos*) on a diet containing zinc concentrations ranging from 3000 to 12 000 mg/kg for 60 days. Food intake and body weight showed decreases as the level of zinc in the diet increased. Zinc caused reductions in pancreas and gonad weight in relation to body weight. The ratio of adrenals and kidney to body weight increased significantly. No significant changes in the liver:body weight ratios were observed. In ducks exposed to zinc there was partial paralysis of the legs, diarrhoea and weight loss within 10 days; severe paralysis was noted in some ducks within 20 days. Slight anaemia was found after 30 days but by day 45 extreme anaemia was observed in most of the exposed birds. High mortality was noted in all groups during the 60-day experiment with only 2 of the 45 exposed ducks surviving the whole time period. Zinc toxicosis consisted of paralysis of the legs, high concentrations of zinc in pancreas and kidney, and yellowish-red kidneys.

Zinc poisoning of birds has been reported as a result of the ingestion of zinc, for instance, from wire mesh cages (Van der Zee et al., 1985; Reece et al., 1986). Grandy et al. (1968) dosed mallard ducks (*A. platyrhynchos*) orally with eight No. 6 zinc shot and observed them for 30 days: three of the 15 birds died within the observation period. The average weight loss among surviving birds was 22%, significantly more than in control birds. Only three of the mallards dosed retained the zinc shot until the end of the study. Signs of intoxication in order of increasing severity were stumbling, an inability to run, complete loss of muscular control of the legs, loss of swimming ability and spasmodic wing movements. However, it should be noted that the zinc pellets were found not to be pure zinc but contained 92% zinc, 0.16% lead, a trace of iron and 7% not determined. French et al. (1987) dosed mallard ducks with five or ten No. 6 zinc shot (99.9% purity) and observed the birds for 28 days. Observation during the experimental period, post-mortem examination and histopathological examination revealed no effects of zinc on the dosed birds.

Mammals can also die from ingestion of zinc. Straube & Walden (1981) reported that 20 of 25 ferrets (*Mustela putorius furo*) being used in an experiment died of renal failure after eating raw meat that was accidentally contaminated with zinc from the wire cages. Zinc poisoning was diagnosed after autopsy and laboratory investigation.

Straube et al. (1980) fed ferrets (*M. putorius furo*) on a basal diet (zinc content 27 mg/kg) with zinc supplements of 500, 1500 or 3000 mg/kg for up to 6 months. The ferrets fed the two highest concentrations showed severe signs of toxicity between weeks 1 and 2, with the animals at the highest exposure dying within 2 weeks. Lesions included diffuse nephrosis, haemorrhages in the intestine and severe macrocytic hypochromic anaemia.

Bleavins et al. (1983) fed mink (*M. vison*) on a diet supplemented with zinc at 500 mg/kg for 2.5 months. No clinical signs of zinc toxicity were observed and the zinc supplement was increased to 1000 mg/kg. Again, no signs of zinc toxicity were observed; however, the offspring of zinc-treated females showed achromotrichia, alopecia, lymphopenia and a reduced rate of growth suggesting copper deficiency, although other signs of this latter condition (anaemia and neurotropenia) were not observed. Aulerich et al. (1991) fed adult and kit male and female mink on a basal diet (zinc content 40 mg/kg) supplemented with zinc at 500, 1000 or 1500 mg/kg for 144 days. No marked adverse effects on food consumption, body weight gains, haematological parameters, fur quality or survival were observed. Histopathological examination of the liver, kidneys and pancreas did not reveal any lesions indicative of zinc toxicosis.

Racey & Swift (1986) housed pipistrelle bats (*Pipistrellus pipistrellus*) in roosting cages treated with zinc octoate. The pregnant female bats (three groups of 10), collected from nursery roosts, were trained to feed on mealworms before transfer to the experimental cages. The cages were metal, lined with plywood and painted with zinc octoate as a solution in white spirit at 0.5 litre/m^2 and containing 8% zinc as metal (as recommended by the manufacturers), or with white spirit. Treatment of the wood was conducted 2 months before introduction of the bats into the cages. During the course of the 142-day experiment, there were 2 deaths in the zinc ocoate group, 2 in the white spirit control group and 3 in the untreated control group.

9.2 Tolerance to zinc

Tolerance to zinc (and other metals) has been documented in a wide variety of plants and animals. Tolerance may occur in two ways (Miller & Hendricks, 1996): through acclimatization at some early stage in the life cycle, or through natural selection. The latter process is heritable, the former is not. However, as noted by Bervoets et al. (1996), few studies have discriminated even the effects of acclimatization on metal uptake or toxicity. Some key studies that have done so are detailed below.

Free-living fungi, as well as those associated in mycorrhiza with higher plants (vesicular-arbuscular mycorrhiza, ectomycorrhiza), have enhanced zinc resistance in zinc-enriched soils (Colpaert & Van Assche, 1992; Gadd, 1993). All tested actinomycetes and non-spore forming bacteria isolated from a site contaminated with metals were zinc tolerant, growing normally in media containing zinc concentrations of 39–130 mg/litre (600–2000 µmol/litre) (Jordan & Lechevalier, 1975). Shehata & Whitton (1981) reported that blue-green algae are often frequent in waters with very high levels of zinc, and laboratory assays have shown that these algae are much more resistant to zinc than most isolates from sites lacking zinc enrichment. Hornor & Hilt (1985) determined the distribution of zinc-tolerant bacteria from three stream sites containing high (3125 µg/litre), medium (291 µg/litre) and low (109 µg/litre) concentrations of zinc. Zinc tolerance was estimated by the ability of bacteria to grow on media amended with zinc (4–512 mg/litre). The presence of zinc-tolerant bacteria was correlated with the degree of heavy metal contamination. Zinc concentrations ranging from 4 to

16 mg/litre were stimulatory to growth of bacteria from contaminated sites while concentrations of 4 mg/litre were inhibitory to bacteria from the control site.

Antonovics et al. (1971) reviewed metal tolerance in plants and found several examples of plant communities that show high tolerance to zinc. Most of the studies that relate to zinc tolerance are associated with soils enriched with zinc either naturally or by metal mining activity. All plants growing for a long time on zinc-enriched soils have evolved a zinc resistance regardless of whether they are fungi, mosses, ferns or angiosperms. Shaw (1990) noted that the development of metal tolerance in plants is among the best observed examples of evolution related to natural and anthropogenic stress.

Acclimatization and adaptation to zinc have been demonstrated by Miller & Hendricks (1996) for *Chironomus riparius*. Other examples of zinc resistance in aquatic organisms are provided by Klerks & Weis (1987), Klerks & Levinton (1989) and Klerks (1990).

Pre-exposure of rainbow trout to zinc at a concentration of 2 mg/litre for more than 5 days significantly decreased the acute toxicity of the metal (Bradley & Sprague, 1985). Anadu et al. (1989) found that acclimatization to a zinc concentration of 50 µg/litre increased the tolerance of juvenile rainbow trout by a factor of 3–5. Maximum tolerance was achieved within 7 days with no further change noted after 2–3 weeks. Further studies revealed that there was no increase in tolerance after acclimatization periods of less than 3 days. Tolerance to zinc was rapidly lost following return to control water, with almost complete reversion to control tolerance after only 7 days. Hobson & Birge (1989) found that tolerance to zinc in 96-h acute toxicity tests increased significantly after 14 days of exposure to zinc at 0.6 mg/litre but decreased significantly following exposure to 1.8 mg/litre for 7 and 14 days. After 21 days of pre-exposure to 0.6 or 1.8 mg/litre there was no significant effect on acute toxicity compared with controls. The authors found no correlation between changes in tolerance and observed changes in metallothionein-like proteins. Hogstrand & Wood (1995) reported that acclimatization to zinc in rainbow trout adapted to fresh water can develop without any detectable increase in zinc accumulation in the gills or liver. Acclimatization to zinc does not necessarily involve induction of metallothionein. The inhibition of the calcium influx by zinc is mainly competitive in its nature, and persists during chronic exposure, indicating that zinc and calcium compete for the same uptake sites. Zinc-adapted fish have a decreased rate of zinc influx compared to controls. The authors therefore speculated that fish are able to regulate the uptake of zinc separately from calcium so that, in zinc-adapted fish, zinc influx can be markedly reduced without altering the influx of calcium.

Joosse et al. (1984) found that a tolerant population of the terrestrial woodlouse (*Porcellio scaber*) from a contaminated site regulated its body content of zinc at a higher level than a control population. Another population of *P. scaber* collected from a zinc-contaminated area was found to be adapted to high zinc and cadmium concentrations (Van Capelleveen, 1987). However, individuals from the contaminated site produced larger quantities of metalloproteins and showed lower growth efficiencies and drought resistance than individuals from a control site.

Differences in assimilation rates for zinc in two populations of centipede (*Lithobius variegatus*) were found to be related to the degree of contamination of the site from which the population was collected (Hopkin & Martin, 1984). Centipedes from a contaminated site survived longer than those from an uncontaminated site when both populations were fed on woodlice hepatopancreas with high concentrations of metals. A review of resistance in terrestrial invertebrates is provided by Posthuma & Van Straalen (1993).

9.3 Interactions with other metals

Zinc can behave antagonistically in combination with copper (Ahsanullah et al., 1988; Vranken et al., 1988; Kraak et al., 1994b) and synergistically with lead or iron alone or in combination (Ahsanullah et al., 1988; Konar & Mullick, 1993), and with mercury or nickel (Vranken et al., 1988). Zinc and cadmium can show additive toxicity (Negilski et al., 1981; Kraak et al., 1994b). Metal interactions can vary, however, depending on physicochemical conditions. For instance, Tomasik et al. (1995) found that zinc toxicity in soft water (50 mg/litre, $CaCO_3$) was lowered by magnesium and molybdenum and increased by cobalt or selenium. However, in hard water (100 mg/litre, $CaCO_3$) zinc was either inhibited by all metals (at a concentration of 1 mg/litre) or there was a weak synergism (at 0.5 mg/litre). Similarly, Biesinger et al. (1986) found that different zinc concentrations resulted in different interactions (with cadmium and mercury).

10. EVALUATION OF HUMAN HEALTH RISKS AND EFFECTS ON THE ENVIRONMENT

10.1 Homeostatic model

Zinc is an essential trace element that can cause symptoms of deficiency and can be toxic when exposures exceed physiological needs. This relationship is described by a homeostatic model that takes the form of a U-shaped curve; the arms of the curve express risk of deficiency or excess, with the portion of the curve between the arms expressing the range of exposure (intake) that is related to optimal function (good health) (Fig. 1). The relationship between intake and health is affected by physiological factors (homeostasis) and by extrinsic factors that affect the availability of zinc for absorption and utilization or that interfere with the metabolism of zinc and biochemical processes that require zinc. In nature these relationships are not necessarily symmetrical.

The homeostatic model defines the principle of an acceptable range of exposures for an essential trace element like zinc. In the acceptable range, zinc, which is necessary for various metabolic processes, embryonic development, cellular differentiation and cell proliferation, provides the substrates for expression of the genetic potential of the individual, i.e., optimum growth, health, reproduction and development. Environmental levels of zinc providing exposures or intakes within the acceptable range do not produce adverse effects among the general human population or the environment. However, there are individuals or groups with imbalances in relation to other trace elements, or with disorders in homeostatic mechanisms that experience effects, of either deficiency or toxicity, from exposures within the acceptable range. These disorders may be acquired or of genetic origin.

10.2 Evaluation of risks to human health

People are exposed to zinc primarily from food, although oral exposure can become excessive through non-dietary sources. Certain occupational exposures can be hazardous.

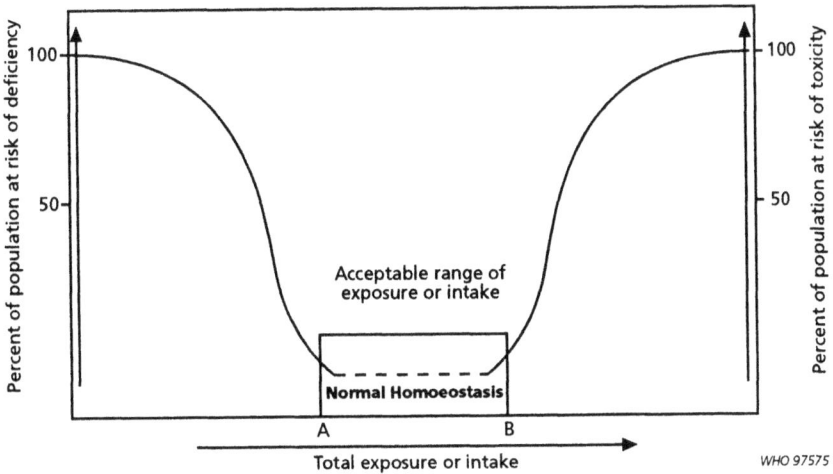

Fig. 1. Percent of population subjected to deficiency and toxicity effects according to exposure/intake. As intake drops below A risk for deficiency increases; at extremely low exposures or intakes all subjects will manifest deficiency. As exposure or intakes increase beyond B a progressively larger proportion of subjects will exhibit effects of toxicity.

10.2.1 *Exposure of general population*

The estimated average daily dietary zinc intakes range from 5.6 to 13 mg/day in infants and children from 2 months up to 19 years and from 8.8 to 14.4 mg/day in adults aged 20–50 years. Flesh foods (i.e., meat, poultry, fish and other seafood) are rich sources of readily available zinc, while fruits and vegetables contain relatively low zinc concentrations. For omnivorous adults, more than one-third of dietary zinc can be provided by flesh foods, whereas for vegetarians, plant-based foods are the major dietary source. Mean daily intake of zinc from drinking-water is estimated to be < 0.2 mg/day.

Intakes via dermal and inhalation routes are generally insignificant in the general population. Consumption of dietary supplements of zinc as well as prolonged treatment with pharmaceuticals containing zinc may result in high exposure to zinc.

The absorption of zinc from diets based on solid foods generally ranges from < 15% to 55% (about 20–30% from an omnivorous diet), depending on the composition of the diet, as well as the nutritional (especially in relation to zinc), physiological and health status of the individual. When major dietary zinc sources are unrefined cereals, nuts and legumes, absorption is low, owing primarily to the inhibitory effect on zinc absorption of phytate and, to a lesser extent, Maillard browning products and dietary fibre.

The major excretory route for ingested zinc is via the faeces.

10.2.2 *Occupational exposure*

Occupational exposure to dusts and fumes of zinc and zinc compounds can occur in a variety of settings in which zinc is produced, or in which zinc and zinc-containing materials are used. Typical airborne exposures observed include 0.19–0.29 mg/m^3 during the smelting of zinc-containing iron scrap, 0.90–6.2 mg/m^3 at non-ferrous foundries and 0.076–0.101 mg/m^3 in hot-dip galvanizing facilities. Far higher exposures are possible during particular job activities, such as welding of zinc-coated steels in the absence of appropriate respiratory protection and/or fume extraction engineering controls.

Occupational exposure to high levels of zinc oxide and/or non-ferrous metals is associated with metal-fume fever. This is usually a short-term, self-limiting syndrome, characterized by fever and chills. Induction of metal-fume fever is most common with ultra-fine particles capable of deep lung penetration under conditions of exposure. Studies on volunteers conducted under short-term exposure conditions (77–153 mg/m^3 for 15–30 min) have detected pulmonary inflammation responses (including cytokine induction) which are consistent with manifestations of metal-fume fever and support an immunological etiology for this acute reversible syndrome.

Based on the available information, it is not possible to define a no-effect level for pulmonary inflammation from exposure to zinc oxide fume.

10.2.3 *Risks of zinc deficiency*

Zinc is an ubiquitous and essential element. Dietary reference values for zinc for adults range from 6 to 15 mg/day (depending upon the bioavailability factor used). However, large numbers of people are believed to ingest insufficient bioavailable zinc. The effects of zinc deficiency are well documented and may be severe. They include impaired neuropsychological functions, oligospermia, growth retardation, impaired reproduction, immune disorders, dermatitis and impaired wound healing. Most of these effects are treatable with adequate amounts of zinc.

Because of the lack of data on zinc kinetics and inadequate measures of internal dose, there is limited understanding of how site-specific zinc concentration relates to manifestations of deficiency (or excess). The estimated absolute absorbed amount of zinc for adults is 2.5 mg daily. This implies a dietary need at 20% bioavailability of 12.5 mg daily. As bioavailability increases, the amount needed in the diet will decrease.

10.2.4 *Risks of zinc excess*

Toxic effects in humans are most obvious from accidental or occupational inhalation exposure to high concentrations of zinc compounds, such as from smoke bombs, or metal-fume fever. Modern occupational health and safety measures can significantly reduce potential exposure. Intentional or accidental ingestion of large amounts of zinc leads to gastrointestinal effects, such as abdominal pain, vomiting and diarrhoea. In the case of long-term intakes of large amounts of zinc at pharmacological doses (150–2000 mg/day), the effects (sideroblastic anaemia, leukopenia and hypochromic microcytic anaemia) are reversible upon discontinuation of zinc therapy and/or repletion of copper status, and are largely attributed to zinc-induced copper deficiency.

High levels of zinc may disrupt the homeostasis of other essential elements. For example, in adults, subtle effects of zinc on copper utilization may occur at doses of zinc near the recommended level of intake of 15 mg/day and up to about 50 mg/day. Copper requirements may be increased and copper utilization may be impaired with changes in clinical chemistry parameters, but these effects are not consistent and depend largely upon the dietary intake of copper. Distortion of lipoprotein metabolism and concentrations associated with large doses of zinc are inferred to be a result of impaired copper utilization. In groups with adequate copper intake, no adverse effects, with the exception of reduced copper retention, have been seen at daily zinc intakes of < 50 mg/day.

There is no convincing evidence that excess zinc plays an etiological role in human carcinogenesis. The weight of evidence supports the conclusion that zinc is not genotoxic or teratogenic. At high concentrations zinc can be cytotoxic.

10.3 Evaluation of effects on the environment

10.3.1 *Environmental risk assessment*

The science of performing environmental risk assessments has evolved rapidly in recent years with standardized techniques being adopted in both the USA and Europe (US EPA, 1992 and OECD, 1995). The key components of environmental risk assessment are: problem formulation, analysis, and risk characterization.

Problem formulation (i.e., hazard identification) consists of identifying and defining the risk problem, assessing the population, community or ecosystem at risk, establishing the model for evaluating the potential for risk, and selecting the biological end-points and environmental media to be analysed. The analysis phase consists of assessments of exposure and effects. The exposure assessment involves detailed studies to characterize the spatial and temporal concentrations of the chemical of interest. Effects assessment involves a series of standardized laboratory and, in some cases, field studies, and is performed to evaluate the dose–response curve for selected toxic end-points and species of interest. In the risk characterization phase, the exposure and effects data are integrated, the potential for co-occurrence of organism and contaminant is determined, and a conclusion is drawn about the potential for risk. The risk statement can be made in terms of a probability statement, frequency or time effects are expected to occur, or number of species to be affected. Risk is assessed by determination of the adequacy of the margin of safety between effects and exposure concentrations and expert judgement is typically used to determine the acceptability of the perceived margin of safety. General consensus exists that the larger the margin of safety the lower is the environmental risk. Margins of safety of < 1.0 are usually indicative of a higher potential for risk and may require further evaluation.

10.3.2 Components of risk assessment for essential elements

For essential elements, the principal components of risk assessment are exposure and effects assessments. Environmental exposure has been assessed by reviewing the fate (transport, distribution and behaviour) of the element from the point of release into and through the environmental compartments of air, water soil/sediment and biota. Effects assessment involves toxicity tests conducted on representative species of the trophic levels in the ecological community of interest, including algae and plants (primary producers), aquatic and terrestrial invertebrates (secondary producers) and fish and terrestrial animals (consumers).

For essential micronutrients, a lower limit exists below which deficiency will occur as illustrated in Fig. 1. This well-established concept is applicable at both the population and the community level. Deficiency as well as toxicity must therefore be considered when assessing environmental risk Within the homeostatic zone, organisms regulate their uptake and/or compartmentation of zinc to maintain optimal life conditions without any stress occurring. Outside this zone, adaptation may occur.

There is a diversity of habitat types in both aquatic and terrestrial environments with different optimal concentration ranges. Because zinc is ubiquitous, it is unlikely, except in some terrestrial regions, particularly agricultural regions where zinc concentrations are very low or where antagonistic nutrient interactions occur, that deficiency in the environment will be a significant issue. However, the use of large safety factors in procedures to limit exposures to below toxic levels might result in target concentrations below essential levels.

One of the key questions in ecotoxicology is the extent to which laboratory tests under controlled conditions are predictive of effects that will be seen in the environment. Traditional toxicity testing has, in the past, focused on the acute and chronic effects of chemicals on the life stages of aquatic and terrestrial organisms. The integration of environmental chemistry and toxicology has allowed a better prediction of the effects of metals on organisms in the environment. This has led to the now accepted view that the total concentration of an essential element such as zinc in an environmental compartment is not a good predictor of its bioavailability (section 4.2). Since site-specific parameters control the bioavailability of essential elements, these parameters must be included in the risk assessment.

Organisms may also develop tolerance on a local scale by acclimatization (physiological behaviour) and adaptation (genetic changes) towards higher as well as lower concentrations. Because of such tolerance, the test-derived toxicity values may be lower or higher than the values for the same species from an adapted population.

10.3.3 *Environmental risk assessment for zinc*

There are limited data available for performing a detailed assessment of the potential risk for zinc for each environmental medium (air, water, soil, sediment). The largest data set is reported for the aquatic environment. The aim of this section is to provide an evaluation of the available biological effects and exposure data for various organisms and media consistent with this risk paradigm, and describe ranges of concentrations for which the potential for risk increases.

10.3.3.1 *Environmental concentrations*

Zinc is released to the environment from natural and anthropogenic sources. On a global scale, emissions from these sources are similar in magnitude. However, on a local scale anthropogenic sources may dominate.

The fate of zinc in the environment is largely determined by sorption processes. Zinc bioavailability is determined by a set of physicochemical and biological parameters as discussed in Chapter 4. Zinc occurs ubiquitously in environmental samples, although the concentration is determined by several factors, such as the local geological and anthropogenic influences. Natural background total zinc concentrations are as follows:

air	up to 300 ng/m^3
fresh water	< 0.1–50 µg/litre
seawater	< 0.002–0.1 µg/litre
sediments	up to 100 mg/kg dw
soil	10–300 mg/kg dw

Higher zinc concentrations can be attributed to anthropogenic contamination, although natural processes (both abiotic and biotic) can contribute to localized high zinc concentrations. Total zinc concentrations in fresh and estuarine waters have been reported to be up to 3900 and 15 µg/litre, respectively. Zinc concentrations of up to 35 000 mg/kg dw have been reported in soil. In mineralized areas these values can be exceeded.

10.3.3.2 *Overview of toxicity data*

The toxicity of zinc to aquatic organisms is affected by factors such as temperature, hardness, pH and dissolved organic carbon. Overall the data are very variable (see chapter 9). Table 45 summarizes minimum effects threshold data. Only data that met the acceptability criteria given in chapter 9 are included. It should be noted that none of the aquatic plant toxicity data contained in chapter 9 met those criteria. Thus, aquatic plants are not included in the table. A dissolved zinc concentration of 20 µg/litre has been shown to have adverse effects on freshwater organisms in soft water (hardness < 100 mg/litre, $CaCO_3$). In hard water (hardness > 100 mg/litre, $CaCO_3$), adverse effects have been reported at dissolved zinc concentrations of 90 µg/litre. In the marine environment, dissolved zinc concentrations of 100 µg/litre have been shown to have adverse effects. Zinc deficiency in aquatic organisms in the open ocean has been reported.

The toxicity of zinc to terrestrial organisms is similarly dependent upon its bioavailability, which in turn is determined by various factors such as the speciation of zinc, and the physicochemical and biological characteristics of the soil. The bioavailable fraction of zinc in soil has been calculated to range from < 1% to 10% of the total zinc concentration. Zinc has to be in a

Table 45.   Minimum thresholds for adverse effects of
dissolved zinc on aquatic organisms

Concentration of dissolved zinc (µg/litre)	Effects in fresh waters	Effects in marine waters
20–50	chronic effects on cladocerans in soft water[a]	
50–100	chronic effects on cladocerans in hard water; acute effects on cladocerans in soft water; acute and chronic effects on fish in soft water; chronic effects on freshwater insects	acute effects on mysids
100–200	acute effects on algae; acute effects on cladocerans in hard water; chronic effects on fish in hard water; chronic effects on molluscs	acute effects on fish
200–1000		acute effects on amphipods; acute effects on decapods
1000–10000	acute effects on molluscs; acute effects on copepods; acute effects on amphipods; acute effects on fish in hard water	acute effects on polychaetes; acute effects on molluscs

[a]   For the purposes of this document, soft water is defined as having a hardness ($CaCO_3$) of < 100 mg/litre, hard water as having a hardness of > 100 mg/litre.

soluble form to be taken up by plants. Plants may also take up zinc that is deposited on the surface of leaves following aerial deposition or application of fertilizer. Symptoms of zinc toxicity in plants differ from those of zinc deficiency, with coralloid rather than extended roots and curled rather than mottled leaves. In the case of zinc toxicity, zinc replaces other metals (e.g., iron, manganese) in the active centres of enzymes (e.g., hydrolases and haem enzymes).

Among terrestrial invertebrates, earthworms (*Eisenia andrei*) and woodlice (*Porcellio scaber*) show adverse effects on reproduction starting at total zinc concentrations of 560 and

1600 mg/kg dw, respectively. For terrestrial vertebrates, only data on zinc toxicity through dietary zinc intake are available (see section 9.1.3.3).

The adverse effects of zinc must be balanced against its essentiality. Zinc is important in enzymes, in the metabolism of proteins and nucleic acids, and in the stabilization of biological membranes. Zinc deficiency has been reported in a wide variety of agricultural plants and animals with severe effects on all stages of reproduction, growth and tissue proliferation. An evaluation of zinc must therefore take into account the adverse effects of low zinc concentrations as well as those attributed to excess zinc. There are ranges of optimal concentrations for essential elements such as zinc, which are dependent upon species and habitat. The setting of an absolute toxic threshold value or minimum toxic concentration is not appropriate for such a chemical. Any risk assessment of the potential effects of zinc on organisms must also take into account the local environmental conditions.

11. CONCLUSIONS AND RECOMMENDATIONS FOR PROTECTION OF HUMAN HEALTH AND THE ENVIRONMENT

11.1 Human health

- There is a decreasing trend in anthropogenic zinc emissions.

- Many pre-1980 environmental samples, in particular in water samples, may have been subject to contamination with zinc during sampling and analysis and, for this reason, zinc concentration data for such samples should be viewed with extreme caution.

- In countries where staple diets are based on unrefined cereals and legumes, and intakes of flesh foods are low, dietary strategies should be developed to improve the content and bioavailability of zinc.

- Preparations intended to increase the zinc intake above that provided by the diet should not contain zinc levels that exceed dietary reference values, and should contain sufficient copper to ensure a ratio of zinc to copper of approximately 7, as is found in human milk.

- There is a need for better documentation of actual exposures to zinc oxide fume in occupational settings. Workplace concentrations should not result in exposure levels as high as those known to have given rise to inflammatory responses in the lungs of volunteers.

- The essential nature of zinc, together with its relatively low toxicity in humans and the limited sources of human exposure, suggests that normal, healthy individuals not exposed to zinc in the workplace are at potentially greater risk from the adverse effects associated with zinc deficiency than from those associated with normal environmental exposure to zinc.

11.2 Environment

- Zinc is an essential element in the environment. The possibility exists both for a deficiency and for an excess of this metal. For this reason it is important that regulatory criteria for zinc, while protecting against toxicity, are not set so low as to drive zinc levels into the deficiency area.

- There are differences in the responses of organisms to deficiency and excess.

- Zinc bioavailability is affected by biotic and abiotic factors, e.g., organism age and size, prior history of exposure, water hardness, pH, dissolved organic carbon and temperature.

- The total concentration of an essential element such as zinc, alone, is not a good predictor of its bioavailability or toxicity.

- There is a range of optimum concentrations for essential elements such as zinc.

- The toxicity of zinc will depend on environmental conditions and habitat types. Thus any risk assessment of the potential effects of zinc on organisms must take into account local environmental conditions.

12. RECOMMENDATIONS FOR FURTHER RESEARCH

12.1 Zinc status

There is a need to improve techniques to assess zinc status. Current methods such as measurements of serum, plasma and leukocyte zinc concentrations are insufficient. Newer approaches for consideration might include the use of zinc kinetics as related to function, or the identification of enzymes that are dependent on zinc and whose changes reflect zinc status as it relates to function.

12.2 Functional indices of zinc status

There is a need for functional indices of zinc status to be measured before and after treatment, and in the absence of other limiting nutrients. Areas of interest include immune function, neuropsychology, dark adaptation, ethanol tolerance, intestinal permeability, and growth and body composition.

12.3 Interactions with other trace elements

Zinc homeostasis requires much clearer definition in respect of both its individual role, and the concurrent role of other ions which may be affected by zinc deficiency or excess. In particular, biological effects at various ratios of zinc and copper require further investigation. Specifically, attention needs to be given to life-cycle issues in which the relationships between concentrations of the metals in organs such as the liver should be related to the dietary ratios of these same metals. It is also important to determine whether the biological end-point is a matter of clinical significance. Data defining the amount of zinc that might interfere with copper utilization in infants, children and pregnant women are needed. The zinc:copper ratio has been shown to induce dyslipidaemias, and studies on the cardiovascular consequences of these conditions need further investigation.

12.4 Supplementation

As excess zinc impairs utilization of copper and other trace elements, research on the potential benefits of adding copper and/or iron to oral zinc preparations is needed.

12.5 Occupational medicine

There is a need to define the NOEL for zinc exposure in occupational medicine.

12.6 The molecular mechanism

Data on zinc transport by proteins, both into and out of the cell, are poorly developed. Further studies on zinc homeostasis and the role of metal-binding proteins (including metallothionein) are required.

12.7 Environment

Work is required to develop and standardize procedures for toxicity testing in terrestrial organisms.

Work is required to understand the homeostatic abilities of sensitive indicator organisms, and to determine the boundaries of the optimal range for different species in different habitat types.

Techniques for measuring and predicting bioavailability in different media, including robust analytical methods for measuring speciation, need to be developed, improved and validated against bioassays.

Important data gaps have been identified including, in particular, chronic marine toxicity data and toxicity data for terrestrial plants.

REFERENCES

Aamodt RL, Rumble WF, Johnston GS, Foster O, & Henkin RI (1979) Zinc metabolism in humans after oral and intravenous administration of Zn-69m. Am J Clin Nutr, **32**: 559–569.

Aamodt RL, Rumble WF, Badcock AK, Foster MD, & Henkin RI (1982) Effects of oral zinc loading on zinc metabolism in humans – I: Experimental studies. Metabolism, **31**(4): 324–326.

Abbasi SA & Soni R (1986) An examination of environmentally safe levels of zinc(II), cadmium(II) and lead(II) with reference to impact on channelfish *Nuria denricus*. Environ Pollut, **A40**: 37–51.

Abu-Hamdan DK, Migdal SD, Whitehouse R, Rabbani P, Prasad AS, & McDonald FD (1981) Renal handling of zinc: effect of cycteine infusion. Am J Physiol, **241**: F487–F494.

Adams MJ, Blundell TL, Dodson EJ, Dodson GG, Vijayan M, Baker EN, Harding MM, Hodgkin DC, Rimmer B, & Sheat S (1969) Structure of rhombohedral 2 zinc insulin crystals. Nature, **224**: 491–495.

Adeniyi FA & Heaton FW (1980) The effect of zinc deficiency on alkaline phosphatase (EC 3.1.3.1) and its isoenzymes. Br J Nutr, **43**: 561–569.

Adriano DC (1986) Zinc. In: Adriano DC ed. Trace elements in the terrestrial environment. New York, Springer, pp 421–469.

Aggett PJ (1989) Severe zinc deficiency. In: Mills B ed. Zinc in human biology. London, Springer, pp 259–279.

Aggett PJ (1991) The diagnostic value of measurements for trace elements in biological material. Beitr Infusionsther, **27**: 53–65.

Aggett PJ (1994) Aspects of neonatal metabolism of trace metals. Acta Paediatr Suppl, **402**: 75–82.

Ågren MS (1990) Percutaneous absorption of zinc from zinc oxide applied topically to intact skin in man. Dermatologica, **180**: 36–39.

Ågren MS (1991) Influence of two vehicles for zinc oxide on zinc absorption through intact skin and wounds. Acta Derm Venereol, **71**: 153–156.

Ahlers WW, Reid MR, Kim JP, & Hunter KA (1990) Contamination-free sample collection and handling protocols for trace elements in natural fresh water. Aust J Mar Freshwater Res, **41**: 713–720.

Ahsanullah M (1976) Acute toxicity of cadmium and zinc to seven invertebrate species from Western Port, Victoria. Australian. J Mar Freshwater Res, **27**: 187–196.

260

Ahsanullah M, Mobley MC, & Rankin P (1988) Individual and combined effects of zinc, cadmium and copper on the marine amphipod *Allorchestes compressa*. Aust J Mar Freshwater Res, **39**: 33–37.

Alaimo K, McDowell MA, Briefel RR, Bischof AM, Caughman CR, Loria CM, & Johnson CL (1994) Dietary intake of vitamins, minerals, and fiber of persons ages 2 months and over in the United States: Third National Health and Nutrition Examination Survey, Phase 1, 1988–91. US Department of Health and Human Services ((PHS) 95-1250 4-2511).

Alam MK & Maughan OE (1992) The effect of malathion, diazinon, and various concentrations of zinc, copper, nickel, lead, iron, and mercury on fish. Biol Trace Elem Res, **34**(3): 225–236.

Alexander J, Aaseth J, & Refsvik T (1981) Excretion of zinc in rat bile – a role of glutathione. Acta Pharmacol Toxicol, **49**: 190–194.

Allen HE (1996) Standards for metals should not be based on total concentration. SETAC News, **16**: 18–19.

Allen DK, Hassel CA, & Lei KY (1982) Function of pituitary-thyroid axis in copper-deficient rats. J Nutr, **112**(11): 2043–2046

Allen JG, Masters HG, Peet RL, Mullins KR, Lewis RD, Skirrow SZ, & Fry J (1983) Zinc toxicity in ruminants. J Comp Pathol, **93**: 363–377.

Allen JG, Morcombe PW, Masters HG, Petterson DS, & Robertson TA (1986) Acute zinc toxicity in sheep. Aust Vet J, **63**(3): 93–95.

Allen HE, Fu GM, & Deng BL (1993) Analysis of acid-volatile sulfide (AVS) and simultaneously extracted metals (SEM) for the estimation of potential toxicity in aquatic sediments. Environ Toxicol Chem, **12**(8): 1441–1453.

Alvarez C, Garcia JE, & Lopez MA (1989) Exocrine pancreas in rabbits fed a copper-deficient diet: Structural and functional studies. Pancreas, **4**(5): 543–549.

Amacher DE & Paillet SC (1980) Induction of trifluorothymidine-resistant mutants by metal ions in L5178Y/TK +/- cells. Mutat Res, **78**: 279–288.

Amundsen CE, Hanssen JE, Semb A, & Steinnes E (1992) Long-range atmospheric transport of trace elements to Southern Norway. Atmos Environ, **26A**: 1309–1324.

Anadu DI, Chapman GA, Curtis LR, & Tubb RA. (1989) Effect of zinc exposure on subsequent acute tolerance to heavy metals in rainbow trout. Bull Environ Contam Toxicol **43**: 329–336.

Anderson PR & Christensen TH (1988) Distribution coefficients of Cd, Co, Ni, and Zn in soils. J Soil Sci, **39**: 15–22.

Anderson MA, Morel FMM, & Guillard RRL (1978) Growth limitation of a coastal diatom by low zinc ion activity. Nature, **276**: 70–71.

Anderson RL, Walbridge CT, & Fiandt JT (1980) Survival and growth of *Tanytarsus dissimilis* (Chironomidae) exposed to copper, cadmium, zinc, and lead. Arch Environ Contam Toxicol, 9: 329–335.

Anderson GH, Peterson RD, & Beaton GH (1982) Estimating nutrient deficiencies in a population from dietary records: The use of probability analyses. Nutr Res, 2: 409–415.

Anderson JR, Aggett FJ, Buseck PR, Germani MS, & Shattuck TW (1988) Chemistry of individual aerosol particles from Chandler, Arizona, an arid urban environment. Environ Sci Technol, 22: 811–818.

Andrews SM, Johnson MS, & Cooke JA (1989) Distribution of trace element pollutants in a contaminated grassland ecosystem established on metalliferous fluorspar tailings. 2: Zinc. Environ Pollut, 59: 241–252.

Ansari MS, Miller WJ, Lassiter JW, Neathhery MW, & Gentry RP (1975) Effects of high but nontoxic dietary zinc on zinc metabolism and adaptations in rats (39072). Proc Soc Exp Biol Med, 150: 534–536.

Ansari MS, Miller WJ, Neathery MW, Lassiter JW, Gentry RP, & Kincaid RL (1976) Zinc metabolism and homeostasis in rats fed a wide range of high dietary zinc levels (39358). Proc Soc Exp Biol Med, 152: 192–194.

Antonovics J, Bradshaw AD, & Turner RG (1971) Heavy metal tolerance in plants. Adv Ecol Res, 7: 1–85.

Apgar J (1970) Effect of zinc deficiency on maintenance of pregnancy in the rat. J Nutr, 100: 470–476.

Apgar J (1973) Effect of zinc repletion late in gestation on parturition in the zinc-deficient rat. J Nutr, 103: 973–981.

Apgar J & Fitzgerald JA (1985) Effects on the ewe and lamb of low zinc intake throughout pregnancy. J Anim Sci, 60: 1530-1538.

Apte SC & Batley GE (1995) Trace metal speciation of labile chemical species in natural waters and sediments: non-electrochemical approaches. In: Tessier A & Turner DA eds. Metal speciation and bioavailability in aquatic systems. Chichester, John Wiley & Sons, pp 259–306.

Apte SC & Gunn AM (1987) Rapid determination of copper, nickel, lead and cadmium in small samples of estuarine and coastal waters by liquid/liquid extraction and electrothermal atomic absorption spectrometry. Anal Chim Acta, 193: 147–156.

Apte SC, Batley GE, Szymczak R, Rendell PS, Lee R, & Waite TD (1998) Baseline trace metal concentrations in New South Wales coastal waters. Marine Freshwater Res, 49: 203–214.

Arakawa T, Tamura T, Igarashi Y, Suzuki H, & Sandstead HH (1976) Zinc deficiency in two infants during parenteral alimentation. Am J Clin Nutr, 29: 197–204.

Arambasic MB, Bjelic S, & Subakov G (1995) Acute toxicity of heavy metals (copper, lead, zinc), phenol and sodium on *Allium cepa* L., *Lepidium sativum* L. and *Daphnia magna* st.: comparative investigations and the practical applications. Water Res, **29**(2): 497–503.

Arnaud J & Favier A (1992) Determination of ultrafiltrable zinc in human milk by electrothermal atomic absorption spectrometry. Analyst, **117**: 1593–1598.

Aslam N & McArdle HJ (1992) Mechanism of zinc uptake by microvilli isolated from human term placenta. J Cell Physiol, **151**: 533–538.

ATSDR (1994) Toxicological profile for zinc. Atlanta, Agency for Toxic Substances and Disease Registry. US Public Health Service, pp1–230.

Aughey E, Grant L, Furman BL, & Dryden WF (1977) The effects of oral zinc supplementation in the mouse. J Comp Pathol, **87**: 1–14.

Aulerich RJ, Bursian SJ, Poppenga RH, Braselton WE, & Mullaney TP (1991) Toleration of high concentrations of dietary zinc by mink. J Vet Diagn Invest, **3**: 232–237.

Aurand K & Hoffmeister H (1980) eds. [Ad hoc field studies on the heavy metal burden of the population in the area of Lker in March 1980.] Berlin, Dietrich Reimer Verlag, pp1–72 (in German).

Babcock AK, Henkin RI, Aamodt RL, Foster DM, & Berman M (1982) Effects of oral zinc loading on zinc metabolism in humans II: *In vivo* kinetics. Metabolism, **31**(4): 335–347.

Baer MT & King JC (1984) Tissue zinc levels and zinc excretion during experimental zinc depletion in young men. Am J Clin Nutr, **39**: 556–570.

Baer MT, King JC, Tamura T, Margen S, Bradfield RB, Weston WL, & Daugherty NA (1985) Nitrogen utilisation, enzyme activity, glucose intolerance and leukocyte chemotaxis in human experimental zinc depletion. Am J, **41**: 1220–1235.

Baes CF & Sharp RD (1983) A proposal for estimation of soil leaching and leaching constants for use in assessment models. J Environ Qual, **12**: 17–28.

Baes C, Sharp RD, Sjoreen AL, & Shor RW (1984) A review and analysis of parameters for assessing transport of environmentally released radionuclides through agriculture. Washington DC, US Department of Energy, pp 53–64 (ORNL-5786).

Baeyens W & Dedeurwaerder H (1991) Particulate trace metals above the southern bight of the North Sea – I. Analytical procedures and average aerosol concentrations. Atmos Environ, **25A**: 293–304.

Baeyens W, Dehairs F, & Dedeurwaerder H (1990) Wet and dry deposition fluxes above the North Sea. Atmos Environ, **24A**: 1693–1703.

Baird DJ, Barber I, Bradley M, Soares AMVM, & Calow P (1991) A comparative study of genotype sensitivity to acute toxic stress using clones of *Daphnia magna* Straus. Ecotoxicol Environ Saf, **21**: 257–265.

Ballester OF & Prasad AS (1983) Anergy, zinc deficiency, and decreased nucleoside phosphorylase activity in patients with sickle cell anemia. Ann Int Med, **98**: 180–182.

Balsberg Påhlsson A-M (1989) Toxicity of heavy metals (Zn, Cu, Cd, Pb) to vascular plants. Water Air Soil Pollut, **47**: 287–319.

Barney GH, Orgebin-Crist MC, & Macapinlac MP (1968) Genesis of eosophageal parakeratosis and histological changes in the testes of the zinc-deficient rat and their reversal by zinc repletion. Br J Nutr, **95**: 526–534.

Barr DH & Harris JW (1973) Growth of the P388 leukemia as an ascites tumor in zinc-deficient mice. Proc Soc Exp Biol Med, **144**: 284–287.

Bartoshuk LM (1978) The psychophysics of taste. Am J Clin Nutr, **31**: 1068–1077.

Batchelor RP, Fehnel JW, Thompson RM, & Drinker KR (1926) A clinical and laboratory investigation of the effect of metallic zinc, of zinc oxide, and of zinc sulphide upon health of workmen. J Ind Hyg, **8**: 322–363.

Bates CJ, Evans PH, Dardenne M, Prentice A, Lunn PG, Northrop-Clewes CA, Hoare S, Cole TJ, Horan SJ, & Longman SC (1993) A trial of zinc supplementation in young rural Gambian children. Br J Nutr, **69**(1): 243–255.

Batley GE (1989a) Collection, preparation and storage of samples for speciation analysis. In: Batley, GE ed.Trace element speciation: analytical methods and problems. Boca Raton FL, CRC Press, pp 1–24.

Batley GE (1989b) Physicochemical separation methods for trace element speciation in aquatic samples. In: Trace element speciation: analytical methods and problems. Boca Raton FL, CRC Press, pp 44–76.

Batley GE (1995) Heavy metals and tributyltin in Australian and estuarine waters: The state of the marine environment report for Australia. In: Zann LP & Sutton DC ed. Technical Amer, 2, pp 63–72.

Batley GE & Farrar YJ (1978) Irradiation techniques for the release of bound heavy metals in natural waters and blood. Anal Chim Acta, **99**: 283–292.

Bauchinger M, Schmid E, Einbrodt HJ, & Dresp J (1976) Chromosome aberrations in lymphocytes after occupational exposure to lead and cadmium. Mutat Res, **40**: 57–62.

Baudouin MF & Scoppa P (1974) Acute toxicity of various metals to freshwater zooplankton. Bull Environ Contam Toxicol, **12**(6): 745–751.

Beavington F & Cawse PA (1979) The deposition of trace elements and major nutrients in dust and rainwater in northern Nigeria. Sci Total Environ, **13**: 263–274.

Beckett PHT & Davis RD (1978) The additivity of the toxic effects of Cu, Ni and Zn in young barley. New Phytol, **81**: 155–173.

Becking GC & Morrison AB (1970) Hepatic drug metabolism in zinc deficient rats. Biochem Pharmacol, **19**: 895–902.

Beer WH, Johnson RF, Guentzel MN, Lozano J, Henderson GI, & Schenker S (1992) Human placental transfer of zinc: normal characteristics and role of ethanol. Alcohol Clin Exp Res, **16**: 98–105.

Beisel WR (1976) Trace elements in infectious processes. Med Clinics North Am, **60**(4): 831–847.

Belanger SE & Cherry DS (1990) Interacting effects of pH acclimation, pH, and heavy metals on acute and chronic toxicity to *Ceriodaphnia dubia* (Cladocera). J Crustac Biol, **10**(2): 225–235.

Belanger SE, Farris JL, Cherry DS, & Cairns J (1986) Growth of Asiatic clams (*Corbicula* sp.) during and after long-term zinc exposure in field-located and laboratory artificial streams. Arch Environ Contam Toxicol, **15**: 427–434.

Beliles RP (1994) Zinc, Zn. In: Clayton GD & Clayton FE ed. Patty's industrial hygiene and toxicology, 4th ed. Part C Toxicology. New York, John Wiley & Sons Inc, pp 2332–2342.

Bengtsson B-E (1974a) Effect of zinc on growth of the minnow *Phoxinus phoxinus*. Oikos, **25**: 370–373.

Bengtsson B-E (1974b) The effect of zinc on the ability of the minnow, *Phoxinus phoxinus* to compensate for torque in a rotating water-current. Bull Environ Contam Toxicol, **12**(6): 654–658.

Bengtsson BE & Bergström B (1987) A flowthrough fecundity test with *Nitocra spinipes* (Harpacticoidea crustacea) for aquatic toxicity. Ecotoxicol Environ Saf, **14**: 260–268.

Benoit DA & Holcombe GW (1978) Toxic effects of zinc on fathead minnows *Pimephales promelas* in soft water. J Fish Biol, **13**: 701–708.

Bentley PJ & Grubb BR (1991) Experimental dietary hyperzinceria tissue disposition of excess zinc in rabbits. Trace Elem Med, **8**(4): 202–207.

Berg JM (1990) Zinc finger domains: hypotheses and current knowledge. Ann Rev Biophys Chem, **19**: 405–421.

Berg JM & Shi Y (1996) The galvanization of biology: a growing appreciation for the role of zinc. Science, **271**: 1081-1085.

Berglind R & Dave G (1984) Acute toxicity of chromate, DDT, PCP, TPBS, and zinc to *Daphnia magna* cultured in hard and soft water. Bull Environ Contam Toxicol, **33**: 63–68.

Bergman HL & Dorward-King EJ (1996) Reassessment of metals criteria for aquatic life protection: priorities for research and implementation. Pensacola, FL, SETAC Press.

Berrow ML & Reaves GA (1984) Background levels of trace elements in soils. In: Proceedings of the International Conference of Environmental of Contamination. Edinburgh, CEP Consultants Ltd, pp 333–340.

Bertholf RL (1988) Zinc. In: Seiler HG & Sigel H ed. Handbook on toxicity of inorganic compounds. New York, Marcel Dekker Inc, pp 787–800.

Bertine KK & Goldberg ED (1971) Fossil fuel combustion and the major sedimentary cycle. Science, **173**: 233–235.

Bervoets L, Blust R, & Verheyen R (1996) Uptake of zinc by midge larvae *Chironomus riparius* at different salinities: role of speciation, acclimation, and calcium. Environ Toxicol Chem, **15**: 1423–1428.

Bettger CM & O'Dell BL (1981) A critical physiological role of zinc in the structure and function of biomembranes. Life Sci, **28**: 1425–1438.

Bettger WJ, Fish TJ, & O'Dell BL (1978) Effects of copper and zinc status of rats on erythrocyte stability and superoxide dismutase activity (40188). Proc Soc Exp Biol Med, **158**: 279–282.

Beyer WN (1986) A reexamination of biomagnification of metals in terrestrial food chains. (Letter). Environ Toxicol Chem, **5**: 863–864.

Beyer WN (1988) Damage to the forest ecosystem on the blue mountain from zinc smelting. In: Proceedings of the 22nd Annual Conference on Trace Substances and Environmental Health, pp 249–262.

Beyer WN & Anderson A (1985) Toxicity to woodlice of zinc and lead oxides added to soil litter. Ambio, **14**(3): 173–174.

Beyer WN & Cromartie EJ (1987) A survey of Pb, Cu, Zn, Cd, Cr, As, and Se in earthworms and soil from diverse sites. Environ Monit Assess, **8**: 27–36.

Beyer WN, Miller GW, & Cromartie EJ (1984) Contamination of the O2 soil horizon by zinc smelting and its effect on woodlouse survival. J Environ Qual, **13**(2): 247–251.

Beyer WN, Pattee OH, Sileo L, Hoffmann DJ, & Mulhern BM (1985) Metal contamination in wild-life living near two zinc smelters. Environ Pollut, **38A**: 63-86.

Bhathena SJ, Recant L, Voyles NR, Timmers KI, Reiser S, Smith JC jr, & Powell AS (1986) Decreased plasma enkephalins in copper deficiency in man. Am J Clin Nutr, **43**: 42-46.

Biesinger KE & Christensen GM (1972) Effects of various metals on survival, growth, reproduction, and metabolism of *Daphnia magna*. J Fish Res Board Can, **29**: 1691–1700.

Biesinger KE, Christensen GM, & Fiandt JT (1986) Effects of metal salt mixtures on *Daphnia magna* reproduction. Ecotoxicol Environ Saf, **11**: 9–14.

Bires J, Dianovsky J, Bartko P, & Juhasova Z (1995) Effects of enzymes and the genetic apparatus of sheep after administration of samples from industrial emissions. Biometals, **8**(1): 53–58.

Black MR, Medeiros DM, Brunett E, & Welke R (1988) Zinc supplements and serum lipids in young adult white male. Am J Clin Nutr, **47**: 970–975.

Blamberg DL, Blackwood UB, Supplee WC, & Combs GF (1960) Effect of zinc deficiency in hens on hatchability and embryonic development. Proc Soc Exp Biol Med, **104**: 217–220.

Blanc P, Wong H, Bernstein MS, & Boushey HA (1991) An experimental human model of metal fume fever. Ann Intern Med, **114**: 930–936.

Blanc PD, Boushey HA, Wong H, Wintermeyer SF, & Bernstein MS (1993) Cytokines in metal fume fever. Am Rev Respir Dis, **147**: 134–138.

Bleavins MR, Aulerich RJ, Hochstein JR, Hornshaw TC, & Napolitano AC (1983) Effects of excessive dietary zinc on the intrauterine and postnatal development of mink. J Nutr, **113**: 2360–2367.

Bloomfield C & McGrath SP (1982) A comparison of the extractabilities of Zn, Cu, Ni and Cr from sewage sludges prepared by treating raw sewage with the metal salts before or after anaerobic digestion. Environ Pollut Series B, **3**: 193–198.

Bodar CWM, Zee AVD, Voogt PA, Wynne H, & Zander DI (1989) Toxicity of heavy metals to early life stages of *Daphnia magna*. Ecotoxicol Environ Saf, **17**: 333–338.

Bogden JD, Oleske JM, Munves EM, Lavenhar MA, Bruening KS, Kemp FW, Holding KJ, Denny TN, & Louria DB (1987) Zinc and immunocompetence in the elderly: baseline data on zinc nutriture and immunity in unsupplemented subjects. Am J Clin Nutr, **46**: 101–109.

Bogden KS, Holding KJ, Denny TN, Guaino MA, Krieger LM, & Holland BK (1988) Zinc and immunocompetence in elderly people: effects on zinc supplementation for 3 months. Am J Clin Nutr, **48**: 655–663.

Bordin G, McCourt J, & Rodríguez A (1992) Trace metals in the marine bivalve *Macoma Balthica* in the Westerschelde Estuary (The Netherlands). Part 1: analysis of total copper, cadmium, zinc and iron concentrations – locational and seasonal variations. Sci Total Environ, **127**: 255–280.

Borroni A, Mazza B, & Nano G (1986) Environmental study of a plating room. Met Finish, pp 41–45.

Bourg ACM & Darmendrail D (1992) Effect of dissolved organic matter and pH on the migration of zinc through river bank sediments. Environ Technol, **13**: 695–700.

Boutron CF, Goerlach U, Candelone J-P, Bolshov MA, & Delmas RJ (1991) Decrease in anthropogenic lead, cadmium and zinc in Greenland snows since the late 1960s. Nature, **353**: 153–156.

Boutron CF, Candelone J-P, & Hong S (1995) Greenland snow and ice cores unique archives of large-scale pollution of the trophosphere of the northern hemisphere by lead and other heavy metals. Science of the Total Environment, **160-161**: 233-241.

Bowen HJM (1979) Environmental chemistry of the elements. Academic Press, New York.

Bradfield RB & Hambidge KM (1980) Problems with hair zinc as an indicator of body zinc status. Lancet, **8164**: 363.

Bradley RW & Sprague JB (1985) The influence of pH, water hardness, and alkalinity on the acute lethality of zinc to rainbow trout (*Salmo gairdneri*). Can J Fish Aquat Sci, **42**: 731–736.

Bremner I & Beattie JH (1990) Metallothionein and the trace minerals. Annu Rev Nutr, **10**: 63–83.

Bremner I & Beattie JH (1995) Copper and zinc metabolism in health and disease: speciation and interations. Proc Nutr Soc, **54**: 489–499.

Bremner I, Young BW, & Mills CF (1976) Protective effect of zinc supplementation against copper toxicosis in sheep. Br J Nutr, **36**: 551–561.

Brennan RF, Armour JD, & Reuter DJ (1993) Diagnosis of zinc deficiency. In: Robson AD ed. Zinc in soils and plants. London, Kluwer Academic Publishers, pp 167–181.

Breskin MW, Worthington-Roberts BS, Knopp RH, Brown Z, Plovie B, Mottet NK, & Mills JL (1983) First trimester serum zinc concentration in human pregnancy1-3. Am J Clin Nutr, **38**: 943–953.

Bresnitz EA, Roseman J, Becker D, & Gracely E (1992) Morbidity among municipal waste incinerator workers. Am J Ind Med, **22**: 363–378.

Brewer GJ, Gretchen MH, Prasad AS, Cossack ZT, & Rabbani P (1983) Oral zinc therapy for Wilson's disease. Ann Int Med, **99**: 314–320.

Brewer GJ, Yuzbasiyan-Gurkan V, & Johnson V (1991) Treatment of Wilson's disease with zinc IX: Response of serum lipids. J Lab Clin Med, **118**: 466–470.

Brix H & Lyngby JE (1982) The distribution of cadmium, copper, lead, and zinc in eelgrass (*Zostera marina* L.). Sci Total Environ, **24**: 51–63.

Brocks A, Reid H, & Glazer G (1977) Acute intravenous zinc poisoning. Br Med J, **1**: 1390–1391.

Broderius SJ & Smith LL (1979) Lethal and sublethal effects of binary mixtures of cyanide and hexavalent chromium, zinc, or ammonia to the fathead minnow

(*Pimephales promelas*) and rainbow trout (*Salmo gairdneri*). J Fish Res Board Can, **36**: 164–172.

Broun ER, Greist A, Tricot G, & Hoffman R (1990) Excessive zinc ingestion. A reversible cause of sideroblastic anemia and bone marrow depression. J Am Med Assoc, **264**(11): 1441–1443.

Brown BT & Rattigan BM (1979) Toxicity of soluble copper and other metal ions to *Elodea canadensis*. Environ Pollut, **20**: 303–314.

Brown JJL (1988) Zinc fume fever. Br J Radiol, **61**: 327–329.

Brown MA, Thom JV, Otth GL, Cova P, & Juarez J (1964) Food poisoning involving zinc contamination. Arch Environ Health, **8**: 657–660.

Brown RFR, Marrs TC, Rice P, & Masek LC (1990) The histopathology of rat lung following exposure to zinc oxide/hexachloroethane smoke or instillation with zinc chloride followed by treatment with 70% oxygen. Environ Health Perspect, **85**: 81–87.

Bruère AN, Cooper BS, & Dillon EA (1990) Zinc. Massey University, Palmerston North, New Zealand, Veterinary Continuing Education, pp 75–81 (Veterinary Clinical Toxicology Publication No. 127).

Bruland KW (1993) Trace metals in oceanic environments – interactions between ocean chemistry and biology. In: Allan RJ & Nriagu JO eds. Heavy metals in the environment. Edinburgh, CEP Consultants.

Bruland KW, Knauer GA, & Martin JH (1978) Zinc in north-east Pacific water. Nature, **271**: 741–743.

Brümmer GW (1986) Heavy metal species, mobility and availability in soils. In: Bernhard M, Brinckman FE, & Sadler PJ eds. The importance of chemical "speciation" in environmental processes. Dahlem, 1986. Heidelberg, Springer, pp 169–192.

Brummerstedt E, Basse A, Flagstad T, & Andersen E (1977) Lethal trait A46 in cattle. Am J Pathol, **87**: 725–738.

Bruni B, Barolo P, Gamba S, Grassi G, & Blatto A (1986) Case of generalized allergy due to zinc and protamine in insulin preparation. Diabetes Care, **9**: 552.

Buchauer MJ (1973) Contamination of soil and vegetation near a zinc smelter by zinc, cadmium, copper, and lead. Environ Sci Technol, **7**: 131–135.

Budavari S ed. (1989) The Merck Index. Rahway, NJ, Merck & Co Inc, pp 1597–1598.

Buell SJ, Fosmire GJ, Ollerich DA, & Sandstead HH (1977) Effects of postnatal cerebellar and hippocampal development in the rat. Exp Neurol, **55**: 199–210.

Buhl KJ & Hamilton SJ (1990) Comparative toxicity of inorganic contaminants released by placer mining to early life stages of salmonids. Ecotoxicol Environ Saf, **20**: 325–342.

Bunker VW, Hinks LJ, Lawson MS, & Clayton BE (1984) Assessment of zinc and copper status of healthy elderly people using metabolic balance studies and measurement of leukocyte concentration. Am J Clin Nutr, **40**: 1096–1102.

Bunzl K & Schimmack W (1989) Associations between the fluctuations of the distribution coefficients of Cs, Zn, Sr, Co, Cd, Ce, Ru, Tc and I in the upper two horizons of a podzol forest soil. Chemosphere, **18**(11/12): 2109–2120.

Burke DM, DeMicco FJ, Taper LJ, & Ritchey SJ (1981) Copper and zinc utilization in elderly adults. J Gerontol, **36**(5): 558–563.

Burton DT & Fisher DJ (1990) Acute toxicity of cadmium, copper, zinc, ammonia, 3,3'-dichlorobenzidine, 2,6-dichloro-4-nitroaniline, methylene chloride, and 2,4,6-trichloro-phenol to juvenile grass shrimp and killifish. Bull Environ Contam Toxicol, **44**: 776–783.

Buzina R, Jusic M, Sapunar J, & Milanovic N (1980). Zinc nutrition and taste acuity in school children with impaired growth. Am J Clin Nutr, **33**(11): 2262–2267.

Cakmak J & Marschner H (1988) Increase in membrane permeability and exudation in roots of zinc deficient plants. J Plant Physiol, **132**: 356–361.

Calabrese A & Nelson DA (1974) Inhibition of embryonic development of the hard clam, *Mercenaria mercenaria*, by heavy metals. Bull Environ Contam Toxicol, **11**(1): 92–97.

Calabrese A, Collier RS, Nelson DA, & MacInnes JR (1973) The toxicity of heavy metals to embryos of the American oyster *Crassostrea virginica*. Mar Biol, **18**: 162–166.

Calabrese A, MacInnes JR, Nelson DA, & Miller JE (1977) Survival and growth of bivalve larvae under heavy-metal stress. Mar Biol, **41**: 179–184.

Callahan MA, Slimak NW, May IP, Fowler CF, Freed JR, Jennings P, Durfee RL, Whitmore FC, Maestri B, Mabey WR, Holt BR, & Gould C (1979) Water-related environmental fate of 129 priority pollutants. Vol. 1: Introduction and technical background, metals and inorganics, pesticides and PCBs. Springfield, VA, Versar, Inc. (EPA-440/4-79-029A).

Campbell GC (1995) Interactions between trace metals and aquatic organisms: a critique of the free-ion activity model. In: Tessier A & Turner DR eds. Metal speciation and bioavailability in aquatic systems. New York, John Wiley and Sons.

Campbell JK & Mills CF (1974) Effects of dietary cadmium and zinc on rats maintained on diets low in copper. Proc Nutr Soc, **33**: 15A–17A.

Campbell JK & Mills CF (1979) The toxicity of zinc to pregnant sheep. Environ Res, **20**: 1–13.

Campbell-Brown M, Ward RJ, Haines AP, North WRS, Abraham R, & McFadyen IR (1985) Zinc and copper in Asian pregnacies – is there evidence for a nutritional deficiency? Br J Obstet Gynaecol, **92**: 875–885.

Canada/EU (1996) Draft report: Technical Workshop on Biodegradation/Persistence and Bioaccumulation/Biomagnification of Metals and Metal Compounds. Held under the auspices of the Canada/European Metals and Minerals Working Group, Brussels, Belgium, December 11–13, 1995.

Canton MC & Cremin MF (1990) The effect of dietary zinc depletion and repletion on rats: Zn concentration in various tissues and activity of pancreatic gamma-glutamyl hydrolase (EC 3.4.22.12) as indices on Zn status. Br J Nutr, 64: 201–209.

Carpenter JM & Ray JH (1969) The effect of zinc chloride on the production of mutations in *Drosophila melanogaster*. Am Zool, 9: 1121.

Carter JW & Cameron IL (1973) Toxicity bioassay of heavy metals in water using *Tetrahymena pyriformis*. Water Res, 7: 951–961.

Carter JP, Grivetti LE, Davis JT, Nasiff S, Mansour A, Mousa WA, Atta AE, Patwardhan VN, Abdel Moneim M, Abdou IA, & Darby WJ (1969) Growth and sexual development of adolescent Egyptian village boys. Effect of zinc, iron, and placebo supplementation. Am J Clin Nutr, 22(1): 59–78.

Castillo-Duran C, Heresi G, Fisberg M, & Uauy R (1987) Controlled trial of zinc supplementation during recovery from malnutrition: effects on growth and immune function1-3. Am J Clin Nutr, 45: 602-608.

Castillo-Duran C, Rodriguez A, Venegas G, Alvarez P, & Icaza G (1995) Zinc supplementation and growth of infants born small for gestational age. J Pediatr, 127(2): 206–211.

Casto BC, Meyers J, & DiPaolo JA (1979) Enhancement of viral transformation for evaluation of the carcinogenic or mutagenic potential of inorganic metal salts. Cancer Res, 39: 193–198.

Cavan KR, Gibson RS, Grazioso CF, Isalgue AM, Ruz M, & Solomons NW (1993) Growth and body composition of periurban Guatemalan children in relation to zinc status: a cross-sectional study. Am J Clin Nutr, 57: 334–343.

CCRX (1991) [Measurement of radioactivity in xenobiotic substances in the biological environment in the Netherlands 1990.] Bilthoven, Coordinating Commission for the measurement of radioactivity in xenobiotic substances, pp 49–51 (Ref A-1) (in Dutch).

CCRX (1994) [Measurement in the environment in the Netherlands 1992.] Bilthoven, Coordinating Commission for measurements in the environment, Ref.A2., Heavy Metals, pp 65–68 (in Dutch).

Centeno JA, Pestaner JP, Nieves S, Ramos M, Mullick FG, & Kaler SG (1996) The assessment of trace element and toxic metal levels in human placental tissues. In: Collery P, Corbella J, Domingo JL, Etienne JC, & Uobel JM eds. Metal Ions Biol Med, 4: 522–524.

Chandra RK (1984) Excessive intake of zinc impairs immune responses. J Am Med Assoc, 252(11): 1443–1446.

Chaney RL (1993) Zinc phytotoxicity. In: Robson AD ed. Zinc in soil and plants, Proceedings of the International Symposium on "Zinc in soils and plants", held at The University of Western Australia, 27–28 September, 1993. Dordrecht, Kluwer, pp 135–150.

Chaney RL, Sterrett SB, & Mielke HW (1984) The potential for heavy metal exposure from urban gardens and soils. In: Preer JR ed. Proceedings of the Symposium on Heavy Metals in Urban Gardens. Washington DC, College of Life Sciences, University of the District of Columbia, pp 37–84.

Chaney RL, Bruins RJF, Baker DE, Korcak RF, Smith JE, & Cole D (1987) Transfer of sludge-applied trace elements to the food chain. In: Page AL, Logan TJ, & Ryan JA eds. Land application of sludge – food chain implications. Chelsea, Lewis, pp 67–99.

Chaney RL, Beyer WN, Gifford CH, & Sileo L (1988) Effects of zinc smelter emissions on farms and gardens at Palmerton, PA. In: Proceedings of the 22nd Annual Conference on Trace Substances and Environmental Health, pp 263–280.

Chang CH, Mann DE, & Gauteri RF (1977) Teratogenicity of zinc chloride, 1,10-phenanthroline in mice. J Pharm Sci, 66: 1755–1758.

Chang AC, Hinesly TD, Bates TE, Doner HE, Dowdy RH, & Ryan JA (1987) Effects of long-term sludge application on accumulation of trace elements by crops. In: Page AL, Logan TJ, & Ryan eds. Land application of sludge – food chain implications. Chelsea, Lewis, pp 53–66.

Chapman GA (1978a) Effects of continuous zinc exposure on sockeye salmon during adult-to-smolt freshwater residency. Trans Am Fish Soc, 107(6): 828–836.

Chapman GA (1978b) Toxicities of cadmium, copper, and zinc to four juvenile stages of Chinook salmon and steelhead. Trans Am Fish Soc, 107(6): 841–847.

Chapman G & Dunlop S (1981) Detoxification of zinc and cadmium by the freshwater protozoan *Tetrahymena pyriformis*: I .The effect of water hardness. Environ Res, 26(1): 81–86.

Chapman GA & Stevens DG (1978) Acutely lethal levels of cadmium, copper, and zinc to adult male Coho salmon and steelhead. Trans Am Fish Soc, 107(6): 837–840.

Chapman PM, Allen HE, Godtfredsen K, & Z'Graggen MN (1996) Evaluation of BCFs as measures for classifying and regulating metals. Environ Sci Technol, 30: 448A–452A.

Cherian L & Gupta VK (1992) Spectrophotometric determination of zinc using 4-carboxyphenyldiazoaminoazobenzene and its application in complex materials. Chem Anal, 37: 69–72.

Cherian MG (1994) The significance of the nuclear and cytoplasmic localization of metallothionein in human liver and tumour cells. Environ Health Perspect, 102(Suppl 3): 131–135.

Cherry FF, Sandstead HH, Rojas P, Johnson LK, Batson HK, & Wang XB (1989) Adolescent pregnancy: associations among body weight, zinc nutrition, and pregnancy outcome. Am J Clin Nutr, **50**: 945–954.

Chester R & Bradshaw GF (1991) Source control on the distribution of particulate trace metals in the North sea atmosphere. Marine Pollut Bull, **22**: 30–36.

Chesters JK & Quarterman J (1970) Effects of zinc deficiency on food intake and feeding patterns of rats. Br J Nutr, **24**: 1061–1069.

Chesters JK & Will M (1978) The assessment of zinc status of an animal from the uptake of ^{65}Zn by the cells of whole blood *in vitro*. Br J Nutr, **38**: 297–306.

Cho CH, Chen SM, Ogle CW, & Young TK (1989) Effects of zinc and cholesterol/choleate on serum lipoproteins and the liver in rats. Life Sci, **44**: 1929–1936.

Clark NJ, O'Brien D, & Edmonds MA (1992) Health hazard report. In: US Department of Health and Human Services; Public Health Control, Centers for Disease Control, National Institute for Occupational Safety and Health ed. Health hazard evaluation. Springfield, US Department of Commerce, pp 1–22 (HETA 91-092-2190).

Clegg MS, Keen CL, & Hurley LS (1989) Biochemical pathologies of zinc deficiency. In: Mills B ed. Zinc in human biology. London, Springer, pp 129–145.

Clements WH & Kiffney PM (1995) The influence of elevation on benthic community responses to heavy metal in Rocky Mountain streams. Can J Fish Aqua Sci, **52**(9): 1966–1977.

Cleven RFMJ, Janus JA, Annema JA, & Slooff W eds. (1993) Integrated criteria document zinc. Bilthoven, National Institute of Public Health and Environmental Protection, pp1–180 (Report No. 710401028).

Codina JC, Pérez-García A, Romero P, & De Vicente A (1993) A comparison of microbial bioassays for the detection of metal toxicity. Arch Environ Contam Toxicol, **25**: 250–254.

Cohen HJ & Powers BJ (1994) A study of respirable versus nonrespirable copper and zinc oxide exposures at a nonferrous foundry. Am Ind Hyg Assoc J, **55**: 1047–1050.

Cole RJ, Frederick RE, Healy RP, & Rolan RG (1984) Preliminary findings of the priority pollutant monitoring project of the nationwide urban runoff program. J Water Pollut Control Fed, **56**: 898–908.

Coleman JE (1992) Zinc proteins: enzymes, storage proteins, transcription factors, and replication proteins. Annu Rev Biochem, **61**: 897–946.

Colin MA, Taper LJ, & Ritchey SJ (1983) Effect of dietary zinc and protein levels on the utilization of zinc and copper by adult females1. J Nutr, **113**: 1480–1488.

Colin JL, Jaffrezo JL, & Gros JM (1990) Solubility of major species in precipitation: factors of variation. Atmos Environ, **24A**(3): 537–544.

Collyard SA, Ankley GT, Hoke RA, & Goldenstein T (1994) Influence of age on the relative sensitivity of *Hyalella azteca* to diazinon, alkylphenol ethoxylates, copper, cadmium, and zinc. Arch Environ Contam Toxicol, **26**: 110–113.

Colpaert J & Van Assche JA (1992) Zinc toxicity in the ectomycorrhizal *Pinus sylvestris* L. Plant Soil, **143**: 201–211

Colwell FS, Hornor SG, & Cherry DS (1989) Evidence of structural and functional adaptation in epilithon exposed to zinc. *Hydrobiologia*, **171**: 79–90.

Conner MM, Flood WH, Rogers AE, & Amdur MO (1988) Lung injury in guinea pigs caused by multiple exposures to ultrafine zinc oxide: changes in pulmonary lavage fluid. J Toxicol Environ Health, **25**(1): 57–69.

Cook-Mills JM & Fraker PJ (1993) Functional capacity of the residual lymphocytes in zinc deficient mice. Br J Nutr, **69**: 835–848.

Copa-Rodríguez FJ & Basadre-Pampín MI (1994) Determination of iron, copper and zinc in tinned mussels by inductively coupled plasma atomic emission spectrometry (ICP-AES). Fresenius J Anal Chem, **348**: 390–395.

Coppen DE & Davies NT (1987) Studies on the effect of dietary zinc dose on [65]Zn absorption in vivo and the effects of zinc status on [65]Zn absorption and body loss in young rats. Br J Nutr, **57**: 35–44.

Coppoolse J, van Bentum F, Schwartz M, Annema JA, & Quarles van Ufford C (1993) Heavy metals in surface waters, sources and measures. RIVM Report 7733003001. Bilthhoven (ISBN 903690 3424) (in Dutch).

Cornejo N, Afailal F, García F, & Palacios M (1994) Determination of zinc in ammoniacal ore leaching solutions by X-ray fluorescence spectrometry using a radioactive source. Fresenius J Anal Chem, **350**: 122–126.

Cossack ZT (1986) Somatomedin-C and zinc status in rats as affected by Zn, protein and food intake. BrJ Nutr, **56**: 163–169.

Cottenie A, Kang B, Kiekens L, & Sajjapongse A (1981) Micronutrient status. In: Greenland DJ ed. Characterization of soils. Oxford, Clarendon Press, pp 149–163.

Couillard Y, Ross P, & Pinel-Alloul B (1989) Acute toxicity of six metals to the rotifer *Brachionus calyciflorus*, with comparisons to other freshwater organisms. Toxic Assess, **4**: 451–462.

Courpron C (1967) Détermination des constantes de stabilité des complexes organo-métalliques des sols. Ann Agron, **18**(6): 623–638.

Cousins RJ (1986) Toward a molecular understanding of zinc metabolism. Clin Physiol Biochem, **4**: 20–30.

Cousins RJ (1989) Theoretical and practical aspects of zinc uptake and absorption. Adv Exp Med Biol, **249**: 3–12.

Cox DH & Harris DL (1960) Effect of excess dietary zinc on iron and copper in the rat. J Nutr, **70**: 514–520.

Cox DH, Schlicker SA, & Chu RC (1969) Excess dietary zinc for the maternal rat, and zinc, iron, copper, calcium, and magnesium content and enzyme activity in maternal and fetal tissues. J Nutr, **98**: 459–466.

Coyle P, Zalewski PD, Philcox JC, Forbes IJ, Ward AD, Lincoln SF, & Rofe AM (1994) Measurement of zinc in hepatocytes by using a fluorescent probe, Zinquin: relationship to metallothionein and intracellular zinc. Biochem J, **303**: 781–786.

Crofton RW, Gvozdanovic D, Gvozdanovic S, Khin CC, Brunt PW, Mowat N, & Aggett PJ (1989) Inorganic zinc and the intestinal absorption of ferrous iron. Am J Clin Nutr, **50**: 141–144.

Cunningham BC, Bass S, Fuh G, & Wells JA (1990) Zinc mediation of the binding of human growth hormone to the human prolactin receptor. Science, **250**: 1709–1712.

Czerwinski AW, Clark M, Serafetinides EA, Perrier C, & Huber W (1974) Safety and efficacy of zinc sulfate in geriatric patients. Clin Pharmacol Ther, **15**: 436–441.

Dasch JM & Wolff GT (1989) Trace inorganic species in precipitation and their potential use in source apportionment studies. Water Air Soil Pollut, **43**: 401–412.

Dastych M (1990) [Determination of copper and zinc in feces.] Aerztl Lab, **36**: 330–333 (in German).

Dave G, Damgaard B, Grande M, Martelin JE, Rosander B, & Viktor T (1987) Ring test of an embryo-larval toxicity test with zebrafish (*Brachydanio rerio*) using chromium and zinc as toxicants. Environ Toxicol Chem, **6**: 61–71.

Davies BE (1993) Radish as an indicator plant for derelict land: Uptake of zinc at toxic concentrations. Commun Soil Sci Plant Anal, **24**(15/16): 1883–1895.

Davies NT (1980) Studies on the absorption of zinc by rat intestine. Br J Nutr, **43**: 189–203.

Davies BE & Roberts LJ (1975) Sci Total Environ, **4**: 249–261; cited in: Adriano DC (1986) Zinc. In: Adriano DC ed. Trace elements in the terrestrial environment. New York, Springer-Verlag, pp 421–469.

Davies NT & Nightingale R (1975) The effects of phytate on intestinal absorption and secretion of zinc, and whole-body retention of Zn, copper, iron and manganese in rats. Br J Nutr, **34**: 243–258.

Davies KL, Davies MS, & Francis D (1991) The influence of an inhibitor of phytochelatin synthesis on root growth and root mertistimatic activity in *Festuca rubra* L. in response to zinc. New Phytol, **118**: 565–570.

Davis RD & Beckett PHT (1978) Upper critical levels of toxic elements in plants. II. Critical levels of copper in young barley, wheat, rape, lettuce and ryegrass, and of nickel and zinc in young barley and ryegrass. New Phytol, **80**: 23–32.

Davis RD & Carlton-Smith CH (1981) The preparation of sewage sludges of controlled metal content for experimental purposes. Environ Pollut (Series B), **2**:167–177.

Dawson DA, Stebler EF, Burks SL, & Bantle JA (1988) Evaluation of the developmental toxicity of metal-contaminated sediments using short-term fathead minnow and frog embryo-larval assays. Environ Toxicol Chem, **7**: 27–34.

Dean RR & Smith JE Jr (1973) Proceedings Joint Conference Recycling of Municipal Sludges, Effluents, Land. Washington DC, National Association State Universities and Land Grant Colleges, pp 39–43.

Dean CE, Hargis BM, & Hargis PS (1991) Effects of zinc toxicity on thyroid function and histology in broiler chicks. Toxicol Lett, **57**(3): 309–18.

Dedeurwaerder HL, Dehairs FA, Decadt GG, & Baeyens W (1982) Estimates of wet dry deposition and response fluxes of several trace metals in the Southern Bight of the North Sea. In: Pruppacher HR, Semonin RC, & Slinn WGN eds. Precipitation scavenging, dry deposition and resuspension. Elsevier, New York, pp 1219–1231.

Deknudt G & Deminatti M (1978) Chromosome studies in human lymphocytes after *in vitro* exposure to metal salts. Toxicology, **10**: 67–75.

Deknudt G & Gerber GB (1979) Chromosomal aberrations in bone marrow cells of mice given a normal or a calcium-deficient diet supplemented with various heavy metals. Mutat Res, **68**: 163–168.

Deknudt GH & Leonard A (1975) Cytogenetic investigations on leucocytes of workers from a cadmium plant. Environ Physiol Biochem, **5**: 319–327.

Depledge MH (1989) Re-evaluation of metabolic requirements for copper and zinc in decapod crustaceans. Mar Environ Res, **27**: 115–126.

Desor JA & Maller O (1975) Taste correlates of disease states: cystic fibrosis. J. Pediatrics, **87**: 93–96.

Devi YU (1987) Heavy metal toxicity to fiddler crabs, *Uca annulipes* Latreille and *Uca triangularis* (Milne Edwards): Tolerance to copper, mercury, cadmium, and zinc. Bull Environ Contam Toxicol, **39**: 1020–1027.

Dewar WA, Wight PAL, Pearson RA, & Gentle MJ (1983) Toxic effects of high concentrations of zinc oxide in the diet of the chick and laying hen. Br Poult Sci, **24**: 397–404.

Dijkshoorn W, Van Broekhoven LW, & Lampe JEM (1979) Phytotoxicity of zinc, nickel, cadmium, lead, copper and chromium in three pasture plant species supplied with graduated amounts from the soil. Neth J Agric Sci, **27**: 241–253.

Dinnel PA, Link JM, Stober QJ, Letourneau MW, & Roberts WE (1989) Comparative sensitivity of sea urchin sperm bioassays to metals and pesticides. Arch Environ Contam Toxicol, **18**: 748–755.

Di Paolo JA & Casto BC (1979) Quantitative studies of *in vitro* morphological transformation of Syrian hamster cells by inorganic metal salts. Cancer Res, **39**: 1008–1013.

Dirilgen N & Inel Y (1994) Cobalt-copper and cobalt-zinc effects on duckweed growth and metal accumulation. J Environ Sci Health, **A29**: 63–81.

Doelman P & Haanstra L (1984) Short-term and long-term effects of cadmium, chromium, copper, nickel, lead and zinc on soil microbial respiration in relation to abiotic soil factors. Plant Soil, **79**: 317–327.

Doelman P & Haanstra L (1986) Short- and long-term effects of heavy metals on urease activity in soils. Biol Fertil Soils, **2**: 213–218.

Doig AT & Challen PJR (1964) Respiratory hazards in welding. Ann Occup Hyg, **7**: 223–231.

Domingo JL, Llobet JM, Patermain JL, & Corbella J (1988) Acute zinc intoxication: comparison of the antidotal efficacy of several chelating agents. Vet Hum Toxicol, **30**(3): 224–228.

Donald CM & Prescott JA (1975) Trace elements in Australian crop and pasture production 1924–1974. In: Nicholas DJD & Egan ER eds. Trace elements in soil-plant-animal systems. Sydney, Academic Press, pp 7–37.

Dorn CR, Pierce JO, Phillips PE, & Chase GR (1976) Airborne Pb, Cd, Zn and Cu concentration by particle size near a Pb smelter. Atmos Environ, **10**: 443–446.

Doshi GR, Krishnamoorthy TM, Sastry VN, & Sarma TP (1973) Sorption behavior of trace nuclides in sea water on manganese dioxide. Indian J Chem, **11**: 158–161.

Dreosti IE & Hurley LS (1975) Depressed thymidine kinase activity in zinc deficient embryos. Proc Soc Exp Biol Med, **150**: 161–165.

Dreosti IE, Grey PC, & Wilkins PJ (1972) Deoxribonucleic acid synthesis, protein synthesis and teratogenesis in zinc-deficient rats. S Afr Med J, **46**: 1585–1588.

Drinker P, Thomson RM, & Sturgis CC (1927a) Metal fume fever: I. Clinical observations on the effect of the experimental inhalation of zinc oxide by 2 apparently normal persons. J Indust Hyg, **9**: 88–97.

Drinker P, Thomson RM, & Finn JL (1927b) Metal fume fever: II. Resistance acquired by inhalation of zinc oxide on 2 successive days. J Indust Hyg, **9**: 98–105.

Drinker P, Thomson RM, & Finn JL (1927c) Metal fume fever: III. The effects of inhaling magnesium oxide fume. J Indust Hyg, **9**: 187–192.

Drinker P, Thomson RM, & Finn JL (1927d) Metal fume fever: IV. Threshold doses of zinc oxide, preventative measures and the chronic effects of repeated exposures. J Indust Hyg, **9**: 331–345.

Duce RA, Hoffman GL, & Zoller WH (1975) Atmospheric trace metals at remote northern and southern hemisphere sites: pollution or natural? Science, **187**: 59–61.

Dudka S & Chlopecka A (1990) Effect of solid-phase speciation on metal mobility and phytoavailability in sludge-amended soil. Water Air Soil Pollut, **51**: 153–160.

Duinker JC & Nolting RF (1982) Dissolved copper, zinc and cadmium in the southern bight of the North sea. Mar Pollut Bull, **13**: 93–96.

Duncan GD, Gray LF, & Daniel LJ (1953) Effect of zinc on cytochrome oxidase activity. Proc Soc Exp Biol Med, **83**: 625–627.

Dvergsten CL, Fosmire GJ, Ollerich DA, & Sandstead HH (1983) Alterations in the postnatal development of the cerebellar cortex due to zinc deficiency. I. Impaired acquisition of granule cells. Brain Res, **271**: 217–226.

Dvergsten CL, Fosmire GJ, Ollerich DA, & Sandstead HH (1984) Alterations in the postnatal development of the cerebellar cortex due to zinc deficiency. II. Impaired maturation of Purkinje cells. Brain Res, **16**: 11–20.

Eckel WP & Jacob TA (1988) Preprint extended abstract. Ambient levels of 24 dissolved metals in U.S. surface and ground waters. In: Reprints of papers presented at the ACS, National Meeting, Los Angeles, September 25–30, 1988, pp 371–372.

Eduljee G, Badsha K, & Scudamore N (1986) Environmental monitoring for PCB and trace metals in the vicinity of a chemical waste disposal facility – II. Chemosphere, **15**: 81–93.

Eisler R (1993) Zinc hazards to fish, wildlife, and invertebrates: a synoptic review. In: Contaminant hazard reviews. Report 26. Zinc hazards to fish, wildlife, and invertebrates: a synoptic review. Washington DC, US Department of the Interior, Fish and Wildlife Service, pp 1–106 (Biological report No. 10).

Eisler R & Hennekey RJ (1977) Acute toxicities of Cd^{2+}, Cr^{+6}, Hg^{2+}, Ni^{2+} and Zn^{2+} to estuarine macrofauna. Arch Environ Contam Toxicol, **6**: 315–323.

El Bassam N, Dambroth M, & Loughman BC (1990) Genetic aspects of plant mineral nutrition. In: Shaw AJ ed. Heavy metal tolerance in plants: evolutionary aspects, Boca Raton FL, CRC Press, pp 21–37.

Elinder C-G & Piscator M (1979) Zinc. In: Friberg L, Nordberg GF, & Vouk VB eds. Handbook on the toxicology of metals. Amsterdam, Elsevier/North-Holland, pp 675–685.

Elinder C-G, Kjellstroem T, Linnman L, & Pershagen G (1978) Urinary excretion of cadmium and zinc among persons from Sweden. Environ Res, **15**: 473–484.

Endre L, Beck FWJ, & Prasad AS (1990) The role of zinc in human health. J Trace Elem Exp Med, **3**: 337–375.

Enserink EL, Maas-Diepeveen JL, & Van Leeuwen CJ (1991) Combined effects of metals; an ecotoxicological evaluation. Water Res, **25**(6): 679–687.

Ernst W (1972) Zinc and cadmium emission in the vicinity of a zinc smelter. Ber Deutsch Bot Ges, **85**: 295–300.

Ernst WHO (1995) Sampling of plant material for chemical analysis. Sci Total Environ, **176**: 15–24.

Ernst WHO (1996) Bioavailability of heavy metals and decontamination by plants. Appl Geochem, **11**: 163–167.

Ernst WHO, Verkleij JAC, & Schat H (1990) Evolutionary biology of metal resistance in Silene vulgaris. Evol Trends Plants, **4**: 45–51.

Erten J, Areasey A, Çavdar AO, & Cin S (1978) Hair zinc levels in healthy and malnourished children. Am J Clin Nutr, **31**(7): 1172-4.

Etxeberria M, Sastre I, Cajaraville MP, & Marigómez I (1994) Digestive lysosome enlargement induced by experimental exposure to metals (Cu, Cd, and Zn) in mussels collected from a zinc-polluted site. Arch Environ Contam Toxicol, **27**: 338–345.

EU (1993) Guidelines of the Scientific Committee on Food for the development of tolerable upper intake levels for vitamins and minerals. Brussels, European Commission (SCF/NUT/UPPLEV/11).

EU (1996) Report of Committee for Veterinary Medicinal Products, Zinc salts, Summary Report. Brussels, The European Agency for the Evaluation of Medicinal Products, Veterinary Medicines Evaluation Unit (EMEA/MRL/113/96).

Evans EH (1945) Casualties following exposure to zinc chloride smoke. Lancet, **2**: 368–370.

Evans RM (1988) The steroid and thyroid hormone receptor superfamily. Science, **240**: 889–895.

Evenson DP, Emerick RJ, Jost LK, Kayongo-Male H, & Stewart SR (1993) Zinc-silicon interactions influencing sperm chromatin integrity and testicular cell development in the rat as measure by flow cytometry. J Anim Sci, **71**: 955–962.

Everall NC, Macfarlane NAA, & Sedgwick RW (1989) The interactions of water hardness and pH with the acute toxicity of zinc to the brown trout, Salmo trutta L. J Fish Biol, **35**: 27–36.

EZI (1996) Zinc recycling: history, present position and prospects. Brussels, European Zinc Institute.

Failla ML & Cousins RJ (1978) Zinc accumulation and metabolism in primary cultures of adult rat liver cells. Regulation by glucocorticoids. Biochem Biophys Acta, **543**: 293–304.

Faintuch J, Faintuch JJ, Toledo M, Nazario G, Machado MCC, & Raia AA (1978) Hyperamylasemia associated with zinc overdose during parenteral nutrition. J Parenter Enteral Nutr, **2**: 640–645.

Fairweather-Tait SJ, Jackson MJ, Fox TE, Whart SG, Eagles J, & Croghan PC (1993) The measurement of exchangeable pools of zinc using the stable isotope [70]Zn. Br J Nutr, **70**: 221–234.

Fairweather-Tait SJ (1995) Iron-zinc and calcium-Fe interactions in relation to Zn and Fe absorption. Symposium on Micronutrient Interactions. University of Southampton, August 1994. Proc Nutr Soc, **54**: 465–473.

Falahi-Ardakani A, Gouin FR, Bouwkamp JC, & Chaney RL (1987) Growth response and mineral uptake of vegetable transplants growing in composted sewage sludge amended medium. II. Influenced by time of application of N and K. J Environ Hortic, **5**(3): 112–115.

Falchuk KH (1988) Zinc deficiency and the *E. gracilis* chromatin. In: Prasad AS ed. Essential and toxic trace metals in human health and disease. New York, Alan R Liss, pp 75–91.

Falchuk KH, Mazus B, Ulpino L, & Vallee BL (1976) Euglena gracilis DNA Dependent RNA Polymerase II: A Zinc Metalloenzyme. Biochemistry, **15**(20): 4468-4475.

Falchuk KH, Ulpino L, Mazus B, & Vallee BL (1977) *E. gracilis* RNA polymerase I: a zinc metalloenzyme. Biochem Biophys Res Commun, **74**(3): 1206–1212.

Faraji B & Swenseid ME (1983) Growth rate, tissue zinc levels and activities of selected enzymes in rats fed a zinc-deficient diet by gastric tube. J Nutr, **113**(2): 447–455.

Farrah H & Pickering WF (1976) The sorption of zinc species by clay minerals. Aust J Chem, **29**: 1649–1656.

Farris JL, Belanger SE, Cherry DS, & Cairns J (1989) Cellulolytic activity as a novel approach to assess long-term zinc stress to *Corbicula*. Water Res, **23**(10): 1275–1283.

Feinglos MN & Jegasothy BV (1979) "Insulin" allergy due to zinc. Lancet, **1**(8108): 122–124.

Ferguson EL, Gibson RS, Thompson LU, & Ounpuu (1989) Dietary calcium, phytate, and zinc intakes and the calcium, phytate, and zinc molar ratios of the diets of a selected group of East African children. Am J Clin Nutr, **50**: 1450–1456.

Ferm B & Carpenter S (1968) The relationship of cadmium and zinc in experimental mammalian teratogenesis. Lab Invest, **18**: 429–432.

Fernandes G, Nair M, Onoe K, Tonaka T, Floyd R, & Good RA (1979) Impairment of cell-mediated immunity functions by dietary zinc deficiency in mice. Proc Natl Acad Sci USA, **76**: 457–461.

Festa MD, Anderson HL, Dowdy RP, & Ellersieck MR (1985) Effect of zinc intake of copper excretion and retention in men. Am J Clin Nutr, **41**: 285–292.

Fiedler HJ & Roesler HJ (1988) [Trace elements in the environment.] Stuttgart, Enke, pp 97–118 (in German).

Finlayson BJ & Verrue KM (1982) Toxicities of copper, zinc, and cadmium mixtures to juvenile chinook salmon. Trans Am Fish Soc, **111**: 645–650.

Fischer PWF, Giroux A, & L'Abbe MR (1983) Effects of zinc on mucosal copper binding and on the kinetics of copper absorption. J Nutr, **113**: 462–469.

Fischer PWF, Giroux A, & L'bbe MR (1984) Effect of zinc supplementation on copper status in adult man. Am J Clin Nutr, **40**: 743–746.

Fishbein L (1981) Sources, transport and alterations of metal compounds: an overview. I. Arsenic, beryllium, cadmium, chromium, and nickel. Environ Health Perspect, **40**: 43–64.

Fisher NS & Jones GJ (1981) Heavy metals and marine phytoplankton: correlation of toxicity and sulfhydryl-binding. J Phycol, **17**: 108–111.

Flanagan PR, Hais J, & Valberg LS (1983) Zinc absorption, intraluminal zinc and intestinal metallothionein levels in zinc-deficient and zinc-replete rodents. J Nutr, **113**: 962–972.

Flanagan P, Cluett J, Chamberlain M, & Valberg L (1985) Dual-isotope method for determination of human zinc absorption: the use of a test meal of turkey meat. J Nutr, **115**: 111–122.

Florence TM (1977) Trace metal species in fresh waters. Water Res, **11**: 681–687.

Florence TM (1980) Speciation of zinc in natural waters. In: Nriagu JO ed. Zinc in the environment. Part 1: Ecological cycling. New York, John Wiley, pp 199–227.

Florence TM (1989) Electrochemical techniques for trace element speciation in waters. In: Batley GE, ed. Trace element speciation: analytical methods and problem. Boca Raton FL, CRC Press, pp 77–116.

Florence TM (1992) Trace element speciation by anodic stripping voltammetry. Analyst, **117**: 551–554.

Florence TM & Batley GE (1980) Chemical speciation in natural waters. CRC Crit Rev Anal Chem, **9**: 219–296.

Fong LY, Sivak A, & Newberne PM (1978) Zinc deficiency and methylbenzyl-nitrosamine-induced esophageal cancer in rats. J Natl Cancer Inst, **61**(1): 145–150.

Fong LY, Lee JS, Chan WC, & Newberne PM (1984) Zinc deficiency and the development of esophageal and forestomach tumors in Sprague-Dawley rats fed precursors of N-nitroso-N-benzylmethylamine. J Natl Cancer Inst, 72(2): 419–425.

Forman WB, Sheehan D, Cappelli S, & Coffman B (1990) Zinc abuse – an unsuspected cause of sideroblastic anemia. West J Med, 152: 190–192.

Forssén A (1972) Inorganic elements in the human body. I. Occurrence of Ba, Br, Ca, Cu, K, Mn, Ni, Sn, Sr, Y and Zn in the human body. Ann Med Exp Fenn, 50: 99–162.

Forth W & Rummel W (1973) Iron absorption. Physiol Rev, 53: 724–792.

Fosmire G, Greely S, & Sandstead HH (1977) Maternal and fetal response to various suboptimal levels of zinc intake during gestation in the rat. J Nutr, 107: 1543–1550.

Foster DM, Aamodt RL, Henkin RI, & Berman M (1979) Zinc metabolism in humans: a kinetic model. Am J Physiol, 237: R340–R349.

Foster PL (1982) Metal resistance of Chlorophyta from rivers polluted by heavy metals. Freshwater Biol, 12: 41–61.

Foulkes EC (1993) Metallothionein and glutathione as determinants of cellular retention and extrusion of cadmium and mercury. Life Sci, 52: 1617–1620.

Foulkes EC & McMullen DM (1987) Kinetics of transepithelial movement of heavy metals in rat jejunum. Am J Physiol, 253: G134–G138.

Fraker PJ, DePasqual-Jardieu P, Zwick CM, & Luecke RW (1978) Regeneration of T-cell helper function in zinc-deficient adult mice. Proc Natl Acad Sci USA, 75(11): 5660–5664.

Fraker PJ, Gershwin ME, Good RA, & Prasad A (1986) Interrelationships between zinc and immune function. Fed Proc, 45: 1474–1479.

Francis JC & Harrison FW (1988) Copper and zinc toxicity in Ephydatia fluviatilis Porifera Spongillidae. Trans Am Microsc Soc, 107: 67–78.

Francis AR, Shetty TK, & Bhattacharya RK (1988) Modifying role of dietary factors on the mutagenicity of aflatoxin B1: In vitro effect of trace elements. Mutat Res, 199: 85–93.

Frank R, Ishida K, & Suda P (1976) Metals in agricultural soils of Ontario. Can J Soil Sci, 56: 181–196.

Frankel AD, Berg JM, & Pabo CO (1987) Metal-dependent folding of a single zinc finger from transcription factor IIIA. Proc Natl Acad Sci USA, 84: 4841–4845.

Frederickson CJ (1989) Neurobiology of zinc and zinc-containing neurons. Int Rev Neurobiol, 31: 145–238.

Freedman LP, Luisi BF, Korszun ZR, Basavappa R, Sigler PB, & Yamamoto KR (1988) The function and structure of the metal coordination sites within the glucocorticoid receptor DNA binding domain. Nature, **334**: 543–546.

Freeland-Graves JH, Han WH, Friedman BJ, & Shorey RL (1980) Effect of dietary Zn/Cu ratios on cholesterol and HDL-cholesterol levels in women. Nutr Rep Int, **22**: 285–293.

Freeland-Graves JH, Hendrickson PJ, Ebangit ML, & Snowden JY (1981) Salivary zinc as an index of zinc status in women fed a low-zinc diet. Am J Clin Nutr, **34**(3): 312–321.

French MC, Haines CW, & Cooper J (1987) Investigation into the effects of ingestion of zinc shot by mallard ducks (*Anas platyrhynchos*). Environ Pollut, **47**: 305–314.

Frostegård Å, Tunlid A, & Bååth E (1993) Phospholipid fatty acid composition, biomass, and activity of microbial communities from two soil types experimentally exposed to different heavy metals. Appl Environ Microbiol, **59**(11): 3605–3617.

Furey WF, Robbins AH, Clancy LL, Winge DR, Wang BC, & Stout CD (1986) Crystal structure of Cd, Zn Metallothionein. Science, **231**: 704–710.

Gadd GM (1993) Interactions of fungi with toxic metals. New Phytol, **124**(1): 25–60.

Gadde RR & Laitinen HA (1974) Studies of heavy metal adsorption by hydrous iron and manganese oxides. Anal Chem, **46**(13): 2022–2026.

Gallery EDM, Blomfield J, & Dixon SR (1972) Acute zinc toxicity in haemodialysis. Br Med J, **4**: 331–333.

Gasaway WC & Buss IO (1972) Zinc toxicity in the mallard duck. J Wildl Manage, **36**(4): 1107–1117.

Geering HR & Hodgson JF (1969) Micronutrient cation complexes in soil solution: III. Characterization of soil solution ligands and their complexes with Zn^{2+} and Cu^{2+}. Soil Sci Soc Am Proc, **33**: 54–59.

Genter RB, Colwell FS, Pratt JR, Cherry DS, & Cairns J (1988) Changes in epilithic communities due to individual and combined treatments of zinc and snail grazing in stream mesocosms. Toxicol Ind Health, **4**(2): 185–201.

Gerritse RG, Vriesema R, Dalenberg JW, & De Roos HP (1982) Effect of sewage sludge on trace element mobility in soils. J Environ Qual, **11**(3): 359–364.

Gibson RS (1980) Hair as a biopsy material for the assessment of trace element status in infancy. A review. J Hum Nutr, **34**(6): 405–416.

Gibson R (1990) Probability approach to evaluating nutrient data. In: Principles of nutritional assessment, 1st ed. Oxford University Press, New York, pp 148–152.

Gibson RS (1994) Zinc nutrition in developing countries. Nutr Res Rev, **7**: 151–173.

Gibson RS & Scythes CA (1982) Trace element intakes of women. Br J Nutr, 48: 241–248.

Gibson RS, Vanderkooy PDS, MacDonald AC, Goldman A, Ryan BA, & Berry M (1989a) A growth-limiting, mild zinc-deficiency syndrome in some Southern Ontario boys with low height percentiles. Am J Clin Nutr, 49: 1266–1273.

Gibson RS, Ferguson EF, Vanderkooy PDS, & MacDonald AC (1989b) Seasonal variations in hair zinc concentrations in Canadian and African children. Sci Total Environ, 84: 291–298.

Giesler E, Nehl H, & Munk R (1983) [Zinc compounds.] In: Bartholomé E, Biekert E, Hellmann H, Ley H, Weigert WM, & Weise E eds. (Ullmann's Encyclopedia of technical chemistry) 4th revised and expanded ed. Weinheim, Verlag Chemie, pp 633–640 (in German).

Gintenreiter S, Ortel J, & Nopp HJ (1993a) Bioaccumulation of cadmium, lead, copper, and zinc in successive developmental stages of Lymantria dispar L. (Lymantriidae, Lepid) - a life cycle study. Arch Environ Contam Toxicol, 25(1): 55–61.

Giroux EL, Durieux M, & Schechter PJ (1976) A study of zinc distribution in human serum. Bioinorganic Chem, 5: 211–218.

Giugliano R & Millward DJ (1984) Growth and zinc homeostasis in the severely Zn-deficient rat. Br J Nutr, 52: 545–560.

Gocke E, King MT, Echardt K, & Wild D (1981) Mutagenicity of cosmetics ingredients licensed by the European Communities. Mutat Res, 90: 91–109.

Goethals K (1991) [Evolution of the quality R. Schelde water.] Water, 56: 12–18 (in Dutch).

Goldenberg RL, Tamura T, Neggers Y, Copper RL, Johnston KE, DuBard MB, & Hauth JC (1995) The effect of zinc supplementation on pregnancy outcome. In: Zinc supplementation in pregnancy. J Am Med Assoc, 274(6): 463–468.

Golovina LP, Lysenko MN, & Kisel' TI (1980) Content and distribution of zinc in the soil of the Ukrainian Poles'ye. Sov Soil Sci, 12: 73–80.

Golub MS, Gershwin ME, Hurley LS, Hendrickx AG, & Saito WY (1985) Studies of marginal zinc deprivation in rhesus monkeys: infant behavior. Am J Clin Nutr, 42: 1229–1239.

González J, Hernández LM, Hernán A, & Baluja G (1985) Multivariate analysis of water contamination by heavy metals at Doñana National Park. Bull Environ Contam Toxicol, 35: 266–271.

Goodwin JS, Hunt WC, Hooper P, & Garry PJ (1985) Relationship between zinc intake, physical activity and blood levels of high-density lipoprotein cholesterol in a healthy elderly population. Metabolism, 34(6): 519–523.

Gordon PR, Woodruff CW, Anderson HL, & O'Dell BL (1982) Effect of acute zinc deprivation on plasma zinc and platelet aggregation in adult males. Am J Clin Nutr, **35**: 113–119.

Gordon T, Chen LC, Fine JM, Schlesinger RB, Su WY, Kimmel TA, & Amdur MO (1992) Pulmonary effects of inhaled zinc oxide in human subjects, guinea pigs, rats, and rabbits. Am Ind Hyg Assoc J, **53**: 503–509.

Graham GA, Byron G, & Norris RH (1986) Survival of *Salmo gairdneri* (rainbow trout) in the zinc polluted Molonglo River near Captains Flat, New South Wales, Australia. Bull Environ Contam Toxicol, **36**: 186–191.

Grandy JW, Locke LN, & Bagley GE (1968) Relative toxicity of lead and five proposed substitute shot types to pen-reared mallards. J Wildl Manage, **32**(3): 483–488.

Grant PT, Coombs TL, & Frank BH (1972) Differences in the nature of the interaction of insulin and proinsulin with zinc. Biochem J, **126**: 433–440.

Greathouse DG & Osborne RH (1980) Preliminary report on nationwide study of drinking water and cardiovascular diseases. J Environ Pathol Toxicol Oncol, **4**: 65–76.

Greaves MW & Skillen AW (1970) Effects of long continued ingestion of zinc sulphate in patients with venous leg ulceration. Lancet, **2**: 889–891.

Greenberg AE, Clesceri LS, & Eaton AD eds. (1992) Standard methods for the examination of water and wastewater, 18th ed. Washington, DC, American Public Health Association.

Greenwood JG & Fielder DR (1983) Acute toxicity of zinc and cadmium to zoeae of three species of partunid crabs (Crustacea: Brachyura). Comp Biochem Physiol, **75C**(1): 141–144.

Greger JL & Geissler AH (1978) Effect of zinc supplementation on taste acuity of the aged. Am J Clin Nutr, **31**: 633–637.

Greger JL & Sickles VS (1979) Saliva zinc levels: potential indicators of zinc status. Am J Clin Nutr, **32**: 1859–1866.

Greger JL, Abernathy RP, & Bennett OA (1978a) Zinc and nitrogen balance in adolescent females fed varying levels of zinc and soy protein. J Clin Nutr, **31**: 112–116.

Greger JL, Baligar P, Abernathy RP, Bennett OA, & Peterson BS (1978b) Calcium, magnesium, phosphorus, copper, and manganese balance in adolescent females. Am J Clin Nutr, **31**: 117–121.

Greger JL, Zaikis SC, Abernathy RP, Bennett OA, & Huffman J (1978c) Zinc, nitrogen, copper, iron, and manganese balance in adolescent females fed two levels of zinc. J Nutr, **108**: 1449–1456.

Gregory J, Foster K, Tyler H, & Wiseman M (1990) The dietary and nutritional survey of British adults.London, HMSO.

Grider A, Bailey LB, & Cousin RJ (1990) Erythrocyte metallothionein as an index of zinc status in humans. Proc Natl Acad Sci, USA, Applied Biological Sciences, **87**: 1259–1262.

Grimshaw DL, Lewin J, & Fuge R (1976) Seasonal and short-term variations in the concentration and supply of dissolved zinc to polluted aquatic environments. Environ Pollut, **11**: 1–7.

Grobler-van Heerden E, Van Vuren JHJ, & Du Preez HH (1991) Bioconcentration of atrazine, zinc and iron in the blood of *Tilapia sparrmanii* (Cichlidae). Comp Biochem Physiol, **100C**(3): 629–633.

GSC (1995) Geological Survey of Canada. National geochemical reconnaissance data. Ottawa, Natural Resources Canada, Government of Canada.

Gubala CP, Landers DH, Monetti M, Heit M, Wade T, Lasorsa B, & Allen-Gil S (1995) The rates of accumulation and chronologies of atmospherically derived pollutants in Arctic Alaska, USA. Sci Total Environ, **160/161**: 347–361.

Gunn SA, Gould TC, & Anderson AD (1963) Cadmium-induced interstitial cell tumors in rats and mice and their prevention by zinc. J Nat Cancer Inst, **31**: 745–759.

Gunshin H, Noguchi T, & Naito H (1991) Effect of calcium on the zinc uptake by brush border membrane vesicles isolated from the rat small intestine. Agric Biol Chem, **55**: 2813–2816.

Gupta T, Talukder G, & Sharma A (1991) Cytotoxicity of zinc chloride in mice in vivo. Biol Trace Elem Res, **30**(2): 95–101.

Guy RD & Chakrabarti CL (1976) Studies of metal–organic interactions in model systems pertaining to natural waters. Can J Chem, **54**: 2600–2611.

Gyorffy EJ & Chan H (1992) Copper deficiency and microcytic anemia resulting from prolonged ingestion of over-the-counter zinc. Am J Gastroenterol, **87**(8): 1054–1055.

Haesaenen E, Lipponen M, Kattainen R, Markkanen K, Minkkinen P, & Brjukhanov P (1990) Elemental concentrations of aerosol samples from the Baltic Sea area. Chemosphere, **21**: 339–347.

Hagan EC, Radomski JL, & Nelson AA (1953) Blood and bone marrow effects of feeding zinc sulfate to rats and dogs. J Am Pharm Assoc, **42**: 700–702.

Haight M, Mudry T, & Pasternak J (1982) Toxicity of seven heavy metals on *Panagrellus silusiae*: the efficacy of the free-living nematode as an *in vivo* toxicological bioassay. Nematologica, **28**: 1–11.

Halas ES, Hanlon MJ, & Sandstead HH (1975) Intrauterine nutrition and aggression. Nature, **257**: 221–222.

Halas ES, Rowe MC, Johnson OR, McKenzie JM, & Sandstead HH (1976) Effects of intrauterine zinc deficiency on subsequent behavior. In: Prasad AS ed. Trace elements in human health and disease. New York, Academic Press, pp 327–343.

Halas ES, Reynolds GM, & Sandstead HH (1977) Intrauterine nutrition and its effects on aggression. Physiol Behav, **19**: 653–661.

Halas ES, Heinrich MD, & Sandstead HH (1979) Long term memory deficits in adult rats due to postnatal malnutrition. Physiol Behav, **22**: 991–997.

Halas ES, Eberhardt MJ, Diers MA, & Sandstead HH (1983) Learning and memory impairment in adult rats due to severe zinc deficiency during lactation. Physiol Behav, **30**: 371–381.

Halas ES, Hunt CD, & Eberhardt MJ (1986) Learning and memory disabilities in young adult rats from mildly zinc deficient dams. Physiol Behav, **37**: 451–458.

Hale JG (1977) Toxicity of metal mining wastes. Bull Environ Contam Toxicol, **17**(1): 66–73.

Hall AC, Young BW, & Bremner I (1979) Intestinal metallothionein and the mutual antagonism between copper and zinc in the rat. J Inorg Biochem, **11**: 57–66.

Hall LW, Ziegenfuss MC, Bushong SJ, Unger MA, & Herman RL (1989) Studies of contaminant and water quality effects on striped bass prolarvae and yearlings in the Potomac river and upper Chesapeake Bay in 1988. Trans Am Fish Soc, **118**: 619-629.

Hallbook T & Lanner E (1972) Serum-zinc and healing of venous leg ulcers. Lancet, **2**(7781): 780–782.

Hallmanns G (1977) Treatment of burns with zinc-tape. A study of local absorption of zinc in humans. Scand J Plast Reconstr Surg, **11**: 155–161.

Halsted JA, Ronaghy HA, Abadi P, Haghshenass M, & Amerhakemi GH (1972) Zinc dificiency in man: the Shiraz experiment. Am J Clin Nutr, **53**: 277–284.

Halsted JA, Smith JC, & Irwin MI (1974) A conspectus of research on zinc requirements of man. J Nutr, **104**: 345–378.

Hambidge KM (1982) Hair analyses: worthless for vitamins, limited for minerals. Am J Clin Nutr, **36**: 943–949.

Hambidge KM (1989) Mild zinc deficiency in human subjects. In: Mills B ed. Zinc in Human Biology. London, Springer, pp 281–296.

Hambidge KM & Walravens PA (1982) Disorders of mineral metabolism. Clin Gastroenterol, **11**(1): 87–117.

Hambidge KM, Hambidge C, Jacobs M, & Baum JD (1972) Low levels of zinc in hair, anorexia, poor growth, and hypogeusia in children. Pediat Res, **6**: 868–874.

287

Hambidge KM, Chavez MN, Brown RM, & Walravens PA (1979) Zinc nutritional status of young middle-income children and effects of consuming zinc-fortified breakfast cereals. Am J Clin Nutr, **32**: 2532–2539.

Hambidge KM, Krebs NF, Jacobs MA, Favier A, Guyette L, & Ikle DN (1983) Zinc nutritional status during pregnancy: a longitudinal study. Am J Clin Nutr, **37**: 429–442.

Hambidge KM, Casey CE, & Krebs NF (1986) Zinc. In: Mertz W ed. Trace elements in human and animal nutrition 5th ed., Orlando FA, Academic Press Inc, pp 1–137.

Hamdi EA (1969) Chronic exposure to zinc of furnace operators in a brass foundry. Br J Ind Med, **26**: 126–134.

Hamer DH (1986) Metallothionein. Annu Rev Biochem, **55**: 913–951.

Hames CAC & Hopkin SP (1991) Assimilation and loss of ^{109}Cd and ^{65}Zn by the terrestrial isopods Oniscus asellus and Porcellio scaber. Bull Environ Contam Toxicol, **47**: 440–447.

Hamilton SJ & Buhl KJ (1990) Safety assessment of selected inorganic elements to fry of chinook salmon (Oncorhynchus tshawytscha). Ecotoxicol Environ Saf, **20**: 307–324.

Hamilton DL, Bellamy JEC, Valberg JD, & Valberg LS (1978) Zinc, cadmium, and iron interactions during intestinal absorption in iron-deficient mice. Can J Physiol Pharmacol, **56**: 384–389.

Hamilton RP, Fox MRS, Fry BVE, Jones AOL, & Jacobs RM (1979) Zinc interference with copper, iron and manganese in young Japanese quail. J Food Sci, **44**: 738–741.

Hanas JS, Hazuda DJ, Bogenhagen DF, Wu FYH, & Wu CW (1983) Xenopus transcription factor a requires zinc for binding to the 5 S RNA gene. J Bio Chem, **258**(23): 14120–14125.

Harding JPC & Whitton BA (1981) Accumulation of zinc, cadmium and lead by field populations of lemanea. Water Res, **15**(3) 301–319.

Hare L, Saouter E, Campbell PG, Tessier A, Ribeyre F, & Boudou A (1991) Dynamics of cadmium, lead, and zinc exchange between nymphs of the burrowing mayfly Hexagenia rigida (Ephemeroptera) and the environment. Can J Fish Aquat Sci, **48**: 39–47.

Harmens H, Gusmao NG, Den Hartog PR, Verkleij JAC, & Ernst WHO (1993a) Uptake and transport of zinc in zinc-sensitive and zinc-tolerant Silene vulgaris. J Plant Physiol, **141**: 309–315.

Harmens H, Den Hartog PR, Ten Bookum WM, & Verkleij JAC (1993b) Increased zinc tolerance in Silene vulgaris (Moench) Garcke is not due to increased production of phytochelatins. Plant Physiol, **103**: 1305–1309.

Harrison RM (1979) Toxic metals in street and household dusts. Sci Total Environ, **11**: 89–97.

Hartwig EE, Jones WF, & Kilen TC (1991) Identification and inheritance of inefficient zinc absorption in soybean. Crop Sci, **31**: 61–63.

Harzer G & Kauer H (1982) Binding of zinc to casein. Am J Clin Nutr, **35**: 981–990.

Heaton RW, Rahn KA, & Lowenthal DH (1990) Determination of trace elements, including regional tracers, in Rhode Island precipitation. Atmos Environ, **24A**(1): 147–153.

Heckel J (1995) Using Barkla polarized x-ray radiation in energy dispersive x-ray fluorescence analysis (EDXRF). J Trace Microprobe Tech, **13**: 97–108.

Helz GR, Huggett RJ, & Hill JM (1975) Behaviour of Mn, Fe, Cu, Zn, Cd and Pb discharged from a wastewater treatment plant into an estuarine environment. Water Res, **9**: 631–636.

Hem JD (1972) Chemistry and occurrence of cadmium and zinc in surface water and ground water. Water Resour Res, **8**: 661–677.

Hempe JM & Cousins RJ (1991) Cysteine-rich intestinal protein binds zinc during transmucosal zinc transport. Proc Natl Acad Sci USA, **88**: 9671–9674.

Hempe JM & Cousins RJ (1992) Cysteine-rich intestinal protein and intestinal metallothionein: an inverse relationship as a conceptual model for zinc absorption in rats. J Nutr, **122**: 89–95.

Henkin RI (1979) Zinc. Baltimore, MD, Univ Park Press.

Henkin RI (1984) Review: zinc in taste function. A critical review. Biol Trace Elem Res, **6**: 263–279.

Henkin RI & Aamodt RL (1983) A redefinition of zinc deficiency. Nutritional bioavailability of zinc. Inglett. Washington, American Chemical Society, pp 83–105.

Henkin RI, Mueller CW, & Wolf RO (1975a) Estimation of zinc concentration of parotid saliva by flameless atomic absorption spectrophotometry in normal subjects and in patients with idiopathic hypogeusia. J Lab Clin Med, **86**: 175–180.

Henkin RI, Patten BM, Re PK, & Bronzert DA (1975b) A syndrome of acute zinc loss. Cerebellar dysfunction, mental changes, anorexia, and taste and smell dysfunction. Arch Neurol, **32**: 751.

Henry CL & Harrison RB (1992) Fate of trace metals in sewage sludge compost. In: Adriano DC ed. Biogeochemistry of trace metals. Boca Raton FL, Lewis Publishers, pp 195–216.

Henstock ME (1996) The recycling of non-ferrous metals. Ottawa, International Council on Metals in the Environment.

Herich R (1969) The effect of zinc on the structure of chromosomes and mitosis. Nucleus, **12**: 81–85.

Herkovits J, Pérez-Coll CS, & Zeni S (1989) Protective effect of zinc against spontaneous malformations and lethality in *Bufo arenarum* embryos. Biol Trace Elem Res, **22**: 247–250.

Hermann R & Neumann-Mahlkau P (1985) The mobility of zinc, cadmium, copper, lead, iron and arsenic in ground water as a function of redox potential and pH. Sci Total Environ, **43**: 1–12.

Hery M, Gerber J-M, Vien I, & Limasset J-C (1991) [Assessment of exposure during catalyst handling in the chemical industry. Preliminary results]. Staub-Reinhalt Luft, **51**: 361–364 (in German).

Hess FM, King JC, & Margen S (1977) Zinc excretion in young women on low zinc intakes and oral contraceptive agents. J Nutr, **107**: 1610–1620.

Heydorn K (1995) Validation of neutron activation analysis techniques. In: Quevauviller PH, Maier EA, & Griepink B eds. Quality assurance for environmental analysis. Amsterdam, Elsevier, pp 89–110.

Hietanen B, Sunila I, & Kristoffersson R (1988) Toxic effects of zinc on the common mussel *Mytilus edulis* L. (Bivalvia) in brackish water. I Physiological and histopathological studies. Ann Zool Fenn, **25**: 341–347.

Hill CH & Matrone G (1970) Chemical parameters in the study on in vivo and in vitro interactions of transition elements. Fed Proc, **29**(4): 1474–1481.

Hill GM, Brewer GJ, Prasad AS, Hydrick CR, & Hartmann DE (1987) Treatment of Wilson's disease with zinc. I. Oral zinc therapy regimens. Hepatology, 7(3): 522–528.

Hilmy AM, El-Domiaty NA, Daabees AY, & Abdel Latife HA (1987) Toxicity in *Tilapia zilli* and *Clarias lazera* (Pisces) induced by zinc, seasonally. Comp Biochem Physiol, **86C**(2): 263–265.

Hjortso E, Qvist J, Bud M, Thomsen JL, Andersen AB, Wiberg-Jorgensen F, Jensen NK, Jones R, Reid LM, & Zapol WM (1988) ARDS after accidental inhalation of zinc chloride smoke. Intensive Care Med, **14**: 17–24.

Hobson JF & Birge WJ (1989) Acclimation-induced changes in toxicity and induction of metallothionein-like proteins in the fathead minnow following sublethal exposure to zinc. Environ Toxicol Chem, **8**: 157–169.

Hodgson JF, Lindsay WL, & Trierweiler JF (1966) Micronutrient cation complexing in soil solution: II. Complexing of zinc and copper in displaced solution from calcareous soils. Soil Sci Soc Am Proc, **30**: 723–726.

Hoell K, Carlson S, Luedemann D, & Rueffer H (1986) [Zinc analysis.] In: Hoell K ed. (Water – Examination, assessment, preparation, chemistry, bacteriology, virology, biology) 7th ed. Berlin, Walter de Gruyter, pp 182–183 (in German).

Hoffman HN, Phyliky RL, & Fleming CR (1988) Zinc induced copper deficiency. Gastroenterology, **94**: 508–512.

Hogan GD & Wotton DL (1984) Pollutant distribution and effects in forests adjacent to smelters. J Environ Qual, **13**: 377–382.

Hogstrand C & Wood CM (1995) Mechanisms for zinc acclimation in freshwater rainbow trout. Mar Environ Res, **39**: 131–135.

Hogstrand C, Lithner G, & Haux C (1989) Relationship between metallothionein, copper and zinc in perch (*Perca fluviatilis*) environmentally exposed to heavy metals. Mar Environ Res, **28**: 179–182.

Hohnadel DC, Sunderman FW Jr, Nechay MW, & McNeely MD (1973) Atomic absorption spectrometry of nickel, copper, zinc, and lead in sweat collected from healthy subjects during sauna bathing. Clin Chem, **19**(11): 1288–1292.

Holbrook JT, Smith JC, & Reiser S (1989) Dietary fructose or starch: effects on copper, zinc, iron, manganese, calcium and magnesium balances in humans. Am J Clin Nutr, **49**: 1290-1294.

Holcombe GW, Benoit DA, & Leonard EN (1979) Long-term effects of zinc exposures on brook trout (*Salvelinus fontinalis*). Trans Am Fish Soc, **108**: 76–87.

Holmgren GGS, Meyer MW, Chaney RL, & Daniels RB (1993) Cadmium, lead, zinc, copper, and nickel in agricultural soils of the United States of America. J Environ Qual, **22**: 335–348.

Holland HD (1978) Physical and chemical transport in river systems. In: Holland HD ed. The chemistry of the atmosphere and oceans. New York, John Wiley & Sons, pp 81–152.

Homma S, Jones R, Qvist J, Zapol WM, & Reid L (1992) Pulmonary vascular lesions in the adult respiratory distress syndrome caused by inhalation of zinc chloride smoke: a morphometric study. Hum Pathol, **23**(1): 45–50.

Hoogenraad TU, Koevoet R, & Ruyter Korver EGWM (1979) Oral zinc sulphate as long-term treatment in Wilson's disease (hepatolenticular degeneration). Eur Neurol, **18**: 205–211.

Hoogenraad TU & Van den Hamer CLA (1983) 3 years of continuous oral zinc therapy in 4 patients with Wilson's disease. Acta Neurol Scand, **67**: 356–364.

Hoogenraad TU, Van den Hamer CLA, & Van Hattum J (1984) Effective treatment of Wilson's disease with oral zinc sulphate: two case reports. Br Med J, **289**: 273–276.

Hooper PL, Visconti L, Garry PJ, & Johnson GE (1980) Zinc lowers high-density lipoprotein-cholesterol levels. J Am Med Assoc, **244**(17): 1960–1961.

Hopkin SP & Martin MH (1984) Assimilation of zinc, cadmium, lead and copper by the centipede *Lithobius variegatus* (Chilopoda). J Appl Ecol, **21**: 535–546.

Hopkin SP & Martin MH (1985) Assimilation of zinc, cadmium, lead, copper, and iron by the spider *Dysdera crocata*, a predator of woodlice. Bull Environ Contam Toxicol, **34**: 183–187.

Hopkin SP & Hames CAC (1994) Zinc, among a 'cocktail' of metal pollutants, is responsible for the absence of the terrestrial isopod Porcellio scaber from the vicinity of a primary smelting works. Ecotoxicology, 3: 68–78.

Hornor SG & Hilt BA (1985) Distribution of zinc-tolerant bacteria in stream sediments. Hydrobiologia, 128: 155–160.

Houba C, Remacle J, Bubois D, & Thorez J (1983) Factors affecting the concentrations of cadmium, zinc, copper and lead in the sediments of the Vesdre River. Water Res, 17(10): 1281–1286.

Hove E, Elvehjem CA, & Hart EB (1937) The physiology of zinc in the nutrition of the rat. Am J Physiol, 119: 768–775.

Hove E, Elvehjem CA, & Hart EB (1938) Further studies on zinc deficiency in rats. Am J Physiol, 124: 750–758.

Huang CP, Elliott HA, & Ashmead RM (1977) Interfacial reactions and the fate of heavy metals in soil-water systems. J Water Pollut Control Fed, 49: 745–756.

Huber AM & Gershoff SN (1973) Effect of zinc deficiency in rats on insulin release from the pancreas. J Nutr, 103: 1739–1744.

Huber AM & Gershoff SN (1975) Effects of zinc deficiency on the oxidation of retinol and ethanol in rats. J Nutr, 195(11): 1486–1489.

Huebert DB & Shay JM (1992) Zinc toxicity and its interaction with cadmium in the submerged aquatic macrophyte Lemna trisulcai L. Environ Contam Toxicol, 11: 715–720

Hunt DTE & Wilson AL (1986) The chemical analysis of water, 2nd ed. Cambridge, Royal Society of Chemistry.

Hunt IF, Murphy NJ, Cleaver AE, Faraji B, Swendseid ME, Coulson AH, Clark VA, Browdy BL, Cabalum MT, & Smith JCJ (1984) Zinc supplementation during pregnancy: effects on selected blood constituents and on program and outcome of pregnancy in low income women of Mexican descent. Am J Clin Nutr, 40: 508–521.

Hunt JW & Anderson BS (1989) Sublethal effects of zinc and municipal effluents on larvae of the red abalone Haliotis rufescens. Mar Biol, 101: 545–552.

Hunt JR, Johnson PE, & Swan PB (1987) Dietary conditions influencing relative zinc availability from foods to the rat and correlations with in vitro measurements. J Nutr, 117: 1913–1923.

Hunt IF, Murphy NJ, Cleaver AE, Faraji B, Swenseid ME, Browdy BL, Coulson AH, Clark VA, Settlage RH, & Smith JC Jr (1995) Zinc supplementation during pregnancy in low-income teenagers of Mexican descent: effects on selected blood constituents and on progress and outcome of pregnancy. Am J Clin Nutr, 42(5): 815–828.

Hurley LS & Shrader RE (1972) Congenital malformations of the nervous system in zinc-deficient rats. Int Rev Neurobiol (Suppl), 1: 7–51.

Hurley LS & Shrader RE (1974) Abnormal development of preimplantation rat eggs after three days of maternal dietary zinc deficiency. Nature, **254**: 427–429.

Hurley LS & Swenerton H (1966) Congenital malformation resulting from zinc deficiency in rats. Proc Soc Exp Biol Med, **123**: 692–696.

Hurley LS, Gowan J, & Swenerton H (1971) Teratogenic effects of short term and transitory zinc deficiency in rats. Teratology, **4**: 199–204.

Husain SL (1969) Oral zinc sulfate in leg ulcers. Lancet, **1**: 1069–1071.

Hussein AS, Cantor AH, & Johnson TH (1988) Use of high levels of dietary aluminum and zinc for inducing pauses in egg production of Japanese quail. Poult Sci, **67**: 1157–1165.

Hutchinson TC & Czyrska H (1975) Heavy metal toxicity and synergism to floating aquatic weeds. Verh Int Ver Theor Angew Limnol, **19**: 2102-2111.

Hutchison F & Wai CM (1979) Cadmium, lead, and zinc in reclaimed phosphate mine waste dumps in Idaho. Bull Environ Contam Toxicol, **23**: 377–380.

IARC (1987) Beryllium and beryllium compounds and chromium and chromium compounds. In: Overall evaluations of carcinogenicity: an updating of IARC Monographs, Volumes 1 to 42. Lyon, International Agency for Research on Cancer (IARC Monographs on the Evaluation of the Carcinogenic Risk of Chemicals to Humans, Suppl. 7).

IHE (1991) [Evaluation of heavy metal content in the ambient air in Belgium.] Brussels, Institute of Hygiene and Epidemiology (in Dutch).

ILO (1991) Occupational exposure limits for airborne toxic substances, 3rd ed. Geneva. International Labour Organisation (Occupational Safety and Health Series, No. 37).

Ilyaletdinov AN, Kamalov MR, & Stukanov VA (1977) [Microbial leaching of zinc and lead from ores of the Tekeli deposit.] *Mikrobiologiya*, **46**(5): 857–866 (in Russian).

ILZSG (1994) Capacity changes in lead and zinc 1980-1993. International Zinc and Lead Study Group (Document LZ/EC.94/5).

ILZSG (1995) Lead and zinc statistics. Monthly bulletin of the International Lead and Zinc Study Group. La Tribune du Cebedeau, **35**: 37, 39–41, 54, 67.

ILZSG (1996) Lead and zinc statistics. Monthly bulletin of the International Lead and Zinc Study Group, La Tribune du Cebedeau,Vol. 36.

Ishizaka A, Tsuchida F, & Ishii T (1981) Clinical zinc deficiency during zinc-supplemented parenteral nutrition. J Pediatr, **99**: 339 (letter).

Istfan NW, Janghorbani M, & Young VR (1983) Absorption of stable 70Zn in healthy young men in relation to zinc intake. Am J Clin Nutr, **38**: 187–194.

Iversen T, Halvorsen N, Mylona S, & Sandnes H (1991) Calculated budgets for airborne acidifying components in Europe 1985, 1987, 1988, 1989, 1990. Norwegian Meteorological Institute Technical report nr.91 (EMEP/MSC-W report 1/91).

Jacob RA, Sandstead HH, Munoz JM, & Klevay LM (1979) Whole body surface loss of trace metals in normal males. Fed Proc Fed Am Soc Exp Biol, **38**: 552.

James LF, Lazar VA, & Binns W (1966) Effects of sublethal doses of certain minerals on pregnant ewes and fetal development. Am J Vet Res, **27**: 132–135.

Jameson S (1976) Effects of zinc deficiency in human reproduction. Acta Med Scand Suppl, **593**: 4–89.

Jameson S (1982) Zinc status and pregnancy outcome in humans. In: Prasad AS et al. Clinical application of recent advances in zinc metabolism. New York, Alan R Liss, pp 39–52.

Jhala US & Baly DL (1991) Zinc deficiency results in a post transcriptional impairment in insulin synthesis. FASEB J, **5**: A94.

John W, Kaifer R, Rahn K, & Wesolowski JJ (1973) Trace element concentrations in aerosols from the San Francisco bay area. Atmos Environ, **7**: 107–118.

Johnson FA & Stonehill RB (1961) Chemical pneumonitis from inhalation of zinc chloride. Dis Chest, **40**: 619–623.

Johnson PE (1982) A mass spectrometric method for use of stable isotopes as tracers in studies of iron, zinc, and copper absorption in human subjects. J Nutr, **112**: 1414–1424.

Johnson WT & Evans GW (1982) Tissue uptake of zinc in rats following the administration of zinc dipicolinate or zinc histidinate. J Nutr, **112**: 914–919.

Johnson PE, Hunt CD, Milne DB, & Mullen LK (1993) Homeostatic control of zinc metabolism in men: zinc excretion and balance in men fed diets low in zinc. Am J Clin Nutr, **57**: 557–565.

Jolly JH (1989) Zinc. In: US Bureau of mines D.O.I. ed. Minerals year book: metals and minerals. Washington DC, pp 1145–1174.

Jones R (1983) Zinc and cadmium in lettuce and radish grown in soils collected near electrical transmission (hydro) towers. Water Air Soil Pollut, **19**: 389–395.

Jones R & Burgess MSE (1984) Zinc and cadmium in soils and plants near electrical transmission (hydro) towers. Environ Sci Technol, **18**: 731–734.

Jones PE & Peters TJ (1981) Oral zinc supplements in non-responsive coeliac syndrome: effect on jejunal morphology, enterocyte production, and brush border disaccharidase activities. Gut, **22**(3): 194–198.

Joosse ENG, Van Capelleveen HE, Van Dalen LH, & Van Diggelen J (1984) Effects of zinc, iron and manganese on soil arthropods associated with decomposition processes. In: Williams JH ed. Proceedings of the 4th International Conference on Heavy Metals in the Environment. Heidelberg, September 1983, pp 467–470.

Jordaan HF & Sandler M (1989) Zinc-induced granuloma – a unique complication of insulin therapy. Clin Exp Dermatol, **14**(3): 227–229.

Jordan MJ & Lechevalier MP (1975) Effects of zinc-smelter emissions on forest soil microflora. Can J Microbiol, **21**: 1855–1865.

Juergensen H & Behne D (1977) Variations in trace element concentrations in human blood serum in the normal state investigated by instrumental neutron activation analysis. J Radioanal Chem, **37**: 375–382.

Kabata-Pendias A & Pendias H (1984) Zinc. In: Trace elements in soils and plants. Boca Raton, FL, CRC Press, pp 99–110.

Kada T, Hirano K, & Shirasu Y (1980) Screening of environmental chemical mutagens by the rec-assay system with Bacillus subtilis. Chem Mutagenesis, **6**: 149–173.

Kalbasi M, Racz GJ, & Lewen–Rudgers LA (1978) Reaction products and solubility of applied zinc compounds in some Manitoba soils. Soil Sci, **125**(1): 55–63.

Kampe W (1986) [Plants need zinc.] In: Zinc Advice Inc., Advice on galvanization. (Without zinc no life). Düsseldorf, Institute for applied zinc galvanization, pp 1–2 (in German).

Kapur SP, Bhussry BR, Rao S, & Harmuth-Hoene E (1974) Percutaneous uptake of zinc in rabbit skin (37927). Proc Soc Exp Biol Med, **145**: 932–937.

Karlsson N, Fangmark I, Haggqvist I, Karlsson B, Rittfeldt L, & Marchner H (1991) Mutagenicity testing of condensates of smoke from titanium dioxide/hexachloroethane and zinc/hexachloroethane pyrotechnic mixtures. Mutat Res, **260**(1): 39–46.

Karra MV, Udipi SA, Kirksey A, & Roepke JL (1986) Changes in specific nutrients in breast-milk during extended lactation. Am J Clin Nutr, **43**(4): 495–503.

Kasarskis EJ & Schuna A (1980) Serum alkaline phosphatase after treatment of zinc deficiency in humans. Am J Clin Nutr, **33**(12): 2609–2612.

Kasprzak KS, Kovatch RM, & Poirier LA (1988) Inhibitory effect of zinc on nickel subsulfide carcinogenesis in Fisher rats. Toxicology, **52**: 253–262.

Katyal JC & Ponnamperuma FN (1974) Zinc deficiency. A widespread nutritional disorder of rice in Agusan del Norte. J Philipp Agric, **58**(3,4): 79–89.

Kauder B (1987) [Natural heavy metal levels in soils.] Wasser Luft Betrieb, **6**: 55–56 (in German).

Kay RG, Tasman-Jones C, Pybus J, Whiting R, & Black H (1976) A syndrome of acute zinc deficiency during total parenteral alimentation in man. Ann Surg, **183**: 331–340.

Keating KI & Caffrey PB (1989) Selenium deficiency induced by zinc deprivation in a crustacean. Proc Natl Acad Sci USA, **86**: 6436–6440.

Keen CL & Gershwin ME (1990) Zinc deficiency and immune function. Annu Rev Nutr, **10**: 415–431.

Keen CL & Hurley LS (1977) Zinc absorption through skin: correction of zinc deficiency in the rat. Am J Clin Nutr, **30**: 528–530.

Keen CL & Hurley LS (1989) Zinc and reproduction: effects of deficiency on foetal and postnatal development. In: Mills B ed. Zinc in human biology. London, Springer, pp 183–220.

Keilin D & Mann T (1940) Carbonic anhydrase. Purification and nature of the enzyme. Biochem J, **34**: 1163–1176.

Kenney MA, Ritchey SJ, Culley P, Sandoval W, Moak S, & Schilling P (1984) Erythrocyte and dietary zinc in adolescent females. Am J Clin Nut, **39**(3): 446–451.

Kersten M & Forstner U (1989) Speciation of trace elements in sediments. In: Batley GE ed. Trace element speciation: analytical methods and problems. Boca Raton FL, CRC Press, pp 245–318.

Kersten M, Dicke M, Kreiws M, Naumann K, Schmidt D, Schulz M, Schwikowski M, & Steiger M (1988) Distribution and fate of heavy metals in the North Sea. In: Salomons W ed. Pollution in the North Sea, Berlin, Springer, pp 300–347.

Ketcheson MR, Barron GP, & Cox DH (1969) Relationships of maternal dietary zinc during gestation & lactaction to development and zinc, iron and copper content of the postnatal rat. J Nutr, **98**: 303–311.

Khangarot BS & Ray PK (1987) Sensitivity of toad tadpoles, *Bufo melanostictus* (Schneider), to heavy metals. Bull Environ Contam Toxicol, **38**: 523–527.

Khangarot BS, Sehgal A, & Bhasin MK (1983) "Man and the Biosphere" - studies on Sikkim Himalayas. Part 1: Acute toxicity of copper and zinc to common carp *Cyprinus carpio* (Linn.) in soft water. Acta Hydrochim Hydrobiol, **11**: 667–673.

Khanum S, Alam AN, Anwar I, Akbar Ali M, & Mujibur Rahaman M (1988) Effect of zinc supplementation on the dietary intake and weight gain of Bangladeshi children recovering from protein-energy malnutrition. Eur J Clin Nutr, **42**(8): 709–714.

Kiekens L (1986) [Heavy metals in soils.] Acad Analecta, **48**: 45.

Kiekens L (1995) Zinc. In: Alloway BJ ed. Heavy metals in soils. 2nd ed. Glasgow, Blackie, pp 284–305.

Kienholz EW, Turk DE, Sunde ML, & Hoekstra WG (1961) Effects of zinc deficiency in the diets of hens. J Nutr, **75**: 211–221.

Kiffney PM & Clements WH (1994) Structural responses of benthic macro-invertebrates communities from different stream orders to zinc. Environ Contam Toxicol, **13**(3): 389–395.

King JC (1986) Assessment of techniques for determining human zinc requirements. J Am Diet Assoc, **86**(11): 1523–1528.

King JC (1990) Assessment of zinc status. J Nutr, **120**: 1474–1479.

King LE & Fraker PJ (1991) Flow cytometric analysis of the phenotypic distribution of splenic lymphocytes in zinc-deficient adult mice. J Nutr, **121**: 1433–1438.

King J & Turnland J (1989) Human zinc requirements. In: Mills B ed. Zinc in human biology. London, Springer, pp 335–350.

Kinnamon KE (1963) Some independent and combined effects of copper, molybdenum, and zinc on the placental transfer of zinc-65 in the rat. J Nutr, **81**: 312–320.

Kirk RE & Othmer DF (1982) Properties of refactory metals. In: Kirk RE & Othmer DF eds. Encyclopedia of chemical technology, 3rd ed, vol 19. Powder coatings to recycling. New York, John Wiley & Sons, p 57.

Kissling MM & Kagi HR (1977) Primary structure of human hepatic metallothionein. Febs Letters, **82**(2): 247–250.

Klerks PL (1990) Adaptations to metals in animals. In: Shaw AJ ed. Heavy metal tolerance in plants: evolutionary aspects. Boca Raton FL, CRC Press, pp 313–321.

Klerks PL & Levinton JS (1989) Effects of heavy metals in a polluted aquatic ecosystem. In: Levin SA, Harwell MA, Kelly JR, & Kimball KD eds. Ecotoxicology: problems and approaches. New York, Springer, pp 41–67.

Klerks PL & Weis JS (1987) Genetic adaptation to heavy metals in aquatic organisms: a review. Environ Pollut, **45**: 173–205.

Klevay LM (1973) Hypercholesterolemia in rats produced by an increase in the ratio of zinc to copper ingested. Am J Clin Nutr, **26**: 1060–1068.

Klevay LM (1975) Coronary heart disease: The zinc/copper hypothesis. Am J Clin Nutr, **28**: 764–774.

Klevay LM (1980) Interactions of copper and zinc in cardiovascular disease. Ann New York Acad Sci, **355**: 140–151.

Klevay LM (1983) Copper and ischemic heart disease. Biol Trace Elem Res, **5**: 245–255.

Klevay LM, Bistrian BR, Fleming CR, & Neumann CG (1987) Hair analysis in clinical and experimental medicine. Am J Clin Nutr, **46**: 233–236.

Knapp JF, Kennedy C, Wasserman GS, & Do Lelli JD (1994) Case 01-1994: a toddler with caustic ingestion. Pediatr Emergency Care, **10**(1): 54–58.

Knotkova D & Porter F (1994) Longer life of galvanised steel in the atmosphere due to reduced SO_2 pollution in Europe. Proceedings of Intergalva1994, Paris. Birmingham, EGGA, pp GD 8/1–GD 8/20.

Knotkova D, Kreislova K, & Boschek P (1995) Trends of corrosivity based on corrosion rates and pollution data. UN/ECE international cooperative programme on effects on materials, including historic and cultural monuments, Report 19.

Konar SK & Mullick S (1993) Problems of safe disposal of petroleum products, detergents, heavy metals and pesticides to protect aquatic life. Sci Total Environ, **Suppl. 2**: 989–1000.

Koo SI, Lee CC, & Norvell JE (1987) Effect of marginal zinc deficiency on the apolipoprotein-B content and size of mesenteric lymph chylomicrons in adult rats. Lipids, **22**(12): 1035–1040.

Kopp JF & Kroner RC (1968) Trace metals in waters in the United States. Federal Water Pollution Control Administration Division Pollution Survey. Cincinnati, Ohio, US Environmental Protection Agency.

Korant BD & Butterworth BE (1976) Inhibition by zinc of rhinovirus protein cleavage: Interaction of zinc with capsid polypeptides. J Virol, **18**(1): 298–306.

Korver RM & Sprague JB (1989) Zinc avoidance by fathead minnows (*Pimephales promelas*): computerized tracking and greater ecological relevance. Can J Fish Aquat Sci, **46**: 494–502.

Kowalska-Wochna E, Moniuszko-Jakoniuk J, Kulikowska E, & Miniuk K (1988) The effect of orally applied aqueous solutions of lead and zinc on chromosome aberrations and induction of sister chromatid exchanges in the rat (Rattus sp.). Genetica Pol, **29**(2): 181–189.

Kozik (1981) Neurosecretion of the hypothalalmo-hypophyseal system after intragastric administration of zinc oxide. Folia Histochem Cytochem, pp 115–122.

Kraak MHS, Toussaint M, Lavy D, & Davids C (1994a) Short-term effects of metals on the filtration rate of the zebra mussel Dreissena polymorpha. Environ Pollut, **84**: 139–143.

Kraak MHS, Lavy D, Schoon H, Toussaint M, Peeters WHM, & Van Straalen NM (1994b) Ecotoxicity of mixtures of metals to the zebra mussel Dreissene polymorpha. Environ Toxicol Chem, **13**: 109–114.

Krebs NF, Hambidge KM, Jacobs MA, & Oliva-Rasbach J (1985) The effects of a dietary zinc supplement during lactation on longitudinal changes in maternal zinc status and milk zinc concentrations. Am J Clin Nutr, **41**: 560–570.

298

Kumar S (1976) Effect of zinc supplementation on rats during pregnancy. Nutr Rep Int, **13**: 33–36.

Kumar S & Pant SC (1984) Comparative effects of the sublethal poisoning of zinc, copper and lead on the gonads of the teleost *Puntius conchonius* Ham. Toxicol Lett, **23**: 189–194.

Kynast G & Saling E (1986) Effect of oral zinc application during pregnancy. Gynecol Obstet Invest, **21**: 117–123.

L'Abbe MR & Fischer PWF (1984) The effects of dietary zinc on the activity of copper-requiring metalloenzymes in the rat. J Nutr, **114**: 823–828.

Lahmann E (1987) [Ambient air measurements in the Federal Republic of Germany.] Staub Rein Luft, **47**(3-4): 82–87.

Laitinen R, Vuori E, & Viikari J (1989) Serum zinc and copper: associations with cholesterol and triglyceride levels in children and adolescents. J Am Coll Nutr, **8**(5): 400–406.

Lake DLP, Kirk WW, & Lester JN (1984) Fractionation, characterization, and speciation of heavy metals in sewage sludge and sludge-amended soils: a review. J Environ Qual, **13**: 175–183.

Lam HF, Peisch R, & Amdur MO (1982) Changes in lung volumes & diffusing capacity in guinea pigs exposed. Toxicol Appl Pharmacol, **66**: 427–433.

Lam HF, Conner MW, & Rogers AE (1985) Functional & morphologic changes in the lungs of guinea pigs exposed to freshly generated ultrafine zinc oxide. Toxicol Appl Pharmacol, **78**: 29–38.

Lam HF, Chen LC, Ainsworth D, Peoples S, & Amdur MO (1988) Pulmonary function of guinea pigs exposed to freshly generated untrafine zinc oxide with and without spike concentrations. Am Ind Hyg Assoc J, **49**(7): 333–341.

Lane HW, Warren DC, Squyres NS, & Cotham AC (1982) Zinc concentrations in hair, plasma, and saliva and changes in taste acuity of adults supplemented with zinc. Biol Trace Elem Res, **4**: 83–93.

Lansdown ABG (1991) Interspecies variations in response to topical application of selected zinc compounds. Food Chem Toxicol, **29**: 57–64.

Lantzy RJ & Mackenzie FT (1979) Atmospheric trace metals: global cycles and assessment of man's impact. Geochim Cosmochim Acta, **43**: 511–525.

LaPerriere JD, Wagener SM, & Bjerklie A (1985) Gold-mining effects on heavy metals in streams, circle quadrangle, Alaska. Water Resour Bull, **21**: 245–252.

Larson L & Hyland J (1987) Ambient water quality criteria for zinc. Washington DC, Environmental Protection Agency (EPA 440/5-87/003) (Fiche).

Laryea MD, Schnittert B, Kersting M, Willhelm M, & Lombeck I (1995) Macronutrient, copper, and zinc intakes of young German children as determinated duplicate food samples and diet records. Ann Nutr Metab, **39**: 271–278.

Lazo JS, Kondo Y, Dellapiazza D, Michalska AE, Choo KHA, & Pitt BR (1995) Enhanced sensitivity to oxidative stress in cultured embryonic cells from transgenic mice deficient in metallothionein I and II genes. J Biol Chem, **270** (10): 5506–5510.

LeBlanc GA (1982) Laboratory investigations into the development of resistance of *Daphnia magna* (Straus) to environmental pollutants. Environ Pollut, **A27**: 309–322.

Lee RE Jr & Von Lehmden DJ (1973) Trace metal pollution in the environment. J Air Pollut Control Assoc, **23**: 853–857.

Lee MS, Gippert GP, Soman KV, Case DA, & Wright PE (1989a) Three-dimensional solution structure of a single zinc finger DNA-binding domain. Science, **245**: 635–637.

Lee HH, Prasad AS, Brewer GJ, & Owyang C (1989b) Zinc absorption in human small intestine. Am J Physiol, **256**: G87–G91.

Lennard ES (1980) Implications in the burn neutrophil of serum and cellular zinc levels. J Surg Res, **29**: 75–82.

Leonard A & Gerber GB (1989) Zinc toxicity: Does it exist? J Am Coll Toxicol, **8**(7): 1285–1290.

Letts D, Kinnes GM, & Blade L (1991) Health hazard evaluation. In: National Institute for Occupational Safety and Health ed. Hastings MI, US Department of Health and Human Services, Public Health Service, Centers for Disease Control, pp 1–36 (Health hazard evaluation report 89-267-2139).

Levine MB, Hall AT, Barrett GW, & Taylor DH (1989) Heavy metal concentrations during ten years of sludge treatment in an old-field community. J Environ Qual, **18**: 411–418.

Lide DR (1991) Physical constants of inorganic compounds. In: Lide DR ed. CRC handbook of chemistry and physics. A ready-reference book of chemical and physical data, 71st ed. Boca Raton FL, CRC Press, pp 4/116–4/118.

Lin HC & Dunson WA (1993) The effect of salinity on the acute toxicity of cadmium to the tropical, estuarine, hermaphroditic fish, *Rivulus marmoratus:*. Arch Environ Contam Toxicol, **25**: 41–47.

Lin J, Chan WC, Fong YY, & Newberne PM (1977) Zinc levels in serum, hair and tumors from patients with esophageal cancer. Nutr Rep Int, **15**: 635–643.

Lindberg E, Ekholm U, & Ulfvarson U (1985) Extent and conditions of exposure in the Swedish chrome plating industry. Arch Occup Environ Health, **50**: 197–205.

Lindén E, Bengtsson B-E, Svanberg O, & Sundström G (1979) The acute toxicity of 78 chemicals and pesticide formulations against two brackish water organisms, the bleak

(*Alburnus alburnus*) and the harpacticoid *Nitacro spinipes*. Chemosphere, **8**(11/12): 843–851.

Lindqvist L & Block M (1994) Excretion of cadmium and zinc during moulting in the grasshopper *Omocestus viridulus* (Orthoptera). Environ Toxicol Chem, **13**(10): 1669–1672.

Lingle JC & Holmberg DM (1957) The response of sweet corn to foliar and soil zinc application on a zinc deficient soil. Proc Amer Soc Hort Sci, **70**: 308–315.

Linn WS, Kleinman MT, Bailey RM, Medway DA, Spier CE, Whynot JD, Anderson KR, & Hackney JD (1981) Human respiratory responses to an aerosol containing zinc ammonium sulfate. Environ Res, **25**: 404–414.

Lioy PJ, Wolff GT, & Kneip TJ (1978) Toxic airborne elements in the New York metropolitan area. J Air Pollut Control Assoc, **28**: 510–512.

Lipman TO, Diamond A, Mellow MH, & Patterson KY (1987) Esophageal zinc content in human squamous esophageal cancer. J Am Coll Nutr, **6**(1): 41–46.

Liu Z, Zhu QQ, & Tang LH (1983) Microelements in the main soils of China. Soil Sci, **135**: 40–46.

LIZ (1992) Life in New Zealand Activities and Health Research Unit, Technical Report No. 26. Dunedin, University of Otago.

Llobet JM, Domingo JL, Colomina MT, Mayayo E, & Corbella J (1988) Subchronic oral toxicity of zinc in rats. Bull Environ Contam Toxicol, **41**(1): 36–43.

Lloyd TB & Showak W (1984) Zinc and zinc alloys. In: Grayson M ed. Encyclopedia of chemical technology 3rd ed. Vol. 24: Vitamins to zone refining. New York, John Wiley & Sons, pp 835–836.

Loennerdal B, Cederblad Å, Davidsson L, & Sandstroem B (1984) The effect of individual components of soy formula and cow's milk formula on zinc bioavailability. Am J Clin Nutr, **40**: 1064–1070.

Lokken PM, Halas ES, & Sandstead HH (1973) Influence of zinc deficiency on behavior. Proc Soc Exp Biol Med, **144**: 680–682.

Long SE & Martin TD (1991) Determination of trace elements in waters and wastes by inductively coupled plasma mass spectrometry. In: Methods for the determination of metals in environmental samples, EPA 600/4-91-010. Cincinnati, Ohio, Environmental Monitoring Systems Laboratory, US EPA, pp 83–122.

Luecke RW, Charles E, Simonel CE, & Fraker PJ (1978) The effect of restricted dietary intake on the antibody medicated response of the zinc deficient A/J mouse. J Nutr, **108**: 881–887.

Luisi BF, Xu WX, Otwinowski Z, Freedman LP, Yamamoto KR, & Sigler PB (1991) Crystallographic analysis of the interaction of glucocorticoid receptor with DNA. Nature, **352**: 497–505.

Lum KR, Kokotich EA, & Schroeder WH (1987) Bioavailable Cd, Pb and Zn in wet and dry deposition. Sci Total Environ, **63**: 161–173.

Luten JB, Bourquet JB, Burggraaf W, Rauchbaar AB, & Rus J (1986) Trace metals in mussels (*Mytilus edulis*) from the Waddenzee, coastal North Sea and the estuaries of the Ems, western and eastern scheldt. Bull Environ Contam Toxicol, **36**: 770–777.

Lykken GI, Mahalko J, Johnson PE, Miline D, Sandstead HH, Garcia WJ, Dintzis FR, & Inglett GE (1986) Effect of browned and unbrowned corn products intrinsically labelled with zinc on absorption of zinc in humans. J Nutr, **116**: 795–801.

Lynch SM & Klevay LM (1992) Effects of a dietary copper deficiency on plasma coagulation factor activities in male and female mice. J Nutr Biochem, **3**: 387–391.

Macapinlac MP, Pearson WN, & Darby WJ (1966) Some characteristics of zinc deficiency in the albino rat. In: Prasad AS ed. Zinc metabolism. Prasad, Thomas, Springfield IL, pp 142–168.

Macapinlac MP, Barney GH, Pearson WN, & Darby WJ (1967) Production of zinc deficiency in the squirrel monkey. J Nutr, **93**: 499–510.

Macapinlac MP, Pearson WN, Barney GJ, & Darby WJ (1968) Protein and nucleic acid metabolism in the testes of zinc-deficient rats. J Nutr, **95**: 569–577.

Macdonald LD, Gibson RS, & Miles JE (1982) Changes in hair zinc and copper concentrations of breast fed and bottle fed infants during the first six months. Acta Paediatr Scand, **71**: 785–789.

Macdonald JM, Shields JD, & Zimmer-Faust RK (1988) Acute toxicities of eleven metals to early life-history stages of the yellow crab *Cancer anthonyi*. Mar Biol, **98**: 201–207.

Machholz R & Lewerenz HJ (1989) [Food toxicology.] Berlin, Springer-Verlag (in German).

MacLean AJ (1974) Effects of soil properties and amendments on the availability of zinc in soils. Can J Soil Sci, **54**: 369–378.

Madoni P, Esteban G, & Gorbi G (1992) Acute toxicity of cadmium, copper, mercury, and zinc to ciliates from activated sludge plants. Bull Environ Contam Toxicol, **49**: 900–905.

Madoni P, Davoli D, & Gorbi G (1994) Acute toxicity of lead, chromium, and other heavy metals to ciliates from activated sludge plants. Bull Environ Contam Toxicol, **53**: 420–425.

Magee AC & Matrone G (1960) Studies on growth, copper metabolism and iron metabolism on rats fed high levels of zinc. J Nutr, **72**: 233–242.

Magliette RJ, Doherty FG, McKinney D, & Venkataramani ES (1995) Need for environmental quality guidelines based on ambient freshwater quality criteria in natural waters – case study "zinc". Bull Environ Contam Toxicol, **54**: 626–632.

302

Mahloudji M, Reinhold JG, Haghasenass M, Ronaghy HA, Spivey-Fox MRS, & Halsted JA (1975) Combined zinc and iron compared with iron supplementation of diets of 6- to 12-year-old village schoolchildren in southern iron. Am J Clin Nutr, **28**: 721–725.

Mahomed K, James D, Golding J, & McCade R (1989) Zinc supplementation during pregnancy: a double blind randomized trial. Br Med J, **299**: 826–830.

Maita K, Hirano M, Mitsumori K, Takahashi K, & Shirasu Y (1981) Subacute toxicity studies with zinc sulfate in mice and rats. J Pestic Sci, **6**: 327–336.

Malagrino W & Mazzilli B (1994) Use of ^{65}Zn as a radiotracer in the bioaccumulation study of zinc by *Poecilia reticulata*. J Radioanal Nucl Chem, **183**(2): 389–393.

Malecki MR, Neuhauser EF, & Loehr RC (1982) The effect of metals on the growth and reproduction of *Eisenia foetida* (Oligochaeta, Lumbricidae). Pedobiologia, **24**: 129–137.

Malle K-G (1992) [Zinc in the environment.] Acta Hydrochim Hydrobiol, **20**: 196–204 (in German).

Malo JL, Malo J, Cartier A, & Dolovich J (1990) Acute lung reaction due to zinc inhalation. Eur Respir J, **3**(1): 111–114.

Malo JL, Cartier A, & Dolovich J (1993) Occupational asthma due to zinc. Eur Respir J, **6**(3): 447–450.

Maltby L & Naylor C (1990) Preliminary observations on the ecological relevance of the *Gammarus* 'scope for growth' assay: effect of zinc on reproduction. Funct Ecol, **4**: 393–397.

Mann H, Fyfe WS, Kerrich R, & Wiseman M (1989) Retardation of toxic heavy metal dispersion from nickel-copper mine tailings, Sudbury district, Ontario: Role of acidophilic microorganisms I. Biological pathway of metal retardation. Biorecovery, **1**: 155–172.

Mantoura RFC & Riley JP (1975) The use of gel filtration in the study of metal binding by humic acids and related compounds. Anal Chim Acta, **78**: 193–200.

Marigomez JA, Angulo E, & Saez V (1986) Feeding and growth responses to copper, zinc, mercury and lead in the terrestrial gastropod *Arion ater* (Linne). J Mol Stud, **52**: 68–78.

Markowitz ME, Rosen JF, & Mizruchi M (1985) Circadian variations in serum zinc (Zn) concentrations: correlation with blood ionized calcium, serum total calcium and phosphate in humans. Am J Clin Nutr, **41**: 680–696.

Marquart H, Smid T, Heederik D, & Visschers M (1989) Lung function of welders of zinc-coated mild steel: cross-sectional analysis and changes over five consecutive work shifts. Am J Ind Med, **16**: 289–296.

Marrs TC, Clifford WE, & Colgrave HF (1983) Pathological changes produced by exposure of rabbits and rats to smokes from mixture of hexachloroethane and zinc oxide. Toxicol Lett, **19**: 247–252.

Marrs TC, Colgrave HF, Edginton JAG, Brown RFR, & Cross NL (1988) The repeated dose toxicity of a zinc oxide/hexachloroethane smoke. Arch Toxicol, **62**: 123–132.

Marschner H (1995) Mineral nutrition of higher plants, 2nd ed. London, Academic Press, pp 347–364.

Marshall JS, Parker JI, Mellinger DL, & Lei C (1983) Bioaccumulation and effects of cadmium and zinc in a Lake Michigan plankton community. Can J Fish Aquat Sci, **40**: 1469–1479.

Mart L (1979) Prevention of contamination and other occurring risks in voltammetric trace metal analysis of natural waters. I. Preparatory steps of filtration and storage of water samples. Fresenius Z Anal Chem, **296**: 350–356.

Martin TR & Holdich DM (1986) The acute lethal toxicity of heavy metals to peracarid crustaceans (with particular reference to fresh-water asellids and gammarids). Water Res, **20**(9): 1137–1147.

Martin M, Hunt JW, Anderson BS, & Palmer FH (1989) Experimental evaluation of the mysid *Holmesimysis costata* as a test organism for effluent toxicity testing. Environ Toxicol Chem, **8**: 1003–1012.

Martin TD, Martin ER, Lobring LB, & McKee GD (1991) Determination of metals in fish tissue by inductively coupled plasma atomic emission spectrometry. In: Methods for the determination of metals in environmental samples. Cincinnati, Ohio, Environmental Monitoring Systems Laboratory, US EPA, pp 177–209 (EPA 600/4-91-010).

Marzin DR & Vo PH (1985) Study of the mutagenicity of metal derivatives with Salmonella typhimurium TA102. Mutat Res, **155**: 49–51.

Mason AZ (1988) The kinetics of zinc accumulation by the marine prosobranch gastropod *Littorina littorea*. Mar Environ Res, **24**: 135–139.

Masters BA, Kelly EJ, Quaife CJ, & Brinster RL (1994) Targeted disruption of metallothionein I and II genes increases sensitivity to cadmium. Proc Natl Acad Sci, **91**: 584–588.

Matarese SL & Matthews JI (1986) Zinc chloride (smoke bomb) inhalation lung injury. Chest, **89**: 308–309.

Mathys W (1977) The role of malate, oxalate, and mustard oil glucosides in the evolution of zinc-resistance in herbage plants. Physiol Plant, **40**: 130–136.

Matseshe JW, Phillips SF, Malagelada J-R, & McCall JT (1980) Recovery of dietary iron and zinc from the proximal intestine of healthy man: studies of different meals and supplements. Am J Clin Nutr, **33**: 1946–1953.

Mayer FL (1987) Zinc sulfate. In: Acute toxicity handbook of chemicals to estuarine organisms. Gulf Breeze FL, Environmental Research Laboratory, US EPA (PB87-188686).

Mayer FL & Ellersieck MR (1986) Zinc sulphate. In: Manual of acute toxicity: interpretation and data base for 410 chemicals and 66 species of freshwater animals. Washington DC, US Department of the Interior, Fish & Wildlife Service (Resource Publication 160).

McCord CP & Friedlander MD (1926) An occupational disease among zinc workers. Arch Intern Med, **37**: 641–659.

McKeague JA & Wolynetz MS (1980) Background levels of minor elements in some Canadian soils. Geoderma, **24**: 299–307.

McKenna IM, Chaney RL, Tao SH, Leach RM, & Williams FM (1992) Interactions of plant zinc and plant species on the bioavailability of plant cadmium to Japanese quail fed lettuce and spinach. Environ Res, **57**: 73–87.

McKenzie JM, Fosmire GJ, & Sandstead HH (1975) Zinc deficiency during the late third of pregnancy: Effects on fetal rat brain, liver, and placenta. J Nutr, **105**: 1466–1475.

McLusky DS & Hagerman L (1987) The toxicity of chromium, nickel and zinc: effects of salinity and temperature, and the osmoregulatory consequences in the mysid *Praunus flexuosus*. Aquat Toxicol, **10**: 225–238.

McNall AD, Etherton TD, & Fosmire GJ (1995) The impaired growth induced by zinc deficiency in rats is associated with decreased expression of the hepatic insulin-like growth factor I and growth hormone receptor genes. J Nutr, **125**(4): 874–879.

Mehrotra A, Mehrotra I, & Tandon SN (1989) Speciation of copper and zinc in sewage sludge. Environ Technol Lett, **10**(2): 195–200.

Meinel W & Krause R (1988) [The correlation between zinc and various pH values in its toxic effect on several groundwater organisms]. Z Angew Zool, **75**(2): 159–182 (in German).

Meisner JD & Quan Hum W (1987) Acute toxicity of zinc to juvenile and subadult rainbow trout, *Salmo gairdneri*. Bull Environ Contam Toxicol, **39**: 898–902.

Melin A & Michaelis H (1983) [Zinc.] In: Bartholomé E, Biekert E, Hellmann H, Ley H, Weigert WM, & Weise E eds. (Ullmanns Encyclopedia of technical chemistry) 4th ed. Weinheim, Verlag Chemie, pp 593–626 (in German).

Memmert U (1987) Bioaccumulation of zinc in two freshwater organisms (*Daphnia magna*, Crustacea and *Brachydanio rerio*, Pisces). Water Res, **21**(1): 99–106.

Menard MP & Cousins RJ (1983) Zinc transport by brush border membrane vesicles from rat intestine. J Nutr, **113**: 1434–1442.

Mench M, Vangronsveld J, Didier V, & Clijsters H (1994) Evaluation of metal mobility, plant availability and immobilization by chemical agents in a limed-silty soil. Environ Pollut, 86(3), 279–286.

Meranger JC, Subramanian KS, & Chalifoux C (1981) Survey for cadmium, cobalt, chromium, copper, nickel, lead, zinc, calcium, and magnesium in Canadian drinking water. J Assoc Off Anal Chem, 64: 44–53.

Mertz W (1987) Trace elements in human and animal nutrition, 15 ed., vol. 2. San Diego, Academic Press.

Messerle BA, Schaffer A, Vasak M, Kagi JHR, & Vuthrich K (1990) Three-dimensional structure of human [^{113}Cd$_7$]metallothionein-2 in solution determined by nuclear magnetic resonance spectroscopy. Mol Biol, 214: 765–779.

Methfessel AH & Spencer H (1973) Zinc metabolism in the rat. I. Intestinal absorption of zinc. J Appl Physiol, 34: 58–62.

MG (1994) The world of metals: zinc. Frankfurt am Main, Metallgesellschaft.

Michalska AE & Choo KHA (1993) Targeting and germ-line transmission of a null mutation at the metallothionein I and II loci in mouse. Proc Natl Acad Sci, 90: 8088–8092.

Millan JL (1987) Promoter structure of the human intestinal alkaline phosphatase gene. Nucleic Acid Res, 15(24): 10599.

Miller MP & Hendricks AC (1996) Zinc resistance in Chironomus riparius: evidence for physiological and genetic components. J N Am Benthol Soc, 15: 106–116.

Milliken J, Waugh D, & Kadish ME (1963) Acute interstitial pulmonary fibrosis caused by a smoke bomb. Can Med Assoc, 88: 36–39.

Mills CF, Quarterman J, Williams RB, Dalgarno AC, & Panic B (1967) The effects of zinc deficiency on pancreatic carboxypeptidase activity and protein digestion and absorption in the rat. Biochem J, 102: 712–718.

Milne DB, Canfield WK, Mahalko JR, & Sandstead HH (1983) Effect of dietary zinc on whole body surface loss of zinc: impact on estimation of zinc retention by balance method. Am J Clin Nutr, 38: 181–186.

Milne DB, Ralston NVC, & Wallwork JC (1985) Zinc content of cellular components of blood: methods for cell separation and analysis evaluated. Clin Chem, 31(1): 65–69.

Milne DB, Canfield WK, Gallagher SK, Hunt JR, & Klevay LM (1987) Ethanol metabolism in postmenopausal women fed a diet marginal in zinc. Am J Clin Nutr, 46: 688–693.

Mirenda RJ (1986) Acute toxicity and accumulation of zinc in the crayfish, Orconectes virilis (Hagen). Bull Environ Contam Toxicol, 37: 387–394.

Misra SG & Tiwari RC (1966) Retention and release of copper and zinc by some Indian soils. Soil Sci, **101**(6): 465–471.

Mitchell SD & Fretz TA (1977) Cadmium and zinc toxicity in white pine, red maple, and Norway spruce. J Am Soc Hortic Sci, **102**(1): 81–84.

Miyaki M, Murata I, Osabe M, & Ohno T (1977) Effect of metal cations on mis-incorporation by E. coli DNA polymerases. Biochem Biophys Res Commun, **77**: 854–860.

Morrison SA, Russell RM, Carney EA, & Oaks EV (1978) Zinc deficiency: a cause of abnormal dark adaptation in cirrhotics. Am J Clin Nutr, **31**(2): 276–281.

Mortvedt JJ & Gilkes RJ (1993) Zinc fertilizers. In: Robson AD ed. Developments in plant and soil sciences, vol. 55. Zinc in soils and plants. Dordrecht, Klewer, pp 33–44.

Moser-Veillon PB & Reynolds RD (1990) A longitudinal study of pyridoxine and zinc supplementation of lactating women. Am J Clin Nutr, **52**(1): 135–141.

Mount DI & Norberg TJ (1984) A seven-day life-cycle cladoceran toxicity test. Environ Toxicol Chem, **3**: 425-434.

Mount DR, Barth AK, Garrison TD, Barten KA, & Hockett JR (1994) Dietary and waterborne exposure of rainbow trout (*Oncorhynchus mykiss*) to copper, cadmium, lead and zinc using a live diet. Environ Toxicol Chem, **13**(12): 2031–2041.

Mueller EJ & Seger DL (1985) Metal fume fever: A review. J Emerg Med, **2**: 271–274.

Mueller G & Furrer R (1994) [The burden of heavy metals in the Elbe.] Naturwiss, **81**: 401–405 (in German).

Mukherjee MD, Sandstead HH, Ratnaparkhi MV, Johnson LK, Milne DB, & Stelling HP (1984) Maternal zinc, iron, folic acid, and protein nutriture and outcome of human pregnancy. Am J Clin Nutr, **40**: 496–507.

Mulhern SA, Stroube WBJ, & Jacobs RM (1986) Alopecia induced in young mice by exposure to excess dietary zinc. Experientia, **42**(5): 551–553.

Münzinger A & Guarducci ML (1988) The effect of low zinc concentrations on some demographic parameters of *Biomphalaria glabrata* (Say), mollusca: gastropoda. Aquat Toxicol, **12**: 51–61.

Münzinger A & Monicelli F (1991) A comparison of the sensitivity of three *Daphnia magna* populations under chronic heavy metal stress. Ecotoxicol Environ Saf, **22**: 24–31.

Murata K, Araki S, & Aono H (1987) Effects of lead, zinc and copper absorption on peripheral nerve conduction in metal workers. Int Arch Occup Environ Health, **59**: 11–20.

Murphy JV (1970) Intoxication following ingestion of elemental zinc. J Am Med Assoc, **212**(12): 2119–2120.

Murray MJ & Flessel CP (1976) Metal-polynucleotide interactions. A comparison of carcinogenic and non-carcinogenic metals in vitro. Biochim Biophys Acta, **425**: 256–261.

Nakatsu C & Hutchinson TC (1988) Extreme metal and acid tolerance of Euglena mutabilis and an associated yeast from Smoking Hills, Northwest Territories, and their apparent mutualism. Microbiol Ecol, **16**: 213–231.

Nanji AA & Anderson FH (1983) Relationship between serum zinc and alkaline phosphatase. Hum Nutr Clin Nutr, 37(6): 461–462.

Naylor C, Pindar L, & Calow P (1990) Inter- and intraspecific variation in sensitivity to toxins; the effects of acidity and zinc on the freshwater crustaceans *Asellus aquaticus* (L.) and *Gammarus pulex* (L.) Water Res, **24**(6): 757–762.

Nebeker AV, Savonen C, Baker RJ, & McCrady JK (1984) Effects of copper, nickel and zinc on the life cycle of the caddisfly *Clistoronia magnifica* (Limnephilidae). Environ Toxicol Chem, **3**: 645–649.

Neggers YH, Cutter GR, Acton RT, Alvarez JO, Bonner JL, Goldenberg RL, Go R, & Roseman JM (1990) A positive association between maternal serum zinc concentration and birth weight. Am J Clin Nutr, **51**: 678–684.

Negilski DS, Ahsanullah M, & Mobley MC (1981) Toxicity of zinc, cadmium and copper to the shrimp *Callianassa australiensis*. II. Effects of paired and triad combinations of metals. Mar Biol, **64**: 305–309.

Nelson DA, Miller JE, & Calabrese A (1988) Effect of heavy metals on bay scallops, surf clams, and blue mussels in acute and long-term exposures. Arch Environ Contam Toxicol, **17**: 595–600.

Nemcsók J, Németh Á, Buzás Z, & Boross L (1984) Effects of copper, zinc and paraquat on acetylcholinesterase activity in carp (*Cyprinus carpio* L.). Aquat Toxicol, **5**: 23–31.

Neuhauser EF, Malecki MR, & Loehr RC (1984) Growth and reproduction of the earthworm *Eisenia fetida* after exposure to sublethal concentrations of metals. Pedobiologia, **27**: 89–97.

Neuhauser EF, Loehr RC, Milligan DL, & Malecki MR (1985) Toxicity of metals to the earthworm *Eisenia fetida*. Biol Fertil Soils, 1: 149–152.

Neumueller O-A (1983) [Roempp's chemical dictionary.] 8th ed. Stuttgart, Franckh'sche Verlagshandlung, pp 1671, 1364, 4376 (in German).

Nève J, Hanocq M, Peretz A, Abi Khalil F, Pelen F, Famaey JP, & Fontaine J (1991) Pharmacokinetic study of orally administered zinc in humans: evidence for an enteral recirculation. Eur J Drug Metab Pharmacokinet, **16**: 315–323.

Newsome DA, Swartz M, Leone NC, Elston RC, & Miller E (1988) Oral zinc in macular degeneration. Arch Ophthalmol, **106**: 192–198.

Niederlehner BR & Cairns J (1993) Effects of previous zinc exposure on pH tolerance of periphyton communities. Environ Toxicol Chem, **12**: 743–753.

Niethammer KR, Atkinson RD, Baskett TS, & Samson FB (1985) Metals in riparian wildlife of the lead mining district of southeasthern Missouri. Arch Environ Contam Toxicol, **14**: 213–223.

Nigam PK, Tyagi S, Saxena AK, & Misra RS (1988) Dermatitis from zinc pyrithione. Contact Dermatitis, **19**(3): 219.

Ninh NX, Thissen JP, Collette L, Gerard G, Khoi HH, & Ketelslegers JM (1996) Zinc supplementation increases growth and circulating insulin-like growth factor I (IGF-I) in growth-retarded Vietnamese children. Am J Clin Nutr, **63**(4): 514–519.

NIOSH (1984) Zinc and compounds, as Zn. In: Eller PM & US Department of Health and Human Service ed. NIOSH manual of analytic methods 3rd ed. Vol. 2 Zinc and compounds, as Zn, Method 7030. Cincinnati OH, National Institute for Occupational Safety and Health, pp 1–3.

Nishioka H (1975) Mutagenic activities of metal compounds in bacteria. Mutat Res, **31**: 185-189.

Norberg AB & Molin N (1983) Toxicity of cadmium, cobalt, uranium and zinc to *Zoogloea ramigera*. Water Res, **17**(10): 1333–1336.

Norberg TJ & Mount DI (1985) A new fathead minnow (*Pimephales promelas*) subchronic toxicity test. Environ Toxicol Chem, **4**: 711–718.

Notenboom J, Cruys K, Heokstra J, & Van Beelen P (1992) Effect of ambient oxygen concentration upon the acute toxicity of chlorophenols and heavy metals to the groundwater copepod *Parastenocaris germanica* (Crustacea). Ecotoxicol Environ Saf, **24**: 131–143.

NPRI (1994) National Pollutant Release Inventory (Canada). Ottawa, Environment Canada.

Nriagu JO (1989) A global assessment of natural sources of atmospheric trace metals. Nature, **338**: 47–49.

Nriagu JO & Davidson CI (1980) Zinc in the atmosphere. In: Nriagu JO ed. Zinc in the environment, Vol. 1. New York, John Wiley & Sons.

Nriagu JO & Pacyna JM (1988) Quantitative assessment of worldwide contamination of air, water and soils by trace metals. Nature, **333**: 134–139.

Nriagu JO, Larson G, Wong HKT, & Cheam V (1996) Dissolved trace metals in Lake superior, Eire, and Ontario. Environ Sci Technol, **30**: 178–187.

Nugegoda D & Rainbow PS (1987) The effects of temperature on zinc regulation by the decapod crustacean *Palaemon elegans* Rathke. Ophelia, **27**: 17–30.

Oberleas D, Muhrer ME, & O'Dell BL (1962) Effects of phytic acid on zinc availability and parakeratoses in swine. J Anim Sci, **21**: 57–61.

O'Dell BL (1968) Trace elements in embryonic development. Fed Proc, **27**(1): 199–204.

O'Dell BL & Savage JE (1960) Effect of phytic acid on zinc availability. Proc Soc Exp Biol Med, **104**: 304–306.

O'Dell BL, Newberne PM, & Savage J (1959) Significance of dietary zinc for the growing chicken. J Nutr, **65**: 508–512.

OECD (1995) Report of the OECD Workshop on environmental hazard/risk assessment. OECD Environment Monograph No. 105. Environment Directorate, Paris, Organisation for the Economic Co-operation and Development, pp 110.

Oestreicher P & Cousins RJ (1985) Copper and zinc absorption in rat: mechanism of mutual antagonism. J Nutr, **115**: 159–166.

Oestreicher P & Cousins RJ (1989) Zinc uptake by basolateral membrane vesicles from rat small intestine. J Nutr, **119**: 639–646.

Ohnesorge FK & Wilhelm M (1991) Zinc. In: Merian E ed. Metals and their compounds in the environment. Occurrence, analysis, and biological relevance. Weinheim, VCH, pp 1309–1342.

Ohno H, Doi R, Yamamura K, Yamashita K, Lizuka S, & Taniguchi N (1985) A study of zinc distribution in erythrocytes of normal humans. Blut, **50**: 113–116.

Oikari A, Kukkonen J, & Virtanen V (1992) Acute toxicity of chemicals to Daphnia magna in humic waters. Sci Total Environ, **117/118**: 367–377.

OSPARCOM (1994) Draft description of best available techniques for the primary production of non-ferrous metals (zinc, copper, lead and nickel works).

Ottley CJ & Harrison RM (1993) Atmospheric dry deposition flux of metallic species to the North Sea. Atmos Environ, **27A**: 685–695.

Pacyna JM, Bartonova A, Cornille P, & Maenhaut W (1989) Modelling of long-range transport of trace elements. A case study. Atmos Environ, **23**(1): 107–114.

Pal N & Pal B (1987) Zinc feeding and conception in the rat. Int J Vit Nutr Res, **57**: 437–440.

Palmiter RD, Sandgren EP, Koeller DM, & Brinster RL (1993) Distal regulatory elements from the mouse metallothionein locus stimulate gene expression in transgenic mice. Mol Cell Biol, **13**(9): 5266–5275.

310

Pare CM & Sandler M (1954) Smoke bomb pneumonitis: description of a case. J Royal Army Med Corps, **100**: 320–322.

Patterson JW, Allen HE, & Scala JJ (1977) Carbonate precipitation for heavy metals pollutants. J Water Pollut Control Fed, **49**: 2397–2410.

Paulauskis JD & Winner RW (1988) Effects of water hardness and humic acid on zinc toxicity to *Daphnia magna* Straus. Aquat Toxicol, **12**: 273–290.

Pavletich NP & Pabo CO (1991) Zinc finger-DNA recognition: Crystal structure of a Zif268-DNA complex at 2.1 A. Science, **252**: 809–817.

Pécoud A, Donzel P, & Schelling JL (1975) Effect of foodstuffs on the absorption of zinc sulfate. Clin Pharmacol Ther, **17**: 469–474.

Pedroli GBM, Maasdam WAC, & Verstraten JM (1990) Zinc in poor sandy soils and associated groundwater. A case study. Sci Total Environ, **91**: 59–77.

Peirson DH, Cawse PA, Salmon L, & Cambray RS (1973) Trace elements in the atmospheric environment. Nature, **241**: 252–256.

Pekarek RS, Sandstead HH, Jacob RA, & Barcome DF (1979) Abnormal cellular immune response during acquired zinc deficiency. Am J Clin Nutr, **32**: 1466–1471.

Penland JG (1991) Cognitive performance effects of low zinc (Zn) intakes in healthy adult men. FASEB J, **5**: A938.

Penland JG, Sandstead HH, Alcock NW, Dayal HH, Chen XC, Li JS, Zhao F, & Yang JJ (1997) A preliminary report: effects of zinc and micronutrient repletion on growth and neurophsychological function of urban Chinese children. J Am Coll Nutr, **16**(3): 268–272.

Pennington JAT & Young BE (1991) Total diet study nutritional elements, 1982–1989. J Am Diet Assoc, **91**: 179–183.

Pennington JAT, Schoen SA, Salmon GD, Young B, Johnson RD, & Marts RW (1995) Composition of core foods of the US Food Supply, 1982-1991. J Food Comp Anal, **8**: 129–169.

Perdue EM & Lytle CR (1983) Distribution model for binding of protons and metal ions by humic substances. Environ. Sci Technol, **17**: 654–660.

Perwak J, Goyer M, Nelken L, Schimke G, Scow K, Walker P, Wallace D, & Delos C (1980) An exposure and risk assessment for zinc. Washington DC, Environmental Protection Agency (EPA/440/4-81/016).

Petering HG, Buskirk HH, & Crim JA (1967) The effect of dietary mineral supplements of the rat on the antitumor activity of 3-ethoxy-2-oxobutyraldehyde bis (thiosemi-carbazone). Cancer Res, **27**: 1115–1121.

Petering HG, Johnson MA, & Stemmer KL (1971) Studies of zinc metabolism in the rat. Arch Environ Health, 23: 93–101.

Petrie JJ & Row PG (1977) Dialysis Anaemia caused by subacute zinc toxicity. Lancet, 1(8023): 1178–1180.

Philcox JC, Coyle P, Michalska A, Choo KHA, & Rofe AM (1995) Endotoxin-induced inflammation does not cause hepatic zinc accumulation in mice lacking metallo-thionein gene expression. Biochem J, 308: 543–546.

Phillips JL & Sheridan PJ (1976) Effect of zinc administration on the growth of L1210 and BW5147 tumors in mice. J Natl Cancer Inst, 57: 361.

Phillips JM, Ackerley CA, Superina RA, Roberts EA, Filler RM, & Levy GA (1996) Excess zinc associated with severe progressive cholestastis in Cree and Ojibwa-Cree children. Lancet, 347: 866–868.

Pickering QH & Henderson C (1966) The acute toxicity of some heavy metals to different species of warmwater fishes. Air Water Pollut Int J, 10: 453–463.

Pickston L, Brewerton HV, Drysdale JM, Hughes JT, Smith JM, Love JL, Sutcliffe ER, & Davidson F (1985) The New Zealand diet: A survey of elements, pesticides, colours, and preservatives. New Zealand J Technol, 1: 81–90.

Pollack S, George JN, Reba RC, Kaufman RM, & Crosby WH (1965) The absorption of nonferrous metals in iron deficiency. J Clin Inves, 44(9): 1470–1473.

Porter FC (1995) Corrosion resistance of zinc and zinc alloys. New York, Basel, Hong Kong, Marcel Dekker.

Porter KG, McMaster D, Elmes ME, & Love AHG (1977) Anemia and low serum copper during zinc therapy. Lancet, 2: 774.

Posthuma L & Van Straalen NM (1993) Heavy metal adaptation in terrestrial invertebrates: a review of occurrence, genetics, physiology, and ecological consequences. Compar Biochem Physiol, 106C: 11–38.

Powell MJ, Davies MS, & Francis D (1986a) Effects of zinc on cell, nuclear and nucleolar size, and on RNA and protein content in the root meristem of a zinc-tolerant and a non-tolerant cultivar of Festuca rubra L. New Phytol, 104: 671–679.

Powell MJ, Davies MS, & Francis D (1986b) The influence of zinc on the cell cycle in the root meristem of a zinc-tolerant and a non-tolerant cultivar of Festuca rubra L. New Phytol, 102: 419–428.

Powell MI & White KN (1990) Heavy metal accumulation by barnacles and its implications for their use as biological monitors. Mar Environ Res, 30: 91–118.

Prahalad AK & Seenayya G (1989) Physico-chemical interactions and bioconcen-tration of zinc and lead in the industrially polluted Husainsager Lake, Hyderabad, India. Environ Pollut, 58A: 139–154.

Prasad A (1963) Zinc metabolism in patients with syndrome of iron deficiency anaemia, hepatosplenomegaly, dwarfism and hyogonadism. J Lab Clin Med, **61**: 537–549.

Prasad AS (1966) Metabolism of zinc and its deficiency in human subjects. Zinc metabolism. Springfield IL, Charles C. Thomas, pp 250–303.

Prasad AS ed. (1976) Zinc. Trace elements and iron in human metabolism. New York, Plenum Press, pp 251–346.

Prasad AS (1983) Clinical, biochemical and nutritional spectrum of zinc deficiency in human subjects: an update. Nutr Rev, **41**(7): 197–208.

Prasad AS (1985) Essential trace elements in human health in disease. J Am Coll Nutr **4**(1): 1–2.

Prasad AS (1988) Clinical spectrum and diagnostic aspects of human zinc deficiency. In: Prasad AS ed. Essential and toxic trace elements in human health and disease. New York, Alan R Liss, pp 3–53.

Prasad AS (1996) Zinc: the biology and therapeutics of an ion. Ann Intern Med, **125**(2): 142–144.

Prasad AS & Cossack ZT (1982) Neutrophil zinc: an indicator of zinc status in man. Trans Assoc Am Physicians, **XCV**: 165–176.

Prasad AS & Oberleas D (1970) Binding of zinc to amino acide and serum proteins *in vitro*. J Lab Clin Med, **76**: 416–425.

Prasad AS & Rabbani P (1981) Nucleoside phosphorylase in zinc deficiency. Trans Assoc Am Physicians, **94**: 314-321.

Prasad AS, Halsted JA, & Nadimi M (1961) Syndrome of iron deficiency anemia, hepatosplenomegaly, hypogonadism, dwarfism and geophagia. Am J Med, **31**: 532-546.

Prasad AS, Miale AJ, Farid Z, Sandstead HH, & Schulert AR (1963a) Zinc metabolism in patients with the syndrome of iron deficiency anemia, hepatosplenomegaly, dwarfism and hypogonadism. J Lab Clin Med, **61**: 537–549.

Prasad AS, Schulert AR, Miale AJ, Farid Z, & Sandstead HH (1963b) Zinc and iron deficiencies in male subjects with dwarfism and hypogonadism but without ancylostomiasis, schistosomiasis or severe anemia. Am J Clin Nutr, **12**: 437–444.

Prasad AS, Oberleas D, Wolf P, & Horwitz JP (1967) Studies on zinc deficiency: changes in trace elements and enzyme activities in tissues of zinc-deficient rats. J Clin Invest, **46**: 549–557.

Prasad AS, Oberleas D, Wolf P, Horwitz JP, Miller ER, & Luecke RW (1969) Changes in trace elements and enzyme activities in tissues of zinc-deficiency pigs. Am J Clin Nutr, **22**(5): 628–637.

Prasad AS, Rabbani P, Abbasi A, Bowersox E, & Fox MRS (1978a) Experimental zinc deficiency in humans. Ann Intern Med, **89**: 483–490.

Prasad AS, Brewer GJ, Schoomaker EB, & Rabbani P (1978b) Hypocupremia induced by zinc therapy in adults. J Am Med Assoc, **240**(20): 2166–2168.

Preston A (1973) Heavy metals in British waters. Nature, **242**: 95–97.

Price RKJ & Uglow RF (1979) Some effects of certain metals on development and mortality within the moult cycle of *Crangon crangon* (L.). Mar Environ Res, **2**: 287–299.

Probst T, Zeh P, & Kim J-I (1995) Multielement determinations in ground water ultrafiltrates using inductively coupled plasma mass spectrometry and monostandard neutron activation analysis. Fresenius J Anal Chem, **351**: 745-751.

Pulido P, Kagi JHR, & Vallee BL (1966) Isolation and some properties of human metallothionein. Biochemistry, **5**(5): 1768–1777.

Pye K (1987) Aeolian dusts and dust deposits. London, Academic Press.

Que Hee SS & Boyle JR (1988) Simultaneous multielemental analysis of some environmental and biological samples by inductively coupled plasma atomic emission spectrometry. Anal Chem, **60**: 1033–1042.

Quevauviller PH & Maier EA (1994) Environment and quality of life. Research trends in the field of environmental analysis. Results of scientific workshops held within the measurements and testing program (BCR). Luxembourg, EC-DG XII. EUR 16000 EN, pp 31-40.

Racey PA & Swift SM (1986) The residual effects of remedial timber treatments on bats. Biol Conserv, **35**: 205–214.

Rachlin JW & Farran M (1974) Growth response of the green algae *Chlorella vulgaris* to selective concentrations of zinc. Water Res, **8**: 575–577.

Rachlin JW, Jensen TE, & Warkentine B (1983) The growth response of the diatom *Navicula incerta* to selected concentrations of the metals: cadmium, copper, lead and zinc. Bull Torrey Bot Club, **110**(2): 217–223.

Ragaini RC, Ralston HR, & Roberts N (1977) Environmental trace metal contamination in Kellogg, Idaho, near a lead smelting complex. Environ Sci Technol, **11**(8): 773–781.

Rainbow PS (1985) Accumulation of Zn, Cu and Cd by crabs and barnacles. Estuary Coast Shelf Sci, **21**: 669–686.

Rainbow PS (1987) Heavy metals in barnacles. In: Southward AJ ed. Barnacle biology. Rotterdam, A.A. Balkema, pp 404–417.

Ramadurai J, Shapiro C, Kozloff M, & Telfer M (1993) Zinc abuse and sideroblastic anemia. Am J Hematol, **42**(2): 227–228.

Rasmussen PE (1996) Trace metals in the environment: a geological perspective.Ottawa, Government of Canada (Geological Survey of Canada, Bulletin 429).

Rath FW, Kortge R, Haase P, & Bismarck M (1991) The influence of zinc administration on the development of experimental lung metastases after an injection of tumour cells into the tail vein of rats. Exp Pathol, **41**(4): 215–217.

Rauser WE (1978) Early effects of phytotoxic burdens of cadmium, cobalt, nickel, and zinc in white beans. Can J Bot, **56**: 1744–1749.

Reader JP, Everall NC, Sayer MDJ, & Morris R (1989) The effects of eight trace metals in acid soft water on survival, mineral uptake and skeletal calcium deposition in yolk-sac fry of brown trout, *Salmo trutta* L. J Fish Biol, **35**: 187–198.

Recio A, Marigómez JA, Angulo E, & Moya J (1988) Zinc treatment of the digestive gland of the slug *Arion ater* L. 1. Cellular distribution of zinc and calcium. Bull Environ Contam Toxicol, **41**: 858–864.

Redpath KJ & Davenport J (1988) The effect of copper, zinc and cadmium on the pumping rate of *Mytilus edulis* L. Aquat Toxicol, **13**: 217–226.

Reece RL, Dickson DB, & Burrowes PJ (1986) Zinc toxicity (new wire disease) in aviary birds. Aust Vet J, **63**(6): 199.

Reeves PG & O'Dell BL (1985) An experimental study of the effect of zinc on the activity of angiotensin converting enzyme in serum. Clin Chem, **31**(4): 581–584.

Rehwoldt R, Menapace LW, Nerrie B, & Alessandrello D (1972) The effect of increased temperature upon the acute toxicity of some heavy metal ions. Bull Environ Contam Toxicol, **8**: 91–96.

Rehwoldt R, Lasko L, Shaw C, & Wirhowski E (1973) The acute toxicity of some heavy metal ions toward benthic organisms. Bull Environ Contam Toxicol, **10**(5): 291–294.

Reif JS, Ameghino E, & Aaronson MJ (1989) Chronic exposure of sheep to a zinc smelter in Peru. Environ Res, **49**: 40–49.

Reinhold JG, Nasr K, Lahimgarzadeh A, & Hedayati H (1973) Effects of purified phytate and phytate-rich bread upon metabolism of zinc, calcium, phosphorus, and nitrogen in man. Lancet, **1**: 283–288.

Reinhold JG, Faradji B, Abadi P, & Ismail-Beigi F (1976) Decreased absorption of calcium, magnesium, zinc and phosphorus by humans due to increased fiber and phosphorus consumption as wheat bread. J Nutr, **106**: 493–503.

Reiser S, Smith JC Jr, Mertz W, Holbrook JT, Scholfield DJ, Powell AS, Caufield WK, & Canary JJ (1985) Indices of copper status in humans consuming a typical American diet containing either fructose or starch. Am J Clin Nutr, **42**(2): 242–251.

Reish DJ & Carr RS (1978) The effect of heavy metals on the survival, reproduction, development, and life cycles for two species of polychaetous annelids. Mar Pollut Bull, 9(1): 24–27.

Reiter R & Poetzl K (1985) [Development of aerosol particles with very small diameter during metallurgical high temperature processes.] Staub-Reinhalt Luft, 45: 66–74 (in German).

Rengel Z & Graham RD (1995) Importance of seed zinc content for wheat growth on Zn-deficient soils. Plant Soil, 173: 267–274.

Reuter JG & Morel FMM (1981) The interaction between zinc deficiency and copper toxicity as it affects the silicic acid uptake mechanisms in *Thalassiosira pseudonana*. Limnol Oceanogr, 26(1): 67–73.

Richards MP & Cousins RJ (1975) Mammalian zinc homeostasis: requirement for RNA and metallothionein synthesis. Biochem Biophys Res Commun, 64(4): 1215–1223.

Richards MP & Cousins RJ (1976) Metallothionein and its relationship to the metabolism of dietary zinc in rats. J Nutr, 106: 1591–1599.

Richards RJ, Atkins J, Marrs TC, Brown RF, & Masek L (1989) The biochemical and pathological changes produced by the intratracheal instillation of certain components of zinc-hexachloroethane smoke. Toxicology, 54(1): 79–88.

Richmond VL (1992) Oxytocin-induced parturition in copper-deficient guinea-pigs. Lab Anim Sci, 42(2): 190–192.

Ritchey SJ, Korslund MK, Gilbert LM, Fay DC, & Robinson MR (1979) Zinc retention and losses of zinc in sweat by preadolscent girls. Am J Clin Nutr, 32: 799–803.

RIWA (1993) [Cooperation of Water Works of Rhine and Maas, 1993.] Yearly report, cooperation of water works of Rhine and Maas, Report on the quantity of the Rhine until the year 1991. Amsterdam, RIWA, pp 51–55, 78 (in German).

Robbins AH, McRee DE, Williamson M, Collett SA, Xuong NH, Furey WF, Wang BC, & Stout CD (1991) Refined crystal structure of Cd, Zn metallothionein at 2.0 A resolution. Mol Biol, 221: 1269–1293.

Roberts RD & Johnson MS (1978) Dispersal of heavy metals from abandoned mine workings and their transference through terrestrial food chains. Environ Pollut, 16: 293–310.

Rogers JE & Li SW (1985) Effect of metals and other inorganic ions on soil microbial activity: soil dehydrogenase assay as a simple toxicity test. Bull Environ Contam Toxicol, 34: 858–865.

Rohrs LC (1957) Metal-fume fever from inhaling zinc oxide. AMA Arch Ind Health, 16: 42–47.

Rojas LX, McDowell LR, Cousins RJ, Martin FG, Wilkinson NS, Johnson AB, & Velasquez JB (1995) Relative bioavailability of two organic and two inorganic zinc sources fed to sheep. J Anim Sci, **73**: 1202–1207.

Ronaghy HA, Caughey JE, & Halstead JA (1968) A study of growth in Iranian village children. Am J Clin Nutr, **21**(5) 488–494.

Ronaghy HA, Reinhold JG, Mahloudji M, Ghavami P, Fox MRS, & Halsted JA (1974) Zinc supplementation of malnourished schoolboys in Iran: increased growth and other effects. Am J Clin Nutr, **27**: 112–121.

Rosko JJ & Rachlin JW (1975) The effect of copper, zinc, cobalt and manganese on the growth of the marine diatom *Nitzschia closterium*. Bull Torrey Bot Club, **102**(3): 100–106.

Rossman TG, Zelikoff JT, Agarwal S, & Kneip TJ (1987) Genetic toxicology of metal compounds: an examination of appropriate cellular models. Toxicol Environ Chem, **14**: 251–262.

Roy SK, Behrens RH, Haider R, Akramuzzaman SM, Mahalanabis D, Wahed MA, & Tomkins AM (1992) Impact of zinc supplementation on intestinal permeability in Bangladeshi children with acute diarrhoea and persistent diarrhoea syndrome. J Pediatr Gastroenterol Nutr, **15**: 289–296.

Royal Belgian Federation of Non-Ferrous Metals (1995) [Producing metals in an environmentally conscious manner.] Brussels, Ecolas (in Dutch).

Rudd T, Lake DL, Mehrotra I, Sterritt RM, Kirk PWW, Campbell JA, & Lester JN (1988) Characterisation of metal forms in sewage sludge by chemical extraction and progressive acidification. Sci Total Environ, **74**: 149–175.

Ruz M, Cavan KR, Bettger WJ, Thompson L, Berry M, & Gibson RS (1991) Development of a dietary model for the study of mild zinc deficiency in humans and evaluation of some biochemical and functional indices of zinc status. Am J Clin Nutr, **53**: 1295–1303.

Ruz M, Cavan KR, Bettger WJ, & Gibson RS (1992) Erythrocytes, erythrocyte membranes, neutrophils and platelets as biopsy materials for the assessment of zinc status in humans. Brit J Nutr, **68**: 515–527.

Ruz M, Castillo-Duran C, Lara X, Codocco J, Rebolledo A, & Atalah E (1997) A 14-mo zinc-supplementation trial in apparently healthy Chilean preschool children. Am J Clin Nutr, **66**(6): 1406–1403.

Sachdev HPS, Mittal NK, & Yadav HS (1990) Oral zinc supplementation in persistent diarrhoea in infants. Ann Trop Paediatr, **10**: 63–69.

Sadasivan V (1951) Studies on the biochemistry of zinc. I. Biochem J, **48**: 527–532.

Sadasivan V (1952) Studies on the biochemistry of zinc. Biochem J, **52**: 452–455.

Samman S & Roberts DCK (1988) The effect of zinc supplements on lipoproteins and copper status. Atherosclerosis, **70**: 247–252.

Samman S, Soto C, Cooke L, Ahmad Z, & Farmakalidis E (1996) Is erythrocyte alkaline phosphatase activity a marker of zinc status in humans? Biol Trace Elem Res, **51**: 285–291.

Sanders JR & Adams TM (1987) The effects of pH and soil type on concentrations of zinc, copper and nickel extracted by calcium chloride from sewage sludge-treated soils. Environ Pollut, **43**: 219–228.

Sanders JR & El Kherbawy MI (1987) The effect of pH on zinc adsorption equilibria and exchangeable zinc pools in soils. Environ Pollut, **44**: 165–176.

Sandler M & Jordaan HF (1989) Cutaneous reaction to zinc – a rare complication of insulin treatment. A case report. S Afr Med J, **75**(7): 342–343.

Sandstead HH (1973) Zinc nutrition in the United States. Am J Clin Nutr, **26**: 1251–1260.

Sandstead HH (1981) Zinc in human nutrition. In: Bronner F & Coburn JW eds. Disorders of mineral metabolism volume 1 trace minerals. New York, Academic Press, pp 93–157.

Sandstead HH (1982a) Nutritional role of zinc and effects of deficiency. In: Winicle M ed. Adolescent nutrition, pp 97–124.

Sandstead HH (1982b) Copper bioavailability and requirements. Am J Clin Nutr, **35**: 809–814.

Sandstead HH (1984) Are estimates of trace element requirements meeting the needs of the user? In: Mills C, Bremner I, & Chester J ed. Trace elements in man and animals, vol. 5. Farnham Royal, Commonwealth Agricultural Bureaux, pp 875–878.

Sandstead HH (1985) Requirement of zinc in human subjects. J Am Coll Nutr, **4**: 73–82.

Sandstead HH & Smith C Jr (1996) Deliberations and evaluations of approaches, endpoints, and paradigms for determining zinc dietary recommendations. Proceedings of Workshop – "New Approaches, Endpoints and Paradigms for RDAs of Mineral Elements", Grand Forks, North Dakota, 10-12 September 1995.

Sandstead HH, Prasad AS, Schulert AR, Farid Z, Miale AJ, Basilly S, & Darby WJ (1967) Human zinc deficiency, endocrine manifestations, and response to treatment. Am J Clin Nutr, **20**(5): 422–442.

Sandstead HH, Lanier VC, Shephard GC, & Gillespie DD (1970) Zinc and wound healing. Am J Clin Nutr, **23**(5): 514–519.

Sandstead HH, Gillespie DD, & Brady RN (1972) Zinc deficiency: Effect on brain of suckling rat. Pediatr Res, **6**: 119–125.

Sandstead HH, Strobel DA, Logan GMJ, Marks EO, & Jacob RA (1978) Zinc deficiency in pregnant rhesus monkeys: effects on behavior of infants. J Clin Nutr, 31: 844–849.

Sandstead HH, Dintzis FR, Bogyo T, Milne DA, Jacob RA, & Klevay LM (1990) Dietary factors that can impair calcium and zinc nutriture of the elderly. In: Princeley J & Sandstead HH ed. Nutrition and aging. New York, Alan R Liss, pp 241–262.

Sandstead HH, Penland JG, Alcock NW, Dayal HH, Chen XC, Li JS, Zhao F, & Yang JJ (1998) Effects of repletion with zinc and other micronutrients on neuropsychologic performance and growth of Chinese children. Am J Clin Nutr, 68(2 Suppl): 470S–475S.

Sandstroem B & Abrahamson H (1989) Zinc absorption and achlorhydria. Eur J Clin Nutr, 43: 877–879.

Sandstroem B & Cederblad Å (1980) Zinc absorption from composite meals. II. Influence of the main protein source. Am J Clin Nutr, 33: 1778–1783.

Sandstroem B & Sandberg AS (1992) Inhibitory effects of isolated inositol phosphates on zinc absorption in humans. J Trace Elem Electrolytes Health Dis, 6: 99–103.

Sandstroem B, Davidson L, Cederblad A, & Lonnerdal B (1985) Oral iron dietary ligands and zinc absorption. J Nutr, 115: 411–414.

Sandstroem B, Almgren A, Kivistoe B, & Cederblad Å (1987) Zinc, absorption in humans from meals based on rye, barley, oatmeal, triticale and whole wheat. J Nutr, 117: 1898–1902.

Sauer GR & Watabe N (1984) Zinc uptake and its effect on calcification in the scales of the mummichog, *Fundulus heteroclitus*. Aquat Toxicol, 5: 51–66.

Sawidis T, Chettri MK, Zachariadis GA, & Stratis JA (1995) Heavy metals in aquatic plants and sediments from water systems in Macedonia, Greece. Ecotoxicol Environ Saf, 32: 73–80.

Sax NI & Lewis RJ (1987) Zinc. In: Sax NI & Lewis RJ eds. Hawley's condensed chemical dictionary. New York, Van Nostrand Reinhold, pp1250–1258.

Say PJ, Diaz, BM, & Whitton BA (1977) Influence of zinc on lotic plants II Environmental effects on toxicity of zinc to *Hormidium rivulare*. Freshwater Biol, 7: 377–384

Sayer MDJ, Reader JP, & Morris R (1989) The effect of calcium concentration on the toxicity of copper, lead and zinc to yolk-sac fry of brown trout, *Salmo trutta* L., in soft, acid water. J Fish Biol, 35: 323–332.

Schat H, Vooijs R, & Kuiper E (1996) Identical major gene loci for heavy metal tolerances that have independently evolved in different local populations and subspecies of Silene vulgaris. Evolution, 50(5): 1888-1895.

319

Schenker MB, Speizer FE, & Taylor JO (1981) Acute upper respiratory symptoms resulting from exposure to zinc chloride aerosol. Environ Res, **25**: 317–324.

Scherz H, Kloos G, & Senser F (1986) Food composition and nutrition tables 1986/87. In: German Research Institute for Food Chemistry ed. Food composition and nutrition tables 1986/87. Stuttgart, Wissenschaftliche VerlagsGmbH.

Schiffer RB, Sunderman FW Jr, Baggs RB, & Moynihan JA (1991) The effects of exposure to dietary nickel and zinc upon humoral and cellular immunity in SJL mice. J Neuroimmunol, **34**: 229–239.

Schilirò G, Russo A, Azzia N, Mancuso GR, Di Gregorio F, Romeo MA, Fallico R, & Sciacca S (1987) Leukocyte alkaline phosphatase (LAP). Am J Pediatric Hematology/ Oncology, **9**: 149–153.

Schlesinger L, Arevalo M, Arredondo S, Lonnerdal B, & Stekel A (1993) Zinc supplementation impairs monocyte function. Acta Paediatr, **82**(9): 734–738.

Schmitt CJ & Brumbaugh WG (1990) National contaminant biomonitoring program: Concentrations of arsenic, cadmium, copper, lead, mercury, selenium, and zinc in US freshwater fish, 1976-1984. Arch Environ Contam Toxicol, **19**: 731–747.

Schmitt Y, Moser V, & Kruse-Jarres JD (1993) Zinc, copper and aluminium in the corpuscular components of peripheral blood in patients with pre-terminal and terminal renal failure. Fresenius J Anal Chem, **346**: 852–858.

Schnitzer M & Skinner SIM (1967) Organo-metallic interactions in soils: 7. Stability constants of Pb^{++}-, Ni^{++}-, Mn^{++}-, Co^{++}-, Ca^{++}-, and Mg^{++}-fulvic acid complexes. Soil Sci, **103**(4): 247–252.

Scholl TO, Hediger ML, Schall JI, Fischer RL, & Khoo CS (1993) Low zinc intake during pregnancy: Its association with preterm and very preterm delivery. Am J Epidemiol, **137**: 1115–1124.

Schroeder HL, Hanover NH, & Brattleboro VT (1970) A sensible look at air pollution by metals. Arch Environ Health, **21**: 798–808.

Schwabe JWR, Neuhaus D, & Rhodes D (1990) Solution structure of the DNA-binding domain of the oestrogen receptor. Nature, **348**: 458–461.

Schweiger P (1984) [Effects of heavy metal burden from waste sludge on agriculturally used soils.] Gewässerschutz Wasser Abwasser, **65**: 439–449 (in German).

Seal CJ & Heaton FW (1983) Chemical factors affecting the intestinal absorption of zinc *in vitro* and *in vivo*. Br J Nutr, **50**: 317–324.

Seeger R & Neumann H-G (1985) [Zinc.] Dtsch Apoth Ztg, **125**(4): 168–170 (in German).

Settlemire CT & Matrone G (1967a) In vivo effect of zinc on iron turnover in rats and life span of the erythrocyte. J Nutr, **92**: 159–164.

Settlemire CT & Matrone G (1967b) In vivo interference of zinc with ferritin iron in the rat. J Nutr, **92**: 153–158.

Shamberger RJ (1979) Benefical effects of trace elements. In: Oehme FW ed. Toxicity of heavy metals in the environment Part 2. New York, Marcel Dekker, pp 751–775.

Sharrett AR, Carter AP, Orheim RM, & Feinlieb M (1982a) Daily intake of lead, cadmium, copperr, and zinc from drinking water: the Seattle study of trace metal exposure. Environ Res, **28**: 456–475.

Sharrett AR, Orheim RM, Carter AP, Hyde JE, & Feinlieb M (1982b) Components of variation in lead, cadmium, copper, and zinc concentration in home drinking water: the Seattle study of trace metal exposure. Environ Res, **28**: 476–498.

Shaw AJ (1990) Heavy metal tolerance in plants: evolutionary aspects. CRC Press, Boca Raton, FL.

Shehata FHA & Whitton BA (1981) Fiels and laboratory studies on blue-green algae from aquatic sites with high levels of zinc. Verh Int Verein Limnol, **21**: 1466–1471.

Shenker JM & Cherr GN (1990) Toxicity of zinc and bleached kraft mill effluent to larval English sole (*Parophrys vetulus*) and topsmelt (*Atherinops affinis*). Arch Environ Contam Toxicol, **19**: 680–685.

Sheppard SC, Evenden WG, Abboud SA, & Stephenson M (1993) A plant life-cycle bioassay for contaminated soil, with comparison to other bioassays: mercury and zinc. Arch Environ Contam Toxicol, **25**: 27–35.

Shiller AM & Boyle E (1985) Dissolved zinc in rivers. Nature, **317**: 49–52.

Shiller AM & Boyle EA (1987) Variability of dissolved trace metals in the Mississippi river. Geochim Cosmochim Acta, **51**: 3273–3277.

Shuman LM (1975) The effect of soil properties on zinc adsorption by soils. Soil Sci Soc Am Proc, **39**: 454–458.

Sileo L & Beyer WN (1985) Heavy metals in white-tailed deer living near a zinc smelter in Pennsylvania. J Wildlife Dis, **21**: 289–296.

Simkiss K & Watkins B (1990) The influence of gut microorganisms on zinc uptake in *Helix aspersa*. Environ Pollut, **66**: 263–271.

Simmer K, Lort-Phillips L, James C, & Thompson RP (1991) A double-blind trial of zinc supplementation in pregnancy. Eur J Clin Nutr, **45**(3): 139–144.

Simon SR, Brande RF, Tindle BH, & Burns SL (1988) Copper deficiency and sideroblastic anemia associated with zinc ingestion. Am J Hematol, **28**: 181–183.

Singh BR & Låg J (1976) Uptake of trace elements by barley in zinc-polluted soils: 1. Availability of zinc to barley from indigenous and applied zinc and the effect of excesssive zinc on the growth and chemical composition of barley. Soil Sci, **121**(1): 32–37.

Sinley JR, Goettl JP, & Davies PH (1974) The effects of zinc on rainbow trout (*Salmo gairdneri*) in hard and soft water. Bull Environ Contam Toxicol, **12**(2): 193–201.

Sirover MA & Loeb LA (1976) Infidelity of DNA Synthesis in vitro: Screening for potential metal mu;tagens or carcinogens. Science, **194**: 1434–1436.

Sivadasan CR, Nambisan PNK, & Damodaran R (1986) Toxicity of mercury, copper and zinc to the prawn *Metapenaeus dobsoni* (Mier). Curr Sci, **55**(7): 337–340.

Skog E & Wahlberg JE (1964) A comparative investigation of the percutaneous absorption of metal compounds in the Guinea pig by means of the radioactive isotopes: 31 Cr, 58 Co, 65 Zn, 110m Ag, 115m Cd, 203 Hg. J Invest Dermatol, **43**: 187–192.

Slater JP, Mildvan AS, & Loeb LA (1971) Zinc in DNA Polymerases. Biochem Biophys Res Commun, **44**(1): 37–43.

Slavin W (1984) Graphite Furnace AAS. A Source Book. Perkin Elmer, Ridgefield.

Smith KT & Cousins RJ (1980) Quantitative aspects of zinc absorption by isolated, vascularly perfused rat intestine. J Nutr, **110**: 316–323.

Smith KT, Cousins RJ, Silbon BL, & Faila ML (1978) Zinc absorption and metabolism by isolated, vascularly perfused rat intestine. J Nutr, **108**: 1849–1857.

Smith MJ & Heath AG (1979) Acute toxicity of copper, chromate, zinc, and cyanide to freshwater fish: effect of different temperatures. Bull Environ Contam Toxicol, **22**: 113–119.

Smith SE & Larson EJ (1946) Zinc toxicity in rats (1946). J Biol Chem, **163**: 29–38.

Smit Vanderkooy PD & Gibson RS (1987) Food consumption patterns of Canadian preschool children in relation to zinc growth status Am J Clin Nutr, **45**(3): 609–616

Snell TW, Moffat BD, Janssen C, & Persoone G (1991) Acute toxicity tests using rotifers. IV. Effects of cyst age, temperature, and salinity on the sensitivity of Brachionus calyciflorus. Ecotoxicol Environ Saf, **21**: 308–317.

Sohn D, Heo M, & Kang C (1989) Particle size distribution of heavy metals in the urban air of Seoul, Korea. In: Brasser LJ & Mulder WC ed. Man and his ecosystem. Proceedings of the 8th World Clean Air Congress, The Hague, 11–15 September 1989, vol. 3. Amsterdam, Elsevier, pp 633–638.

Solbé de LG JF (1973) The relation between water quality and the status of fish populations in Willow Brook. Proc Soc Water Treat Exam, **22**: 41–61.

Solbé JFdLG (1977) Water quality, fish and invertebrates in a zinc-polluted stream. In: Alabaster JS ed. Biological monitoring of inland fisheries. London, Applied Science Publishers, pp 97–105.

Solomons NW & Jacob RA (1981) Studies on the bioavailability of zinc in humans: effects of heme and nonheme iron on the absorption of zinc1-4. Am J Clin Nutr, **34**: 475–482.

Solomons NW, Jacob RA, & Pineda O (1979) Studies on the bioavaility of zinc in man. II. Absorption of zinc from organic and inorganic sources. J Lab Clin Med, **94**: 335–343.

Solomons NW, Pineda O, Viteri F, & Sandstead HH (1983) Studies on the bioavailability of zinc in humans: mechanism of the intestinal interaction of nonheme iron and zinc1,2. J Nutr, **113**: 337–349.

Somasundaram B (1985) Effects of zinc on epidermal ultrastructure in the larva of *Clupea harengus*. Mar Biol, **85**: 199–207.

Somasundaram B, King PE, & Shackley S (1984a) The effects of zinc on postfertilization development in eggs of *Clupea harengus* L. Aquat Toxicol, **5**: 167–178.

Somasundaram B, King PE, & Shackley SE (1984b) The effect of zinc on the ultrastructure of the trunk muscle of the larva of *Clupea harengus* L. Comp Biochem Physiol, **79C**(2): 311–315.

Somasundaram B, King PE, & Shackley SE (1984c) The effects of zinc on the ultrastructure of the brain cells of the larvae of *Clupea harengus* L. Aquat Toxicol, **5**: 323–330.

Somasundaram B, King PE, & Shackley SE (1985) The effect of zinc on the ultrastructure of the posterior gut and pronephric ducts of the larva of *Clupea harengus* L. Comp Biochem Physiol, **81C**(1): 29–37.

South TL & Summers MF (1990) Zinc fingers. Adv Inorg Biochem, **8**: 199–248.

Spear PA (1981) Zinc in the aquatic environment: chemistry, distribution, and toxicology. In: National Research Council of Canada, Associate committee on scientific criteria for environmental quality ed. Zinc in the aquatic environment: chemistry, distribution, and toxicology No. NRCC 17589. Ottawa, Publications NRCC/CNRC, pp 1–145.

Spehar RL (1976) Cadmium and zinc toxicity to flagfish, *Jordanella floridae*. J Fish Res Board Can, **33**: 1939–1945.

Spehar RL, Leonard EN, & DeFoe DL (1978) Chronic effects of cadmium and zinc mixtures on flagfish (*Jordanella floridae*). Trans Am Fish Soc, **107**(2): 354–360.

Spence JW & McHenry JN (1994) Development of regional corrosion maps for galvanised steel by linking the RADM engineering model with an atmospheric corrosion model. Atmos Environ, **28**(18): 3033–3046.

Spencer H, Rosoff B, Feldstein A, Cohn SH, & Gusmano E (1965) Metabolism of zinc-65 in man. Radiat Res, **24**: 432–445.

Spencer H, Asmussen CR, Holtzman RB, & Kramer L (1979) Metabolic balances of cadmium, copper, manganese, and zinc in man. Am J Clin Nutr, **32**: 1867–1875.

Spencer H, Kramer L, & Osis D (1983) Zinc balances in humans during different intakes of calcium and phosphorus. In: Inglett F ed. Nutritional bioavailability of zinc. Washington, DC, pp 223–232 (ACS Symposium. Series No. 210).

Spencer H, Rubio N, & Kramer L (1987) Effect of zinc supplementation on the intestinal absorption of calcium. J Am Coll Nutr, **6**: 47–51.

Spencer H, Norris C, & Osis D (1992) Further studies of the effect of zinc on intestinal absorption of calcium in man. J Am Coll Nutr, **11**: 561–566.

Spender QW, Cronk CE, Charney EB, & Stallings VA (1989) Assessment of linear growth of children with cerebral palsy. Dev Med Child Neurol, **31**: 206-214.

Sprague JB (1986) Toxicity and tissue concentrations of lead, zinc, and cadmium for marine mollusks and crustaceans. International Lead Zinc Research Organization, Research Triangle Park, NC, 215 pp.

Sprenger M, McIntosh A, & Lewis T (1987) Variability in concentrations of selected trace elements in water and sediment of six acidic lakes. Arch Environ Contam Toxicol, **16**: 383–390.

Spry DJ, Hodson PV, & Wood CM (1988) Relative contribution of dietary and waterborne zinc in the rainbow trout, *Salmo gairdneri*. Can J Fish Aquat Sci, **45**: 32–41.

Spurgeon DJ, Hopkin SP, & Jones DT (1994) Effects of cadmium, copper, lead and zinc on growth, reproduction and survival of the earthworm *Eisenia fetida* (Savigny): assessing the environmental impact of point-source metal contamination in terrestrial ecosystems. Environ Pollut, **84**: 123-130.

Spurgeon DJ & Hopkin SP (1996) Effects of variations of the organic matter content and pH of soils on the availability and toxicity of zinc to the earthworm Eisenia fetida. Pedobiologia, **40**: 80–96.

SSSA (1990) Soil testing and plant analysis, 3rd ed. Madison, WI, Soil Science Society of America.

Stahl JL, Greger JL, & Cook ME (1989) Zinc, copper and iron utilisation by chicks fed various concentrations of zinc. Br Poult Sci, **30**: 123–134.

Stahl JL, Greger JL, & Cook ME (1990) Breeding-hen and progeny performance when hens are fed excessive dietary zinc. Poult Sci, **69**: 259–263.

Stanners D & Bourdeau P eds (1995) The Dobris Assessment. Europe's Environment. Copenhagen, European Environment Agency.

Starodub ME, Wong PTS, Mayfield CI, & Chau YK (1987) Influence of complexation and pH on individual and combined heavy metal toxicity to a freshwater green alga. Can J Fish Aquat Sci, **44**: 1173–1180.

Stauber JL (1995) Toxicity testing using marine and freshwater unicellular algae. Aust J Ecotoxicol, **1**: 15–24.

Stauber JL & Florence TM (1989) The effect of culture medium on metal toxicity to the marine diatom *Nitzschia closterium* and the freshwater green alga *Chlorella pyrenoidosa*. Water Res, **23**(7): 907–911.

Stauber JL & Florence TM (1990) Mechanism of toxicity of zinc to the marine diatom *Nitzschia closterium*. Marine Biol, **105**(3): 519–524.

Stevenson FJ (1991) In: Mordvedt JJ, Giordano P, & Lindsay WL ed. Micronutrients in agriculture. Madison WI, Soil Science Society of America,

Straube EF & Walden NB (1981) Zinc poisoning in ferrets (*Mustella putoris furo*). Laboratory animals, **15**(1): 45–47.

Straube EF, Schuster NH, & Sinclair AJ (1980) Zinc toxicity in the ferret. J Comp Pathol, **90**: 355–361.

Strobel DA & Sandstead HH (1984) Social and learning changes following prenatal or postnatal zinc deprivation in rhesus monkeys. In: Frederickson et al. Ed. The neurobiology of zinc. Part B: Deficiency, toxicity, and pathology. New York, Alan R. Liss, pp 121–138.

Strömgren T (1982) Effect of heavy metals (Zn, Hg, Cu, Cd, Pb, Ni) on the length growth of *Mytilus edulis.* Mar Biol, **72**: 69–72.

Stuart GW, Searle PF, & Palmiter RD (1985) Identification of multiple metal regulatory elements in mouse metallothionein-I promoter by assaying synthetic sequences. Nature, **317**: 828–831.

Sturgeon RE, Berman SS, Willie SN, & Desaulniers JAH (1981) Preconcentration of trace elements from seawater with silica-immobilised 8-hydroxyquinoline. Anal Chem, **53**: 2337–2340.

Sturniolo GC, Montino MC, Rosetto L, Martin A, D'Ina R, D'Odorico A, & Naccarato R (1991) Inhibition of gastric acid secretion reduces zinc absorption in man. J Am Coll Nutr, **10**: 372–375.

Suedel BC, Boraczek JA, Peddicord RK, Clifford PA, & Dillon TM (1994) Trophic transfer and biomagnification potential of contaminants in aquatic ecosystems. Rev Environ Contam Toxicol, **136**: 21–89.

Sutton & Nelson (1937) Studies on zinc. Proc Soc Exp Biol Med, **36**: 211–213.

Swaine DJ (1955) Zinc. In: Commonwealth Agricultural Bureaux England ed The trace element content of soil. York, Herald printing Works, pp 122–130 (Commonwealth Bureau of Soil Science, Technical communication No. 48).

Swanson CA, Turnlund JR, & King JC (1983) Effect of dietary sources and pregnancy on zinc utilization in adult women fed controlled diets. J Nutr, **113**: 2557–2567.

Swenerton H & Hurley L (1980) Zinc deficiency in rhesus and bonnet monkeys, including effects on reproduction. J Nutr, **110**: 575–583.

Swenerton H, Shrader R, & Hurley L (1969) Zinc deficient embryos: Reduced thymidine incorporation. Science, **166**: 1014–1015.

Tacnet F, Watkins DW, & Ripoche P (1990) Studies of zinc transport into brush-border membrane vesicles isolated from pig small intestine. Biochem Biophysika, **1024**: 323–330.

Takkar PN & Walker CD (1993) The distribution and correction of zinc deficiency. In: Robson DA ed. Zinc in soils and plants, Dordrecht, Kluwer.

Tamura T, Freeberg LE, Johnston KE, & Keen CL (1994) *In vitro* zinc stimulation of angiotensin-converting enzyme activities in various tissues of zinc-deficient rats. Nutr Res, **14**(6): 919–928.

Tamura T & Goldenberg RL (1996) Zinc nutriture and pregnancy outcome. Nutr Res, **16**(1): 139–181.

Taper LJ, Hinners MH, & Ritchey SJ (1980) Effects of zinc intake on copper balance in adult females1. Am J Clin Nutr, **33**: 1077–1082.

Taylor A (1986) Usefulness of measurements of trace elements in hair. Ann Clin Biochem, **23**: 364–378.

Taylor D, Maddock BG, & Mance G (1985) The acute toxicity of nine 'grey list' metals (arsenic, boron, chromium, copper, lead, nickel, tin, vanadium and zinc) to two marine fish species: dab (*Limanda limanda*) and grey mullet (*Chelon labrosus*). Aquat Toxicol, **7**: 135–144.

Teraoka H (1989) Impact of atmospheric trace metals on rice roots. Arch Environ Contam Toxicol, **18**: 269–275.

Thompson RPH (1991) Assessment of zinc status. Proc Nutr Soc, **50**: 19–28.

Thompson ED, McDermott JA, Zerkle TB, Skare JA, Evans BLB, & Cody DB (1989) Genotoxicity of zinc in 4 short-term mutagenicity assays. Mutat Res, **223**: 267–272.

Thornton TW (1996) Metals in the global environment: Facts and misconceptions. Ottawa, International Council on Metals and the Environment.

Tijero J, Guardiola E, Mirada F, & Cortijo M (1991) Effect of Cu^{2+}, Ni^{2+} and Zn^{2+} on an anaerobic digestion system. J Environ Sci Health, **26**(6): 799–811.

Timmermans KR, Peeters W, & Tonkes M (1992a) Cadmium, zinc, lead and copper in *Chironomus-riparius* (Meigen) larvae (Diptera, Chironomidae) – uptake and effects. Hydrobiologia, **241**(2): 119–134.

Timmermans KR, Spijkerman E, & Tonkes M (1992) Cadmium and zinc uptake by two species of aquatic invertebrate predators from dietary and aquatic sources. Can J Fish Sci, **49**: 655–662.

Tipping E (1993) Modelling the competition between alkaline earth cations and trace metal species for binding by humic substances. Environ Sci Technol, **27**: 520–529.

Todd WR, Elvehjem CA, & Hart EB (1934) Zinc in the nutrition of the rat. Am J Physiol, **107**: 146–156.

Tomasik P, Magadza CHD, Mhizha S, & Chirume A (1995) The metal-metal interactions in biological systems. Part III. *Daphnia magna*. Water Air Soil Pollut, **82**: 695–711.

Trefry JH & Presley BJ (1971) Heavy metal transport from the Mississippi river to the Gulf of Mexico. In: Windom HL & Duce BA ed. Marine pollution transfer. Lexington Books, pp 33–76.

Trevisan A, Buzzo A, & Gori GP (1982) [Biological indicators in occupational exposure to low concentrations of zinc.] Med Lavoro, **6**: 614-618 (in Italian).

TRI (1993) US Toxic Release Inventory. Washington DC, Office of Pollution Prevention and Pesticides, US Environmental Protection Agency.

TRI (1995) US Toxic Release Inventory. Washington DC, Office of Pollution Prevention and Pesticides, US Environmental Protection Agency.

Tsai MJ & O'Malley WB (1994) Molecular mechanisms of action of steroid/thyroid receptor superfamily members. Annu Rev Biochem, **63**: 451–459.

Tuovinen PK (1988) In: Solomons I & Forstner U eds. Heavy metals in dredged materials and mine tailings. Berlin, Springer.

Turnlund JR, King JC, Keyes WR, Gong B, & Michel MC (1984) A stable isotope study of zinc absorption in young men. Effects of phytate and alpha-cellulose. Am J Clin Nutr, **40**: 1071–1077.

Tyler LD & McBride MB (1982) Mobility and extractability of cadmium, copper, nickel, and zinc in organic and mineral soil columns. Soil Sci, **134**(3): 198–205.

UBA (Umweltbundesamt) ed. (1992) [Environment data 1990/91.] Berlin, Erich Schmidt, pp 444–454 (in German).

UBA (Umweltbundesamt) ed. (1994) [Environment data 1992/93.] Berlin, Erich Schmidt, pp 200–531 (in German).

Udomkesmalee E, Dhanamitta S, Sirisinha S, Charoenkiatkul S, Tuntipopipat S, Banjong O, Rojroongwasinkul N, Kramer TR, & Smith JC Jr (1992) Effect of vitamin A and zinc supplementation on the nutriture of children in Northeast Thailand. Am J Clin Nutr, **56**(1): 50–57.

UK (1991) Dietary reference values for food energy and nutrients for the United Kingdom. Department of Health Report on Health & Social Subjects. London, HMSO.

UN ECE (1979) Guidelines for the control of emissions from the non-ferrous metallurgical industries: Primary zinc. Geneva, United Nations Economic Commission for Europe, pp 215-257.

Ure AM & Berrow ML (1982) The elemental constituents of soils. In: The Royal Society of chemistry ed. Environmental chemistry, Vol. 2. A review of the literature published up to mid-1980, London, Spottiswoode Ballantyne, pp 94–95, 136–143.

Uriu-Hare JY, Stern JS, & Keen CL (1989) Influence of maternal dietary Zn intake on expression of diabetes-induced teratogenicity in rats. Diabetes, **38**: 1282–1290.

US Bureau of Mines (1994) Mineral commodity summaries: Zinc, pp 194–195.

US DHHS (1994) Toxicological profile for zinc (update) TP-93-15. Atlanta, GA, US Department of Health and Human Services, Public Health Service, Agency for Toxic Substances and Disease Registry.

US EPA (1980) Exposure and risk assessment for zinc. Washington DC, US Environmental Protection Agency, Office of Water Regulations and Standards (WH-553) (EPA 440/4-81-016. PB85-212009).

US EPA (1984) Health effects assessment for zinc (and compounds). Cincinnati, OH, Environmental Protection Agency (EPA/540/1-86/048) (Fiche).

US EPA (1987) Ambient water quality criteria for zinc-1987. US Environmental Protection Agency Report 440/5-87-003, pp 1-207.

US EPA (1992) Framework for ecological risk assessment. Washington DC, US Environmental Protection Agency Risk Assessment Forum (EPA/630/R-92/001).

US National Academy of Sciences (1977) Inorganic solutes. Drink Water Health, **1**: 205–316.

US National Academy of Sciences (1979) Zinc. US National Academy of Sciences, National Research Council, Subcommittee on Zinc. Baltimore MD, University Park Press.

US National Academy of Sciences (1989) Recommended dietary allowances, 10th ed. Washington DC, National Academy of Sciences Press.

Valberg LS, Flanagan PR, & Chamberlain MJ (1984) Effects of iron, tin, and copper on zinc absorption in humans. Am J Clin Nutr, **40**: 536–541.

Vallee BL & Auld DS (1990a) Zinc coordination, function, and structure of zinc enzymes and other proteins. Biochemistry, **29**: 5647–5659.

Vallee BL & Auld DS (1990b) Active-site zinc ligands and activated H_2O of zinc enzymes. Proc Natl Acad Sci USA, **87**: 220–224.

Vallee BL & Falchuk KH (1993) The biochemical basis of zinc physiology. Physiol Rev, **73**(1): 79–118.

Vallee BL, Coleman JE, & Auld DS (1991) Zinc fingers, zinc clusters, and zinc twists in DNA-binding protein domains. Proc Natl Acad Sci USA, **88**: 999–1003.

Van Alsenoy V, Bernard P, & Van Grieken R (1990) [Heavy metals in North Sea and Schelde sediments.] Water **51** - March/April 1990 (in Dutch).

Van Assche F (1995) Environment and health effects of zinc – a risk assessment. European Zinc Institute, pp 237–252.

Van Assche C & Jansen G (1978) [Application of selectively acting cation exchanges on soils contaminated with heavy metals.] Landwirtsch Forsch, **34**: 215–228 (in German).

Van Assche F & Clijsters H (1986) Inhibition of photosynthesis in Phaseolus vulgaris by treatment with toxic concentrations of zinc: effects on electron transport and photophosphorylation. Physiol Plant, **66**: 717–721.

Van Assche C & Clijsters H (1990) A biological test system for the evaluation of the phytotoxicity of metal-contaminated soils. Environ Pollut, **66**: 157–172.

Van Assche F, Cardinaels C, & Clijsters H (1988) Induction of enzyme capacity in plants as a result of heavy metal toxicity: dose–response relations in *Phaseolus vulgaris* L., treated with zinc and cadmium. Environ Pollut, **52**: 103–115.

Van Bladel R, Godfrin JM, & Cloos P (1988) [Influence of the physico-chemical properties of soil on the absorption of the heavy metals copper and zinc.] Landbouwtijdschrift, **3**: 721–736 (in Dutch).

Van Capelleveen HE (1987) Ecotoxicology of heavy metals for terrestrial isopods. PhD Thesis, Free University of Amsterdam.

Van Daalen J (1991) Air quality and deposition of trace elements in the province of South-Holland. Atmos Environ, **25A**: 691–698.

Van den Berg CMG (1986) The determination of trace metals in sea-water using cathodic stripping voltammetry. Sci Total Environ, **49**: 89–99.

Van den Berg CMG, Merks AGA, & Duursma EK (1987) Organic complexation and its control of the dissolved concentrations of copper and zinc in the Scheldt estuary. Estuary Coast Shelf Sci, **24**: 785–797.

Van der Wal JF (1990) Exposure of welders to fumes and gases in Dutch industries: summary of results. Ann Occup Hyg, **59**: 45–54.

Van der Weijden CH & Middelburg JJ (1989) Hydrogeochemistry of the River Rhine: Long term and seasonal variability, elemental budgets, base levels and pollution. Water Res, **23**(10): 1247–1266.

Van der Werff M & Pruyt M.J (1982) Long term effects of heavy metals on aquatic plants Chemosphere, **11**(8): 727–739.

Van der Zee J, Zwart P, & Schotman AJH (1985) Zinc poisoning in a Nicobar pigeon. J Zoo Anim Med, **16**: 68–69.

Van Dokkum W (1995) The intake of selected minerals and trace elements in European countries. Nutr Res Rev, **8**: 271–302.

Van Gestel CAM, Dirven-van Breeman EM, & Baerselman R (1993) Accumulation and elimination of cadmium, chromium and zinc and effects on growth and reproduction in *Eisenia andrei* (Oligochaeta, Annelida). Sci Total Environ, **Suppl. 1**: 585–597.

Van Ginneken I (1994) The effect of zinc oxide on the growth of the unicellular alga *Selenastrum capricornutum*. Report Nr AASc/0022, Janssen Pharmaceutica NV, Beerse, Belgium, 22 pp.

Vangronsveld J & Clijsters H (1992) A biological test system for the evaluation of metal phytotoxicity and immobilization by additives in metal-contaminated soils. In Merian E & Haerdi W eds. Interrelation between chemistry and biology. Northwood, Science and Technology Letters, pp117–125.

Vangronsveld J & Clijsters H (1994) Toxic effects of metals. In: Farago M ed. Plants and the chemical elements. Weinheim, VCH, pp 149–177.

Vangronsveld J, Van Assche F, & Clijsters H (1990) Immobilization of heavy metals in polluted soils by application of a modified alumino-silicate: biological evaluation. In: Barcelo J ed. Environmental contamination. Proceedings of the 4th International Conference, Barcelona, Edinburgh, CEP-Consultants, pp 283–285.

Vangronsveld J, Van Assche F, & Clijsters H (1995a) Reclamation of a bare industrial area contaminated by nonferrous metals – in-situ metal immobilization and revegetation. Environ Pollut **87**: 51–59.

Vangronsveld J, Sterckx J, Van Assche F, & Clijsters H (1995b). Rehabilitation studies on an old nonferrous waste dumping ground – effects of revegetation and metal immobilization by beringite. J Geochem Explor, **52**: 221–229.

Van Reen R (1953) Effects of excessive dietary zinc in the rat and the interrelationship with copper. Arch Biochem Biophys, **46**: 337–344.

Van Steveninck RFM, Van Steveninck ME, Fernando DR, Edwards LB, & Wells AJ (1990) Electron probe X-ray microanalytical evidence for two distinct mechanisms of Zn and Cd binding in a Zn tolerant clone of *Lemma minor* L. C-R-ACAD-SCI-SER-III, **310**(13): 671–678.

Van Tilborg WJM (1996) "A further look at zinc" refuted. Rozendaal, Van Tilborg Business Consultancy.

Van Tilborg W & van Assche F (1994) Industry addendum to the Dutch criteria document zinc. Rozendaal, VRBC, pp1–50.

Van Woensel M (1994) The effect of zinc powder on the growth of the unicellular alga *Selenastrum capricornutum*. Report Nr AASc/0021, Janssen Pharmaceutica NV, Beerse, Belgium, 27 pp.

Vardia HK, Rao PS, & Durve VS (1988) Effect of copper, cadmium and zinc on fish-food organisms *Daphnia lumholtzi* and *Cypris subglobosa*. Proc Indian Acad Sci (Animal Sci), **97**(2): 175–180.

VDI-Nachrichten (1995) [Association of Chemical Industry (VCI), 2nd stage of the Environment agreement between VCI and Rotterdam is signed (unpublished press release).] Chemistry News Press Release, pp 1-4 (in German).

Veillon C & Patterson KY (1995) Selenium, chromium and zinc content of proposed SRM 1846 infant formula. Anal Chem, **352**: 77–79.

Verriopoulos G & Dimas S (1988) Combined toxicity of copper, cadmium, zinc, lead, nickel, and chrome to the copepod *Tisbe holothuriae*. Bull Environ Contam Toxicol, **41**: 378–384.

Verus AP & Samman S (1994) Urinary zinc as a marker of zinc intake: results of a supplementation tri in free-living men. Eur J Clin Nutr, **48**: 219–221.

Vidal F & Hidalgo J (1993) Effect of zinc and copper on preimplantation mouse embryo development in vitro and metallothionein levels. Zygote, **1**(3): 225–229.

Viets FG (1962) Chemistry and availability of micronutrients in soils. J Agric Food Chem, **10**(3): 174–178.

Vilkina GA, Pomerantzeva MD, & Ramaya LK (1978) Lack of mutagenic activity of cadmium and zinc salts in somatic and germ mouse cells. Genetica, **14**: 2212–2214.

Vinogradov AP (1959) The geochemistry of rare and dispersed chemical elements in soils. In: Consultants Bureau Inc. ed. The geochemistry of rare and dispersed chemical elements in soils 2nd ed. London, Chapman and Hall, pp 5–25, 149–154, 182–185.

VMM (1994) [1993 Annual report of the surface water measurements.] Erembodegem, VMM Publishers (in Dutch).

VMM (1996) [1995 Annual report of the surface water measurements.] Erembodegem, VMM publishers (in Dutch).

Vogelmeier G, Konig G, Bencze K, & Fruhmann G (1987) Pulmonary involvement in zinc fume fever. Chest, **92**: 946–948.

Von Wirén N, Marschner H, & Romheld V (1996) Roots of iron-efficient maize also absorb phytodiderophore-chelated zinc. Plant Physiol, **111**: 1119–1125.

Voroshilin SI, Platko EG, Fink T, & Nikiforova VJA (1978) Cytogenetic effects of inorganic and acetate compounds of tungsten, zinc, cadmium and cobalt in animal and human cells. Tsitol Genet, **12**: 241–243.

Vranken G, Tiré C, & Heip C (1988) The toxicity of paired metal mixtures to the nematode *Monhystera disjuncta* (Bastian, 1865). Mar Environ Res, **26**: 161–179.

Vymazal J (1986) Occurrence and chemistry of zinc in freshwaters – its toxicity and bioaccumulation with respect to algae: a review. Part 2: Toxicity and bioaccumulation with respect to algae. Acta Hydrochim Hydrobiol, **14**: 83–102.

Wada L, Thurnland JR, & King JC (1985) Zinc utilization in young men fed adequate and low zinc intakes. J Nutr, **115**: 1345–1354.

Waidmann E, Emons H, & Duerbeck HW (1994) Trace determination of Tl, Cu, Pb, Cd, and Zn in specimens of the limnic environment using isotope dilution mass spectrometry with thermal ionization. Fresenius J Anal Chem, **350**: 293–297.

Wainwright SJ & Beckett PJ (1975) Kinetic studies on the binding of zinc ions by the lichen *Usnea Florida* (L.) web. New Phytol, **75**: 91–98.

Wainwright SJ & Woolhouse HW (1977) Some physiological aspects of copper and zinc tolerance in *Agrostis tenuis* Sibh.: Cell elongation and membrane damage. J Exper Bot, **28**(105): 1029–1036.

Wallenius K, Mathur A, & Abdulla M (1979) Effect of different levels of dietary; zinc on development of chemically induced oral cancer in rats. Int J Oral Surg, **8**: 56–62.

Wallock LM, King JC, Hambidge KM, English-Westcott JE, & Pritts J (1993) Meal-induced changes in plasma, erythrocyte, and urinary zinc concentrations in adult women. Am J Clin Nutr, **58**: 695–701.

Wallwork JC, Fosmire GJ, & Sandstead HH (1981) Effect of zinc deficiency on appetite and plasma amino acid concentrations in the rat. Br J Nutr, **45**: 127–136.

Walravens PA & Hambidge KM (1976) Growth of infants fed a zinc supplemented formula. Am J Clin Nutr, **29**: 1114–1121.

Walravens PA, Krebs NF, & Hambidge KM (1983) Linear growth of low income preschool children receiving a zinc supplement. Am J Clin Nutr, **38**: 195–201.

Walravens PA, Hambidge KM, & Koepfer DM (1989) Zinc supplementation in infants with a nutritional pattern of failure to thrive: a double-blind, controlled study. Pediatrics, **83**(4): 532–538.

Walravens PA, Chakar A, Mokni R, Denise J, & Lemonnier D (1992) Zinc supplements in breastfed infants. Lancet, **340**(8821): 683–685.

Walsh CT, Sandstead HH, Prasad AS, Newberne PM, & Fraker PJ (1994) Zinc: health effects and research priorities for the 1990s. Health Perspect, **102**(Suppl 2): 5–46.

Wandzilak TM & Benson RW (1977) Yeast RNA polymerase III: A zinc metallo-enzyme. Biochem Biophys Res Commun, **76**(2): 247–252.

Ward NI (1987) The future of multi-(ultra-trace) element analysis in assessing human and disease: a comparison of NAA and ICPSMS. In: Ward NI ed. Copenhagen, World Health Organization, pp 118–123 (Environmental Health No. 20).

Warkany J & Petering HG (1972) Congenital malformation of the central nervous system in rats produced by maternal zinc deficiency. Teratology, **5**: 319–334.

Wastney ME, Aamodt RL, Rumble WF, & Henkin RI (1986) Kinetic analysis of zinc metabolism and its regulation in normal humans. Am J Physiol, **251**: R398–R408.

Watson TA & McKeown BA (1976) The effect of sublethal concentrations of zinc on growth and plasma glucose levels in rainbow trout, *Salmo gairdneri* (Richardson). J Wildl Dis, **12**: 263–270.

Watson WS, Mitchell KG, Lyon TDB, Bethel MIF, & Crean GP (1987) A simple blood sample method for measuring oral zinc absorption in clinical practice. Clin Phys Physiol Meas, **8**: 173–178.

Weaver FC (1989) Progressive changes in pancreatic vasculature accompanying copper deficiency-induced glandular atrophy. Int J Pancreatol, **4**(2): 175–186.

Weeks JM (1992) The use of the terrestrial amphipod *Arcitalitrus dorrieni* (Crustacea; Amphipoda; Talitridae) as a potential biomonitor of ambient zinc and copper availabilities in leaf-litter. Chemosphere, **24**(10): 1505–1522.

Weeks JM (1993) Effects of dietary copper and zinc concentrations on feeding rates of two species of talitrid amphipods (Crustacea). Bull Environ Contam Toxicol, **50**: 883–890.

Weeks JM & Rainbow PS (1991) The uptake and accumulation of zinc and copper from solution by two species of talitrid amphipods (Crustacea). J Mar Biol. Assoc United Kingdom, **71**: 811–826.

Weeks JM & Rainbow PS (1993) The relative importance of food and seawater as sources of copper and zinc to talitrid amphipods (Crustacea; Amiphipoda; Talitridae). J Appl Ecol, **30**: 722–735.

Weismann K & Hover H (1985) Serum alkaline phsphatase and serum zinc levels in the diagnosis and exclusion of zinc deficiency in man. Am J Clin Nutr, **41**(6): 1214–1219.

Westbrook GL & Mayer ML (1987) Micromolar concentrations of Zn2+ antagonize NMDA and GABA responses of hippocampal neurons. Nature, **328**: 640–643.

White RT (1988) Open reflux vessels for microwave digestion. In: Kingston HM & Jassie LB eds. Introduction to microwave sample preparation. Washington DC, American Chemical Society, pp 54–77.

White JR & Driscoll CT (1987) Zinc cycling in an acidic Adirondack lake. Environ Sci Technol, **21**: 211–216.

White SL & Rainbow PS (1985) On the metabolic requirements for copper and zinc in molluscs and crustaceans. Mar Environ Res, **16**: 215–229.

Whitehouse RC, Prasad AS, Rabbani PI, & Cossack ZT (1982) Zinc in plasma, neutrophils, lymphocytes, and erythrocytes as determined by flameless atomic absorption spectrophotometry. Clin Chem, **28**: 475–480.

WHO (1973) Trace elements in human nutrition. Report of a WHO Expert Committee. Geneva, World Health Organization (WHO Technical Report Series, No. 532).

WHO (1982) Zinc. In: Evaluation of certain food additives and contaminants. Twenty-sixth Report of the Joint FAO/WHO Expert Committee on Food Additives. Geneva, World Health Organization, pp 32–22 (WHO Technical Report Series No. 683).

WHO (1996a) Guidelines for drinking-water quality, 2nd ed. Vol. 2, Health criteria and other supporting information. Geneva, World Health Organization, pp 382–388.

WHO (1996b) Trace elements in human nutrition and health. Chapter 5. Zinc. Geneva, World Health Organization, pp 72–103.

Wiemeyer SN, Scott JM, Anderson MP, Bloom PH, & Stafford CJ (1988) Environmental contaminants in California condors. J Wildl Man, **52**: 238–247.

Wieslander G, Norbaeck D, & Edling C (1994) Occupational exposure to water based paint and symptoms from the skin and eyes. Occup Environ Med, **51**: 181–186.

Williams DE, Vlamis J, Pukite AH, & Corey JE (1984) Metal movement in sludge-treated soils after six years of sludge addition: 1. Cadmium, copper, lead and zinc. Soil Sci, **137**(5): 351–359.

Willie SN, Sturgeon RE, & Berman SS (1983) Comparison of 8-quinolinol-bonded polymer supports for the preconcentration of trace metals in seawater. Anal Chim Acta, **149**: 59-66.

Willis M (1988) Experimental studies of the effects of zinc on *Ancylus fluviatilis* (Müller) (Mollusca; Gastropoda) from the Afon Crafnant, N.Wales. Arch Hydrobiol, **112**(2): 299-316.

Willis M (1989) Experimental studies on the effects of zinc on *Erpobdella octulata* (L.) (Annelida: Hirudinea) from the Afon Crafnant, N.Wales. Arch Hydrobiol, **116**(4): 449–469.

Wilson DO (1977) Nitrification in three soils amended with zinc sulfate. Soil Biol Biochem, **9**: 277–280.

Windom HL, Byrd JT, Smith RG Jr, & Huan F (1991) Inadequacy of NASQAN data for assessing metal trends in the nations rivers. Environ Sci Technol, **25**: 1137–1142.

Winner RW (1981) A comparison of body length, brood size and longevity as indices of chronic copper and zinc stresses in *Daphnia magna*. Environ Pollut, **A26**: 33–37.

Winner RW & Gauss JD (1986) Relationship between chronic toxicity and bioaccumulation of copper, cadmium and zinc as affected by water hardness and humic acid. Aquat Toxicol, **8**: 149–161.

Wong CK (1992) Effects of chromium, copper, nickel, and zinc on survival and feeding of the cladoceran Moina macrocopa. Bull Environ Contam Toxicol, **49**: 593–599.

Wong CK (1993) Effects of chromium, copper, nickel, and zinc on longevity and reproduction of the cladoceran *Moina macrocopa*. Bull Environ Contam Toxicol, **50**: 633–639.

Wong MH & Bradshaw AD (1982) A comparison of the toxicity of heavy metals, using root elongation of rye grass, *Lolium perenne*. New Phytol, **91**: 255–261.

Woo W (1983) The effect of dietary zinc on high density lipoprotein synthesis. Nutr Rep Int, **33**: 1019–1025.

Wright AL, King JC, Baer MT, & Citron LJ (1981) Experimental zinc depletion and altered taste perception for NaC1 in young adult males. Am J Clin Nutr, **34**: 848–852.

Wu FYH & Wu CW (1987) Zinc in DNA replication and transcription. Annu Rev Nutr, **7**: 251–272.

Xie X & Smart TG (1991) A physiological role for endogenous zinc in rat hippocampal synaptic neurotransmission. Nature, **349**: 521–524.

Xu Q & Pascoe D (1993) The bioconcentration of zinc by *Gammarus pulex* (L.) and the application of a kinetic model to determine bioconcentration factors. Water Res, **27**(11): 1683–1688.

Xu Q & Pascoe D (1994) The importance of food and water as sources of zinc during exposure of *Gammarus pulex* (Amphipoda). Arch Environ Contam Toxicol, **26**: 459–465.

Yaaqub RR, Davies TD, Jickells TD, & Miller JM (1991) Trace elements in daily collected aerosols at a site in southeast England. Atmos Environ, **25A**: 985–996.

Yadrick MK, Kenney MA, & Winterfeld EA (1989) Iron, copper and zinc status: response to supplementation with zinc or zinc and iron in adult females. Am J Clin Nutr, **49**: 145–150.

Yeats PA (1988) The distribution of trace metals in ocean waters. Sci Total Environ, **72**: 131–149.

Yeats PA & Bewers JM (1982) Discharge of metals from the St. Lawrence river. Can J Earth Sci, **19**: 982–992.

Yip CC (1971) A bovine pancreatic enzyme catalyzing the conversion of proinsulin to insulin. Proc Nat Acad Sci USA, **68**(6): 1312–1315.

Yip R, Reeves JD, Lonerdal B, Keen CL, & Dallman PR (1985) Does iron supplementation compromise zinc nutrition in healthy infants. Am J Clin Nutr, **42**: 683–687.

Yokoi K, Alcock NW, & Sandstead HH (1994) Iron and zinc nutriture of premenopausal women: Associations of diet with serum ferritin and plasma zinc disappearance and of serum ferritin with plasma zinc and plasma zinc disappearance. J Lab Clin Med, **124**: 852–861.

Young DR, Jan TK, & Hershelman GP (1980) Cycling of zinc in the nearshore marine environment. In: Nriagu JO ed. Zinc in the environment, Part I: ecological cycling. New York, John Wiley, pp 297-335.

Yukawa M, Amano K, Suzuki-Yasumoto M, & Terai M (1980) Distribution of trace elements in the human body determined by neutron activation analysis. Arch Environ Health, **35**: 36–44.

Zalewski PE, Forbes IJ, & Betts WH (1993) Correlation of apoptosis with change in intracellular labile Zn(II) using Zinquin [(2-methyl-8-p-toluenesulphonamido-6-quinolyl-oxy)acetic acid], a new specific fluorescent probe for Zn(II). Biochem. J, **296**: 403–408.

Zaporowska H & Wasilewski W (1992) Combined effect of vanadium and zinc on certain selected haematological indices in rats. Comp Biochem Physiol C, **103**(1): 143–147.

Zeng J, Vallee BL, & Kagi JHR (1991) Zinc transfer from transcription factor IIIA fingers to thionein clusters. Proc Natl Acad Sci USA, **88**: 9984–9988.

Zhang J & Huang WW (1993) Dissolved trace metals in the Huanghe: the most turbid large river in the world. Water Res, **27**: 1–8.

Zoetemen BCJ (1978) Sensory assessment and chemical composition of drinking water. The Hague, Van der Gang.

Zoller WH, Gladney ES, & Duce RA (1974) Atmospheric concentrations and sources of trace metals at the South Pole. Science, **183**: 198–200.

Zou E & Bu S (1994) Acute toxicity of copper, cadmium, and zinc to the water flea, *Moina irrasa* (Cladocera). Bull Environ Contam Toxicol, **52**: 742–748.

Zschiesche W (1988) [New occupational medicine information on exposures during welding.] Arbeitsmed Sozialmed Praevativmed, **23**: 223–224 (in German).

Zuurdeeg BW (1992) [Natural background levels of heavy metals and some other trace elements in surface water in the Netherlands.] Utrecht, Geochemical Publication (in Dutch).

Zwick D, Frimpong NA, & Tulp OL (1991) Progressive zinc-induced changes in glycemic responses in lean and obese LA/N-cp rats. FASEB J, **5**: A94.

RESUME ET CONCLUSIONS

1. Identité , propriétés physiques et chimiques

Le zinc n'existe pas à l'état métallique dans la nature. Il n'est présent qu'à l'état divalent (Zn II). L'ion zinc peut être solvaté; sa solubilité dépend du pH et de l'anion. Le zinc est un élément de transition capable de former des complexes avec divers ligandes organiques. Les composés organozinciques n'existent pas dans la nature.

2. Méthodes d'analyse

Le zinc étant très répandu dans l'environnement , il faut un soin particulier pour effectuer les prélèvements ainsi que pour préparer et analyser les échantillons afin d'éviter toute contamination. Dans le cas des échantillons solides on a le plus souvent recours à une minéralisation par des acides concentrés assistée par traitement microonde. Dans le cas d'échantillons d'eau, on peut commencer par concentrer le zinc au moyen d'une extraction par solvant en présence d'agents complexants ou d'une séparation sur résine chélatante.

Pour le dosage du zinc, on a couramment recours aux méthodes instrumentales suivantes: spectrométrie d'emission atomique à source plasma à couplage inductif, spectrométrie d'absorption atomique avec four à électrodes de graphite, voltamétrie par redissolution anodique et spectrométrie de masse à source plasma à couplage inductif. Pour le dosage des faibles quantités on donne la préférence à la spectrométrie d'absorption atomique avec four à électrodes de graphite, à la voltamétrie par redissolution anodique et à la spectro-métrie de masse avec source plasma à couplage inductif.

En opérant de manière minutieuse, on parvient à doser le zinc jusqu'à des concentrations de 0,006 µg/litre dans l'eau et de 0,01 mg/kg dans les échantillons solides.

Pour la recherche et le dosage des diverses espèces chimiques présentes dans l'eau (ou *spéciation*), il faut mettre en oeuvre l'une ou l'autre des techniques de séparation indiquées plus haut ou recourir à la différenciation des espèces labiles que permet la voltamétrie par redissolution anodique.

3. Sources d'exposition humaine et environnementale

La plupart des roches et de nombreux minéraux renferment du zinc. Sur le plan commercial, le minerai le plus important est la sphalérite ou blende (ZnS), qui constitue la principale source de zinc pour l'industrie. En 1994, la production mondiale de zinc a été de 7 089 000 tonnes et la consommation de zinc métallique a atteint 6 895 000 tonnes.

Le zinc est largement utilisé pour la protection d'autres métaux (zingage, galvanisation), en moulage sous pression, dans le BTP et pour la confection d'alliages divers. Les dérivés minéraux du zinc ont des applications diverses, notamment dans la fabrication d'équipements automobiles, d'accumulateurs et de piles sèches ou encore dans le domaine dentaire, médical ou pour la confection d'objets ménagers. On utilise les organozinciques comme fongicides,comme antibiotiques pour applications locales et comme lubrifiants.

Vers 100 à 150 °C, le zinc devient malléable et il est alors facile de le travailler. Comme il est capable de réduire la plupart des ions d'autres métaux, on l'emploie pour la confection d'électrodes dans les piles sèches ainsi qu'en hydrométallurgie.

La majeure partie du zinc naturellement présent dans l'eau y est amenée par l'érosion. Le zinc présent dans l'atmosphère par suite de processus naturels provient d'émissions ignées et de feux de forêt. Les émissions d'origine humaine sont du même ordre de grandeur. Les principales sources de zinc dues aux activités humaines sont l'extraction du minerai, la production de zinc, la production de fer et d'acier, la corrosion des structures galvanisées, la combustion du charbon et de carburants divers, l'élimination et l'incinération des déchets et enfin l'utilisation d'engrais et de pesticides à base de zinc.

4. Transport, distribution et transformation dans l'environnement

Le zinc présent dans l'atmosphère est en majeure partie fixé à des particules d'aérosols. La granulométrie de ces particules dépend de la source qui les émet. Une importante proportion du zinc libéré lors de divers processus industriels est adsorbée sur des particules suffisamment petites pour être respirables.

Le transport et la distribution du zinc atmosphérique varient en fonction de la granulométrie des particules et des propriétés des composés en cause. Le zinc s'élimine de l'atmosphère en se déposant par voie humide ou par voie sèche. Le zinc adsorbé sur des particules de faible densité et de petit diamètre peut être en revanche transporté sur de grandes distances.

La distribution et le transport du zinc dans l'eau, les sédiments et le sol dépendent de l'espèce chimique en cause et des caractéristiques de l'environnement. C'est le pH qui détermine principalement la solubilité du zinc. Si le pH est acide, le zinc peut être présent dans l'eau sous forme ionique. Si par contre le pH est supérieur à 8,0, il peut y avoir précipitation du zinc. Il peut également former des complexes organique stables, par exemple avec les acides humiques et fulviques. La formation de ces complexes augmente la mobilité et la solubilité du zinc. Comme il est adsorbé sur les argiles et les matières organiques, il peu probable qu'il puisse se détacher du sol par lessivage. Les sols acides et sablonneux à faible teneur en matières organiques ne sont guère capables de retenir le zinc.

Le zinc est un élément essentiel et sa concentration *in vivo* est donc régulée chez la plupart des êtres vivants. Les animaux aquatiques ont tendance à le prélever dans l'eau plutôt que dans leur nourriture. Pour être biodisponible, le zinc doit obligatoirement être en solution et sa biodisponibilité dépend des caractéristiques physiques et chimiques de l'environnement et des processus biologiques. Toute évaluation d'ordre écologique doit donc se faire site par site.

5. Concentrations dans l'environnement

Le caractère ubiquitaire du zinc le fait se retrouver dans un peu tous les échantillons d'origine biologique ou environnementale. Sa concentration dans les sédiments et dans l'eau douce est fortement influencée par les conditions géologiques et anthropogéniques locales et elle varie donc dans de larges proportions. La concentration totale naturelle du zinc se situe en général autour de

< 0,1–50 µg /litre dans les eaux douces, de 0,002–0,1 µg/litre dans l'eau de mer, de 10–300 mg/kg (poids sec) dans les sols; elle peut aller jusqu'à 300 mg/ kg (poids sec) dans les sédiments et jusqu'à 300 ng/m^3 dans l'air. Des concentrations plus élevées peuvent être attribuées à la présence naturelle de minerais enrichis en zinc, à des sources anthropogéniques ou encore à des processus biotiques ou abiotiques. Les échantillons dont la teneur en zinc s'explique par la présence de sources anthropogéniques, peuvent en contenir jusqu'à 4 mg/litre (eau), 35 g/kg (sol), 15 µg/litre (eaux estuarielles) ou 8 µg/m^3 (air).

Chez les organismes représentatifs, on observe des concentrations de zinc comprises entre 200 et 2000 mg/kg en cas d'exposition au zinc présent dans l'eau.

Chez les végétaux et les animaux, la teneur en zinc est plus élevée à proximité des sources de pollution anthropogéniques. Les variations de teneur sont importantes d'une espèce à l'autre; la teneur peut varier, par exemple, en fonction du stade évolutif, du sexe, de la saison, du régime alimentaire et de l'âge. Dans la plupart des cultures et des pâturages, la concentration normale du zinc se situe dans la fourchette 10–100 mg/kg de poids sec. Certaines plantes accumulent le zinc, mais dans une proportion qui dépend de la nature du sol et des caractéristiques de la plante.

La quantité de zinc inhalée avec l'air ambiant est négligeable, mais sur les lieux de travail, l'exposition aux poussières et vapeurs contenant du zinc peut varier dans d'importantes proportions.

5.1 Apport chez l'Homme

On estime que l'apport de zinc total d'origine alimentaire est de 5,6 à 10 mg par jour pour les nourrissons et les enfants de 2 mois à 11 ans, de 12,3 à 13,0 mg par jour pour les jeunes de 12 à 19 ans et de 8,8 à 14,4 mg par jour pour les adultes de 20 à 50 ans. L'apport journalier moyen par l'eau de boisson est estimé à moins de 0,2 mg.

Les valeurs de référence pour l'apport de zinc d'origine alimentaire varient en fonction des habitudes alimentaires du pays, des hypothèse formulées au sujet de la biodisponibilité du zinc présent dans les aliments et également de l'âge, du sexe et de l'état physiologique des sujets. Ces valeurs se situent dans les fourchettes suivantes: 3,3 à 5,6 mg par jour pour les nourrissons de 0 à 12 mois, 3,8 à 10,0 mg par jour pour les enfants de 1 à 10 ans et 8,7 à 15 mg par jour pour les adolescents de 11 à 18 ans. Pour les adultes, les valeurs vont de 6,7 à 15 mg par jour dans la tranche d'âge 19–50 ans, à 7,3–15 mg par jour chez la femme enceinte- en supposant un régime alimentaire offrant une biodisponibilité moyenne- et à 11,7–19 mg par jour chez la mère allaitante, selon le stade.

6. Cinétique et métabolisme chez l'Homme et les animaux de laboratoire

Des études portant sur l'exposition au zinc par la voie respiratoire (par le nez uniquement) de cobayes, de rats et de lapins ont montré que le taux de rétention pulmonaire était de 5 à 20% après exposition à des aérosols d'oxyde de zinc à la concentration de $5-12 \text{ mg/m}^3$ pendant 3 à 6 heures. Dans l'intestin, l'absorption du zinc s'effectue selon un mécanisme homéostatique qui n'est pas encore complètement élucidé, mais qui est sous la dépendance des sécrétions pancréatiques et intestinales ainsi que de l'excrétion par la voie fécale. Il est possible que cette homéostase fasse intervenir des métalloprotéines comme la métallothionéine et la protéine intestinale riche en cystéine. D'autres mécanismes de nature inconnue peuvent également être à l'oeuvre. Au niveau de la muqueuse intestinale, la résorption peut se faire par transport actif ou passif. Chez l'animal, la résorption du zinc peut osciller entre 10 et 40%, selon l'état nutritionnel et la présence d'autres ligandes dans l'alimentation. La résorption du zinc par la voie percutanée peut s'opérer en présence d'oxyde ou de chlorure de zinc et elle augmente en cas de carence zincique. Une fois absorbé, le zinc se dépose principalement dans les muscles, les os, le foie, le pancréas, les reins et d'autres organes. La demi-vie biologique du zinc est de 4 à 50 jours chez le rat, selon la dose administrée, et elle est d'environ 280 jours chez l'Homme.

7. Effets sur les animaux de laboratoire

Le zinc a une faible toxicité aiguë pour les rongeurs, la valeur de la DL_{50} allant de 30 à 600 mg/kg de poids corporel, en fonction de la nature du sel de zinc administré. Après inhalation ou instillation intratrachéenne de composés du zinc, on a observé des effets aigus tels qu'une détresse respiratoire, un oedème pulmonaire et une infiltration leucocytaire du parenchyme pulmonaire.

Chez des rongeurs brièvement exposés par voie orale à des composés du zinc, on peut observer les effets toxiques suivants : faiblesse, anorexie, anémie, ralentissement de la croissance, dépilation, moindre assimilation des aliments ainsi qu'une modification du taux des enzymes hépatiques et sériques, des anomalies morphologiques et enzymatiques dans l'encéphale et des anomalies histologiques et fonctionnelles au niveau rénal. Le seuil d'apparition des effets indésirables a été fixé à environ 160 mg/kg p.c. chez le rat. Chez des veaux exposés à une forte concentration de zinc dans leur alimentation, on a observé des anomalies pancréatiques. Une exposition de brève durée par la voie respiratoire à des concentrations d'oxyde de zinc $\geq 5,9$ mg/m^3 a provoqué une inflammation et des lésions pulmonaires chez des cobayes et des rats.

Une exposition de longue durée par voie orale à des dérivés du zinc a permis de constater que les organes cibles de l'action toxique de cet élément sont le système hémopoïétique chez le rat, le furet et le lapin; le rein chez le rat et le furet, et le pancréas chez la souris et le furet. La dose sans effet observable (NOEL)- en prenant en compte l'effet sur la croissance et l'anémie imputables à la présence de sulfate de zinc dans l'alimentation- a été trouvée inférieure à 100 mg/kg chez le rat. Chez les animaux de laboratoire exposés au zinc, on peut observer que l'augmentation de la concentration de cet élément dans l'organisme s'accompagne d'une diminution de la concentration du cuivre, ce qui incite à penser que certains signes de toxicité attribués à une exposition excessive au zinc pourraient en fait s'expliquer par une carence cuprique provoquée par le zinc. Par ailleurs, certaines études montrent que l'exposition au zinc modifie la concentration d'autres métaux essentiels dans l'organisme des animaux exposés, notamment celle du fer. De fait, certains signes de toxicité observés chez des animaux exposés à de fortes concentrations de zinc peuvent être atténués par l'adjonction de cuivre ou de fer à l'alimentation.

A très forte dose, le zinc est toxique pour la souris et le hamster gravides. Des rats qui avaient reçu pendant 5 mois une alimentation contenant soit 0,5%, soit 1% de zinc, se sont révélés incapables de se reproduire jusqu'à ce que le zinc soit retiré de leur nourriture. Chez des souris et des rats, on a également constaté qu'une forte teneur en zinc de l'alimentation (2000 mg/kg) s'accompagnait d'une augmentation des résorptions et des mortinaissances; ces effets ont été aussi observés chez des moutons et des hamsters. Dans une étude au cours de laquelle des rats avaient été exposés, pendant toute la durée de la gestation, à des doses de zinc ne dépassant pas 150 mg/kg, on a également constaté une augmentation du nombre de résorptions. Toutefois, dans une autre étude portant également sur des rats, on n'a pas observé d'effets délétères sur le foetus en développement à la dose de 500 mg/kg. Chez des rats, l'administration après le coït d'une nourriture contenant 4000 mg de zinc par kg, a empêché l'implantation des ovules. Chez des ratons exposés à des composés du zinc, l'élévation de la concentration de cet élément s'accompagnait d'une réduction de celle du fer et du cuivre.

Des études de génotoxicité ont été effectuées sur divers systèmes. Si la plupart des résultats se sont révélés négatifs, quelques-uns ont tout de même été positifs.

Chez l'animal, une carence en zinc se caractérise par une moindre croissance, une perturbation de la réplication cellulaire, des effets indésirables sur la reproduction et sur le développement - ces derniers persistant après le sevrage- et une diminution des réponses immunitaires.

8. Effets sur l'Homme

Après exposition unique ou brièvement réitérée à de l'eau ou à des boissons contenant 1000 à 2500 mg de zinc par litre, des symptômes d'intoxication consistant en débâcle gastrointestinale, nausées et diarrhée se sont manifesté. Des symptômes analogues, mais qui ont parfois abouti à la mort ont été signalés après administration accidentelle de fortes doses de zinc par voie intraveineuse. Chez des insuffisants rénaux en dialyse, on a constaté l'apparition de symptômes d'une intoxication par le zinc due à l'utilisation d'eau conservée dans des réservoirs galvanisés. Ces symptômes étaient réversibles puisqu'ils ont disparu lorsque l'eau a été filtrée sur charbon actif.

Chez l'Homme, lorsqu'il y a disproportion entre l'apport de zinc et celui de cuivre, il en résulte une carence cuprique qui accroît les besoins de cuivre, en augmente l'excrétion et en rend le bilan déficitaire. On a observé divers effets consécutifs à des apports de zinc d'origine pharmacologique, effets qui vont d'une leucopénie et d'une anémie microcytaire hypochrome à une diminution du taux sérique des lipoprotéines de haute densité. Ces anomalies disparaissent lorsque cesse l'apport zincique avec supplémentation cuprique concomitante.

Chez l'Homme, les effets d'une carence en zinc sont nombreux et se traduisent notamment par les anomalies suivantes: troubles neurosensoriels, oligospermie, altération des fonctions neuropsychologiques, retard de croissance, allongement de la durée de cicatrisation, troubles de l'immunité et dermatite. Ces troubles sont généralement réversibles et cèdent à une supplémentation zincique.

Il n'existe pas d'indice biochimique du bilan zincique qui soit à la fois spécifique et sensible. La méthode la plus fiable pour diagnostiquer une carence en zinc consiste à chercher à mettre en évidence une réaction positive à une supplémentation zincique dans le cadre d'un essai contrôlé en double insu (en l'absence d'autres carences en nutriments limitants). Cette méthode est longue et souvent impraticable, aussi préfère -t-on généralement recourir à un ensemble d'indices alimentaires, biochimiques et physiologiques. Plusieurs valeurs anormales concordantes sont plus fiables qu'une seule valeur aberrante pour diagnostiquer une carence en zinc. En faisant appel à des indices physiologiques fonctionnels comme la croissance, le sens du goût et l'adaptation à l'obscurité et en les complétant par une épreuve biochimique (par exemple, le dosage du zinc dans le plasma ou les cheveux) on perçoit mieux les conséquences fonctionnelles de la carence que l'on s'efforce de déterminer.

L'exposition par la voie respiratoire au chlorure de zinc contenu dans les bombes fumigènes utilisées par les forces armées a produit les effets suivants: oedème et fibrose interstitiels, pneumonie, oedème de la muqueuse bronchique, ulcérations et même mort en cas d'extrême exposition dans un espace confiné. Ces effets pourraient s'expliquer par la nature hygroscopique et astringente des particules libérées par ces dispositifs.

L'exposition professionnelle aux particules finement dispersées qui se forment lorsque certains métaux, dont le zinc, sont volatilisés, peut conduire à une pathologie aiguë qui porte le nom de "fièvre des fondeurs". Elle se caractérise par des symptômes divers au nombre desquels de la fièvre, des frissons, de la nausée et de la fatigue. C'est une affection généralement grave mais passagère et les sujets ont tendance à acquérir une tolérance. En exposant des volontaires à des concentrations de zinc de 77 à 150 mg/m^3 pendant 15 à 30 minutes, on a pu observer chez certains d'entre eux les symptômes suivants: une réaction inflammatoire marquée liée à la dose avec augmentation du nombre de polynucléaires dans le liquide de lavage broncho-alvéolaire et un accroissement sensible des cytokines. On a signalé des cas d'asthme professionnel chez des ouvriers travaillant avec des fondants pour soudure tendre, mais les données sont insuffisantes pour qu'on puisse invoquer une relation de cause à effet. Un cas rare incitant à penser à une telle relation a été récemment observé: il s'agissait d'un ouvrier travaillant dans un atelier de galvanisation par trempage dans des bains de zinc fondu.

9. Effets sur les autres êtres vivants au laboratoire et dans leur milieu naturel

Le zinc joue un rôle important dans la stabilité de la membrane cellulaire, dans plus de 300 enzymes et dans le métabolisme des protéines et des acides nucléiques. Il faut faire la part entre les effets indésirables du zinc et son caractère d'élément essentiel. Des carences en zinc ont été observées chez toutes sortes de plantes cultivées et d'animaux, avec de graves effets à tous les stades de la reproduction, de la croissance et de la prolifération tissulaire. Un peu partout dans le monde, les carences en zinc touchant diverses cultures ont pour effet d'importantes pertes au niveau des récoltes. Dans l'environnement aquatique, les carences en zinc sont rares, mais on peut les provoquer expérimentalement.

Des facteurs biotiques ou abiotiques tels que l'âge et la taille de l'organisme en cause, une exposition antérieure éventuelle, la dureté de l'eau, le pH, la teneur en carbone organique dissous ou la température peuvent influer sur la toxicité du zinc. Une approche écochimique et écotoxicologique intégrée a permis une meilleure prévision des effets du zinc sur les êtres vivants dans leur milieu naturel. La conséquence en a été que l'on admet désormais que la concentration totale d'un élément essentiel comme le zinc dans tel ou tel compartiment de l'environnement ne peut, à elle seule, suffire pour prévoir correctement sa biodisponibilité.

En ce qui concerne les invertébrés d'eau douce on a obtenu, pour le zinc en solution, des valeurs de la toxicité aiguë qui vont de 0,07 mg/litre chez la daphnie à 575 mg/litre chez une espèce d'isopode. Dans le cas des invertébrés marins, les valeurs obtenues vont de 0,097 mg/litre pour une espèce de mysidé à 11,3 mg /litre pour une espèce de crevette. Les concentrations létales en exposition instantanée pour les poissons d'eau douce se situent entre 0,066 et 2,6 mg/litre; pour les poissons de mer, elles sont comprises entre 0,19 et 17,66 mg/litre.

On a montré que le zinc a des effets génésiques, biochimiques, physiologiques et comportementaux indésirables sur divers organismes aquatiques. Ces effets indésirables se manifestent à partir de 20 µg/litre. Cependant, la toxicité du zinc pour ces organismes dépend d'un grand nombre de facteurs comme la température, la dureté et le pH de l'eau ainsi que des expositions antérieures éventuelle à cet élément.

D'une façon générale, les effets toxiques du zinc sur les végétaux se traduisent par des troubles métaboliques différents de ceux que l'on observe en cas de carence zincique. Dans le tissu foliaire de la plupart des espèces végétales, une concentration de zinc de l'ordre de 200–300 mg/kg de poids sec constitue un seuil critique pour la croissance.

Les études effectuées sur le terrain montrent que les invertébrés aquatiques, les poissons et les végétaux terrestres qui vivent à proximité de sources de pollution zincique en subissent les effets nocifs. Toutefois, les végétaux terrestres, les algues, les micro-organismes et les invertébrés qui vivent non loin des zones fortement polluées par le zinc, ont acquis une tolérance à cet élément.

10. Conclusions

10.1 Santé humaine

- Les émissions de zinc dues à l'activité humaine ont tendance à diminuer.

- Il est possible qu'un grand nombre d'échantillons - notamment des échantillons d'eau - prélevés avant 1980 aient été contaminés par du zinc lors des prélèvements et de l'analyse et c'est pourquoi la plus grande prudence s'impose concernant les valeurs de la concentration.

- Dans les pays où l'alimentation de base est constituée de légumineuses et de céréales non raffinées et où la consommation de produits carnés est faible, il faudrait élaborer des stratégies pour accroître la teneur en zinc des aliments et la bio-disponibilité de cet élément.

- Les préparations qui sont destinées à augmenter l'apport de zinc au-delà de l'apport alimentaire ne doivent pas avoir une teneur en zinc supérieure à la teneur de référence et contenir en outre suffisamment de cuivre pour que le rapport Zn/Cu soit d'environ 7, comme dans le lait humain.

- Il faudrait que l'exposition réelle aux vapeurs d'oxyde de zinc sur les lieux de travail soit mieux documentée. La concentration en oxyde de zinc ne devrait pas y entraîner une exposition aussi intense que celle qui provoque une réaction inflammatoire pulmonaire chez des volontaires.

- Etant donné que le zinc est un élément essentiel, qu'il est relativement peu toxique pour l'Homme et que les sources d'exposition humaine sont limitées, il apparaît que pour des sujets en bonne santé qui ne sont pas professionnellement exposés à cet élément, le risque d'une carence zincique est potentiellement plus important que celui qui découle de l'exposition ordinaire dans l'environnement.

10.2 Environnement

- Le zinc est un élément essentiel. Il peut y avoir carence ou au contraire excès de ce métal. C'est pourquoi il importe que la réglementation relative au zinc, tout en assurant une protection contre ses effets toxiques, n'impose pas des teneurs trop faibles qui pourraient entraîner une carence zincique.

347

- Il existe des différences dans les réactions à une carence zincique ou à un excès de zinc.

- La biodisponibilité du zinc dépend d'un certain nombre de facteurs biotiques ou abiotiques comme par exemple l'âge et la taille de l'organisme en cause, l'éventualité d'une exposition antérieure, la dureté de l'eau, le pH, la teneur en carbone organique dissous et la température.

- La concentration totale d'un élément essentiel tel que le zinc, ne permet pas à elle seule, d'en prévoir correctement la bio-disponibilité ou la toxicité.

- Pour des éléments essentiels comme le zinc, il existe une fourchette de concentration optimale.

- La toxicité du zinc dépend des conditions environnementales et de la nature des biotopes, de sorte que toute évaluation du risque relative aux effets potentiels de cet élément sur les êtres vivants doit tenir compte de l'écologie locale.

RESUMEN Y CONCLUSIONES

1. Identidad y propiedades físicas y químicas

El zinc metálico no se encuentra de forma natural en el medio ambiente. Está presente sólo en estado divalente como Zn (II). El zinc iónico sufre solvatación, dependiendo su solubilidad del pH y de la concentración de aniones. El zinc es un elemento de transición y puede formar complejos con diversos ligandos orgánicos. En la naturaleza no existen compuestos organometálicos de zinc.

2. Métodos analíticos

Debido a la ubicuidad del zinc en el medio ambiente, hay que prestar especial atención durante el muestreo, la preparación y el análisis de las muestras para evitar su contaminación. La preparación de muestras a partir de material sólido normalmente requiere la mineralización con ácidos concentrados y con la ayuda de microondas. Para las muestras acuosas, se ha utilizado la extracción con disolventes en presencia de sustancias capaces de formar iones complejos y la separación mediante resinas quelantes para conseguir una concentración previa del zinc.

Las técnicas instrumentales normalmente utilizadas para la determinación del zinc son: espectrometría de emisión atómica de plasma con acoplamiento inductivo, espectrometría de absorción atómica en horno de grafito, voltametría de arranque anódico y espectrometría de masas de plasma con acoplamiento inductivo. Para el análisis de concentraciones bajas se utilizan la espectrometría de absorción atómica en horno de grafito, la voltametría de arranque anódico y la espectrometría de masas de plasma con acoplamiento inductivo.

Prestando especial atención se pueden detectar concentraciones muy bajas, de hasta 0,006 µg/litro y 0,01 mg/kg en muestras acuosas y sólidas, respectivamente.

Los análisis de especiación en agua requieren la aplicación de técnicas de separación con cualquiera de los métodos antes mencionados o la separación de enlaces lábiles mediante voltametría de arranque anódico.

3. Fuentes de exposición humana y ambiental

La mayoría de las rocas y muchos minerales contienen zinc en cantidades variables. Desde el punto de vista comercial, la esfalerita (SZn) es el mineral más importante y la fuente principal del metal para la industria del zinc. En 1994, la producción de zinc metálico en todo el mundo fue de 7 089 000 toneladas y su consumo de 6 895 000 toneladas.

El zinc se utiliza ampliamente como revestimiento protector de otros metales, en la fundición de color y en la industria de la construcción, así como para aleaciones. Los compuestos de zinc inorgánico tienen varias aplicaciones, por ejemplo de equipamiento para el automóvil, baterías de acumuladores y pilas secas y aplicaciones dentales, médicas y domésticas. Los compuestos orgánicos de zinc se utilizan como fungicidas, antibióticos tópicos y lubrificantes.

El zinc se convierte en maleable cuando se calienta a 100–150 °C y entonces se moldea fácilmente para darle distintas formas. Es capaz de reducir la mayoría de los otros estados metálicos y, por consiguiente, se utiliza como electrodo en pilas secas y en hidrometalurgia.

La emisión natural más importante de zinc al agua procede de la erosión. Las aportaciones naturales al aire se deben fundamental-mente a emisiones ígneas y a incendios forestales. Las fuentes antropogénicas y naturales son de una magnitud semejante. Las principales fuentes antropogénicas de zinc son la extracción, las instalaciones de producción de zinc, la producción de hierro y acero, la corrosión de estructuras galvanizadas, la combustión de carbón y otros combustibles, la eliminación e incineración de desechos y el uso de fertilizantes y plaguicidas con zinc.

4. Transporte, distribución y transformación en el medio ambiente

El zinc se encuentra en la atmósfera fundamentalmente unido a partículas de aerosol. El tamaño de la partícula depende de la fuente de emisión del zinc. Una proporción importante del zinc que se libera en los procesos industriales se adsorbe sobre partículas suficientemente pequeñas para quedar en la gama respirable.

El transporte y la distribución del zinc atmosférico varían en función del tamaño de las partículas y de las propiedades de los compuestos de zinc correspondientes. El zinc se elimina de la atmósfera mediante deposición seca y húmeda. El zinc que se adsorbe sobre partículas con densidad y diámetro bajos puede recorrer largas distancias.

La distribución y el transporte del zinc en el agua, los sedimentos y el suelo dependen de las especies de zinc presentes y de las características del medio ambiente. La solubilidad del zinc está en función sobre todo del pH. Con valores de pH ácido, el zinc puede estar presente en la fase acuosa en su forma iónica. A un pH superior a 8 puede precipitar. También puede formar complejos orgánicos estables, por ejemplo con los ácidos húmico y fúlvico. La formación de estos compuestos puede aumentar la movilidad y/o la solubilidad del zinc. No es probable la lixiviación del zinc a partir del suelo, debido a su adsorción sobre la arcilla y la materia orgánica. Los suelos ácidos y los arenosos con un contenido orgánico bajo tienen menos capacidad de absorción.

El zinc es un elemento esencial, por lo que la mayoría de los organismos regulan sus niveles in vivo. El zinc no se bioamplifica. Los animales acuáticos lo absorben del agua más que de los alimentos. Sólo el zinc disuelto tiende a estar biodisponible, dependiendo dicha biodisponibilidad de las características físicas y químicas del medio ambiente y de los procesos geológicos. En consecuencia, la evaluación en el medio ambiente se debe realizar de manera específica para cada lugar.

5. Concentraciones en el medio ambiente

El zinc se encuentra en todas partes en muestras del medio ambiente y biológicas. Sus concentraciones en los sedimentos del suelo y en el agua dulce dependen en gran medida de influencias geológicas y antropogénicas locales, de manera que son muy variables. La concentración natural total de fondo de zinc suele ser <0,1–50 µg/litro en el agua dulce, 0,002–0,1 µg/litro en el agua marina, 10–300 mg/kg de peso seco en el suelo, hasta 100 mg/kg de peso seco en los sedimentos y hasta 300 ng/m^3 en el aire. Se pueden atribuir niveles superiores a la presencia natural de minerales con concentraciones elevadas de zinc, a fuentes antropogénicas o a procesos abióticos y bióticos. En muestras con contaminación de origen antropogénico, se obtienen concentraciones de zinc de hasta 4 mg/litro en el agua, 35 g/kg en el suelo, 15 µg/litro en el agua de estuarios y 8 µg/m^3 en el aire.

Su concentración en organismos representativos durante la exposición al zinc presente en el agua es del orden de 200–2000 mg/kg.

Las concentraciones en plantas y animales son más altas cerca de fuentes puntuales antropogénicas de contaminación por zinc. Las variaciones interespecíficas en cuanto al contenido de zinc son considerables; los niveles intraespecíficos varían, por ejemplo, con la fase de la vida, el sexo, la estación, la alimentación y la edad. Las concentraciones normales de zinc en la mayor parte de los cultivos y pastos son del orden de 10–100 mg/kg de peso seco. Algunas plantas acumulan zinc, pero la magnitud de la acumulación en los tejidos vegetales varía con las características del suelo y de la propia planta.

Sólo se inhalan cantidades insignificantes de zinc del aire ambiente, pero es posible la exposición a una gran variedad de tipos de polvo y de vapor de zinc y sus compuestos en el entorno ocupacional.

5.1 Ingesta humana

La gama estimada de ingesta diaria total de zinc con los alimentos es de 5,6–10 mg/día para los lactantes y los niños de edades comprendidas entre los dos meses y los 11 años, de 12,3–13,0 mg/día para los niños de 12–19 años y de 8,8–14,4 mg/día para los adultos de 20–50 años. La ingesta diaria media de zinc con el agua de bebida se estima en <0,2 mg/día.

Los valores de referencia del zinc en los alimentos varían según los hábitos alimentarios del país, las hipótesis sobre la biodisponibilidad del zinc en los alimentos y la edad, el sexo y las condiciones fisiológicas. Los valores de referencia en los alimentos oscilan entre 3,3 y 5,6 mg/día para los lactantes de 0–12 meses, entre 3,8 y 10,0 mg/día para niños de 1–10 años y entre 8,7 y 15 mg/día para los adolescentes de 11–18 años. Para los adultos, los valores varían entre 6,7 y 15 mg/día para las edades de 19–50 años, entre 7,3 y 15 mg/día durante la gestación, suponiendo un consumo de alimentos con una disponibilidad moderada de zinc, y entre 11,7 y 19 mg/día durante la lactación, dependiendo de la fase.

6. Cinética y metabolismo en animales de laboratorio y en el ser humano

En estudios de inhalación (solamente nasal) realizados con cobayas, ratas y conejos se observaron valores de retención del 5%–20% en el pulmón tras la exposición a aerosoles de óxido de zinc a una concentración de 5–12 mg/m^3 durante 3–6 horas. La absorción intestinal del zinc está regulada por un mecanismo homeostático que no se conoce del todo, pero que depende fundamentalmente de la secreción pancreática e intestinal y de la excreción fecal. En la homeostasis pueden intervenir proteínas fijadoras de metales, como por ejemplo la metalotioneína y una proteína intestinal rica en cisteína. También pueden existir otros mecanismos desconocidos. La absorción a partir de la mucosa intestinal puede entrañar procesos de transporte tanto activo como pasivo. En los animales, la absorción puede variar en la gama del 10%–40% en función de su situación nutricional y de la presencia de otros ligandos en los alimentos. Se puede producir absorción cutánea de zinc a partir del óxido y el cloruro de zinc, siendo mayor en casos de deficiencia de zinc. El zinc absorbido se deposita principalmente en los músculos, los huesos, el hígado, el páncreas, el riñón y otros órganos. La semivida biológica del zinc es de unos 4–50 días en la rata, en función de la dosis administrada, y de unos 280 días en las personas.

353

7. Efectos en los animales de laboratorio

La toxicidad aguda por vía oral en roedores expuestos al zinc es baja, con valores de la DL_{50} del orden de 30–600 mg/kg de peso corporal, en función de la sal de zinc administrada. Entre los efectos agudos en roedores tras la inhalación o la instilación intratraqueal de compuestos de zinc figuran trastornos respiratorios, edema pulmonar e infiltración de leucocitos en los pulmones.

Los efectos tóxicos del zinc para los roedores tras una exposición breve por vía oral incluyen debilidad, anorexia, anemia, disminución del crecimiento, pérdida de pelo y una utilización menor de los alimentos, así como cambios en las concentraciones de las enzimas del hígado y el suero, cambios morfológicos y enzimáticos en el cerebro y cambios histológicos y funcionales en el riñón. La concentración a la cual el zinc no produce síntomas adversos en los roedores se ha establecido en unos 160 mg/kg de peso corporal en las ratas. Se observaron cambios pancreáticos en crías expuestas a concentraciones altas de zinc en los alimentos. La exposición breve por inhalación de cobayas y ratas a concentraciones de óxido de zinc $\geq 5{,}9$ mg/m^3 provocó inflamación y daño pulmonar.

La exposición prolongada por vía oral al zinc puso de manifiesto que los órganos destinatarios de la toxicidad eran en ratas, hurones y conejos el sistema hematopoyético; en ratas y ratones el riñón; y en ratones y hurones el páncreas. Se notificó una concentración sin efectos observados (NOEL) con respecto al crecimiento y la anemia para el sulfato de zinc en los alimentos < 100 mg/kg en la rata. Al aumentar la concentración de zinc en el organismo de animales experimentales expuestos a él se produce una reducción de la concentración de cobre, lo que parece indicar que algunos de los signos de toxicidad atribuidos a la exposición a concentraciones excesivas de zinc podrían deberse a una deficiencia de cobre inducida por el zinc. Además, diversos estudios han puesto de manifiesto que la exposición al zinc altera las concentraciones de otros metales esenciales en el organismo de los animales expuestos, incluido el hierro. Algunos signos de toxicidad observados en los animales expuestos a concentraciones elevadas de zinc se pueden atenuar añadiendo cobre o hierro a los alimentos.

Las concentraciones muy altas de zinc son tóxicas para los ratones y los hámsteres en gestación. Las ratas expuestas a un 0,5% y un 1% de zinc en los alimentos durante cinco meses no pudieron concebir hasta que se suprimió el zinc de su alimentación. Las concentraciones altas de zinc en los alimentos (2000 mg/kg) se asociaron también con un aumento de las resorciones y de los casos de muerte prenatal en ratones y ratas; este resultado también se ha observado en ovejas y hámsteres. Las resorciones aumentaron en un estudio en el cual las ratas estuvieron expuestas durante todo el período de la gestación a concentraciones bajas de zinc, de apenas 150 mg/kg. Sin embargo, en otro estudio con ratas no se observaron efectos perjudiciales sobre el desarrollo del feto a dosis de 500 mg/kg. Se comprobó que la exposición después del coito de ratas a concentraciones de zinc de 4000 mg/kg en los alimentos interfería con la implantación del óvulo. El aumento de la concentración de zinc en crías de ratas expuestas a él provocó una reducción de las concentraciones de cobre y de hierro.

Se han realizado estudios de genotoxicidad en diversos sistemas. La mayoría de los resultados han sido negativos, pero se han notificado algunos positivos.

La deficiencia de zinc en los animales se caracteriza por una reducción del crecimiento y de la replicación celular, efectos reproductivos adversos y efectos adversos en el desarrollo, que persisten tras el destete, así como una disminución de la respuesta inmunitaria.

8. Efectos en el ser humano

Se han notificado incidentes de intoxicación con síntomas de malestar gastrointestinal, náuseas y diarrea tras una exposición única o breve a concentraciones de zinc en el agua o en bebidas de 1000–2500 mg/litro. Se han descrito síntomas semejantes, que ocasionalmente provocan la muerte, tras la administración intravenosa inadvertida de grandes dosis de zinc. En pacientes de diálisis renal expuestos al zinc mediante el uso de agua almacenada en compartimentos galvanizados se han observado síntomas de toxicidad del zinc que fueron reversibles al filtrar el agua por carbón activado.

355

Se demostrado que una ingesta desproporcionada de zinc en relación con el cobre induce deficiencia de cobre en las personas, provocando un aumento de la necesidad de cobre, mayor excreción de cobre y una situación de deficiencia de cobre. La ingesta farmacológica de zinc se ha asociado con efectos que van desde la leucopenia y/o la anemia microcítica hipocrómica hasta la disminución de la concentración de lipoproteínas de alta densidad en el suero. Estas condiciones fueron reversibles al interrumpir la terapia de zinc y administrar un suplemento de cobre.

Los efectos en la salud humana asociados con la deficiencia de zinc son numerosos y entre ellos figuran cambios neurosensoriales, oligospermia, alteración de las funciones neurosicológicas, retraso del crecimiento, retraso de la cicatrización de las heridas, trastornos inmunitarios y dermatitis. Estas condiciones generalmente son reversibles cuando se corrigen con la administración de zinc suplementario.

No hay un índice bioquímico único, específico y sensible de la situación con respecto al zinc. El método más fidedigno para detectar su deficiencia es observar una respuesta positiva a un suplemento de zinc en ensayos doble ciego controlados (en ausencia de otras deficiencias limitantes de nutrientes). Sin embargo, este método es lento y con frecuencia poco práctico y en general se prefiere la determinación de una combinación de índices alimentarios, bioquímicos y fisiológicos funcionales. Varios valores anormales concordantes son más fidedignos que un valor aislado anómalo en el diagnóstico de un estado de deficiencia de zinc. La inclusión de índices fisiológicos funcionales, como el crecimiento, la agudeza degustativa y la adaptación a la oscuridad con una prueba bioquímica (por ejemplo, la concentración de zinc en el plasma o en el pelo), permite evaluar el alcance de las consecuencias funcionales de la situación de deficiencia de zinc.

La exposición al cloruro de zinc por inhalación tras el uso militar de "bombas de humo" produjo efectos como edema intersticial, fibrosis intersticial, neumonitis, edema de la mucosa bronquial, ulceración e incluso la muerte en condiciones extremas de exposición en espacios cerrados. Estos efectos posiblemente son atribuibles al carácter higroscópico y astringente de las partículas liberadas por tales dispositivos.

La exposición ocupacional a la materia particulada finamente dispersa que se forma cuando se volatilizan ciertos metales, entre ellos el zinc, puede provocar una enfermedad aguda denominada "fiebre de los fundidores", caracterizada por diversos síntomas, como por ejemplo fiebre, escalofríos, disnea, náuseas y fatiga. El trastorno es generalmente grave pero transitorio, y las personas tienden a adquirir tolerancia. La exposición de voluntarios a concentraciones de zinc de 77–150 mg/m^3 durante 15–30 minutos dio lugar a un aumento de los síntomas en algunos de ellos, una acusada respuesta inflamatoria relacionada con la dosis, con aumento de los linfocitos polinucleares en el fluido de lavado bronquioalveolar, y un aumento acentuado de la citoquinesis. Se ha notificado asma ocupacional en personas que trabajan con fundentes para soldadura blanda, pero las pruebas no fueron suficientes para indicar una relación causal. Hace poco se ha diagnosticado un caso raro que parece indicar una relación de ese tipo en un trabajador de una instalación de galvanización por inmersión en caliente (zinc).

9. Efectos sobre otros organismos en el laboratorio y sobre el terreno

El zinc es importante para la estabilidad de las membranas, en más de 300 enzimas y en el metabolismo de las proteínas y los ácidos nucleicos. Hay que encontrar un equilibrio entre los efectos adversos del zinc y su carácter esencial. Se ha notificado deficiencia de zinc en una amplia variedad de plantas cultivadas y animales, con efectos graves en todas las fases de la reproducción, el crecimiento y la proliferación tisular. Las deficiencias de zinc en diversos cultivos han producido pérdidas importantes en todo el mundo. La deficiencia de zinc es rara en los organismos acuáticos del medio ambiente, pero se puede inducir en condiciones experimentales.

En la toxicidad del zinc pueden influir factores tanto bióticos como abióticos, como la edad y el tamaño del organismo, la exposición anterior, la dureza del agua, el pH, el carbono orgánico disuelto y la temperatura. La integración de la química y la toxicología en el medio ambiente ha permitido predecir mejor los efectos en los organismos presentes en él. Esto ha llevado al punto de vista actualmente aceptado de que la concentración total de un elemento esencial como el zinc en un compartimento del medio ambiente no es, considerada de manera aislada, un buen indicador de su biodisponibilidad.

Los valores de la toxicidad aguda del zinc disuelto para los invertebrados de agua dulce oscilan entre 0,07 mg/litro para la pulga de agua y 575 mg/litro para un isópodo. Los valores de la toxicidad aguda para los invertebrados marinos varían entre 0,097 mg/litro para un mísido y 11,3 mg/litro para el camarón de fangal. La concentración letal aguda para los peces de agua dulce es del orden de 0,066–2,6 mg/litro; la gama para los peces marinos es de 0,19–17,66 mg/litro.

Se ha demostrado que el zinc tiene efectos reproductivos, bioquímicos, fisiológicos y de comportamiento adversos en diversos organismos acuáticos. Se ha puesto de manifiesto que las concentraciones de zinc >20 µg/litro tienen efectos adversos en los organismos acuáticos. Sin embargo, la toxicidad del zinc para tales organismos depende de numerosos factores, por ejemplo la temperatura, la dureza y el pH del agua y la exposición anterior al zinc.

En general, la toxicidad del zinc en las plantas provoca trastornos en el metabolismo que son diferentes de los que produce su deficiencia. La concentración crítica de zinc en el tejido foliar para un efecto sobre el crecimiento varía en la mayoría de las especies de 200 a 300 mg/kg de peso seco.

En estudios sobre el terreno se han detectado efectos adversos en los invertebrados acuáticos, los peces y las plantas terrestres en las cercanías de fuentes de contaminación por zinc. Se ha observado la aparición de tolerancia al zinc en plantas terrestres, algas, microorganismos e invertebrados de las cercanías de zonas con concentraciones elevadas de zinc.

10. Conclusiones

10.1 Salud humana

- Hay una tendencia decreciente en las emisiones antropogénicas de zinc.

- Muchas de las muestras del medio ambiente tomadas antes de 1980, en particular las muestras de agua, pueden haber experimentado una contaminación por zinc durante el muestreo y el análisis, por lo que habría que tomar con precaución extremada los datos sobre la concentración de zinc en esas muestras.

- En los países donde los alimentos básicos consisten en cereales y legumbres no refinados y la ingesta de alimentos frescos es baja, se deberían formular estrategias alimentarias para mejorar el contenido y la biodisponibilidad de zinc.

- Las preparaciones orientadas a aumentar la ingesta de zinc por encima de la proporcionada mediante los alimentos no deberían contener concentraciones de zinc que superasen los valores de referencia para la alimentación y deberían contener una concentración suficiente de cobre para garantizar una razón zinc:cobre aproximada de 7, como en la leche humana.

- Es necesario documentar mejor las exposiciones reales al humo de óxido de zinc en el entorno ocupacional. Las concentraciones en el lugar de trabajo no deberían producir niveles de exposición tan altos como los que se sabe que dieron lugar a respuestas inflamatorias en los pulmones de los voluntarios.

- La naturaleza esencial del zinc, junto con su toxicidad relativamente baja para el ser humano y las fuentes limitadas de exposición humana, parecen indicar que las personas normales sanas no expuestas al zinc en el lugar de trabajo tienen potencialmente un riesgo de sufrir los efectos adversos asociados con la deficiencia de zinc superior al que acompaña a una exposición normal al zinc en el medio ambiente.

10.2 Medio ambiente

- El zinc es un elemento esencial. Existe la posibilidad tanto de una deficiencia como de un exceso de este metal. Por este motivo, es importante que los criterios de reglamentación del zinc, al mismo tiempo que protegen de la toxicidad, no se establezcan en un nivel tan bajo que pueda llevar a la deficiencia.

• Hay diferencias en las respuestas a la deficiencia y al exceso.

• La biodisponibilidad del zinc depende de factores bióticos y abióticos, por ejemplo: la edad y el tamaño de los organismos, el historial anterior de exposición, la dureza del agua, el pH, el carbono orgánico disuelto y la temperatura.

• La concentración total de un elemento esencial como el zinc por sí sola no es un buen indicador de su biodisponibilidad o toxicidad.

• Hay una gama de concentraciones óptimas para los elementos esenciales como el zinc.

• La toxicidad del zinc dependerá de las condiciones ecológicas y de los tipos de hábitat, de manera que en cualquier evaluación del riesgo de los efectos potenciales del zinc en los organismos se deben tener en cuenta las condiciones del medio ambiente local.